# Optimization and Energy Maximizing Control Systems for Wave Energy Converters

# Optimization and Energy Maximizing Control Systems for Wave Energy Converters

Editors

**Giuseppe Giorgi**
**Sergej Antonello Sirigu**

MDPI • Basel • Beijing • Wuhan • Barcelona • Belgrade • Manchester • Tokyo • Cluj • Tianjin

*Editors*
Giuseppe Giorgi  Sergej Antonello Sirigu
Politecnico di Torino  Politecnico di Torino
Italy  Italy

*Editorial Office*
MDPI
St. Alban-Anlage 66
4052 Basel, Switzerland

This is a reprint of articles from the Special Issue published online in the open access journal *Journal of Marine Science and Engineering* (ISSN 2077-1312) (available at: https://www.mdpi.com/journal/jmse/special_issues/control_systems_wave_energy).

For citation purposes, cite each article independently as indicated on the article page online and as indicated below:

LastName, A.A.; LastName, B.B.; LastName, C.C. Article Title. *Journal Name* **Year**, *Volume Number*, Page Range.

**ISBN 978-3-0365-2824-3 (Hbk)**
**ISBN 978-3-0365-2825-0 (PDF)**

# Contents

# About the Editors

**Giuseppe Giorgi**, Ph.D., a research fellow, is currently working at the Marine Offshore Renewable Energy Lab (MOREnergy Lab) of Politecnico di Torino, Italy. He holds a Ph.D. in Engineering, obtained in the Centre for Ocean Energy Research (COER) of Maynooth University, Ireland, with major contributions in the nonlinear hydrodynamic modelling of wave energy conversion under controlled conditions. He is a Marie-Curie individual fellow alumni, working on computationally efficient nonlinear Froude–Krylov force calculation methods, applied to control and parametric resonance detection. His current research interests involve offshore renewable energy in general, numerical modelling and experimental testing.

**Sergej Antonello Sirigu**, Ph.D., a research fellow, is currently working at the Marine Offshore Renewable Energy Lab (MOREnergy Lab) of Politecnico di Torino, Italy. He holds a Ph.D. in Mechanical Engineering, with major contributions in design and testing of innovative wave energy converters, based on inertial coupling and resonance water tanks. His current research interests involve holistic techno-economic optimization based on genetic algorithms and evolutionary theory, as well as hardware-in-the-loop and wave tank experimental testing.

Journal of
*Marine Science
and Engineering*

MDPI

Editorial

# Optimization and Energy Maximizing Control Systems for Wave Energy Converters

**Giuseppe Giorgi** *and **Sergej Antonello Sirigu**

Marine Offshore Renewable Energy Lab (MOREnergy Lab), Department of Mechanical and Aerospace Engineering (DIMEAS), Politecnico di Torino, 10129 Turin, Italy; sergej.sirigu@polito.it
* Correspondence: giuseppe.giorgi@polito.it

**Citation:** Giuseppe, G.; Sirigu, S.A. Optimization and Energy Maximizing Control Systems for Wave Energy Converters. *J. Mar. Sci. Eng.* **2021**, *9*, 1436. https://doi.org/10.3390/jmse9121436

Received: 30 November 2021
Accepted: 14 December 2021
Published: 15 December 2021

**Publisher's Note:** MDPI stays neutral with regard to jurisdictional claims in published maps and institutional affiliations.

In recent years, we have been witnessing great interest and activity in the field of wave energy converters' (WECs) development, striving for competitiveness and economic viability via increasing power conversion while decreasing costs and ensuring survivability. In the community, the consensus is that both optimization and control are sine qua non conditions for success, but many challenges, peculiar to WECs, need to be addressed. Unlike other traditional control applications, the control objective for WECs is to exaggerate the motion, potentially inducing strong nonlinearities in the system and stressing the power-conversion chain and mechanical structure. Therefore, it is crucial to include techno-economical constraints in both the optimization and control objective functions. Furthermore, although often considered as consecutive independent phases, optimization and control are mutually dependent and, ideally, should be considered together.

The book *Optimization and Energy Maximizing Control Systems for Wave Energy Converters* includes eleven contributions [1–11] to this Special Issue published during 2020–2021. The overall objective of this Special Issue is to draw the most updated picture of the heterogeneous challenges that still need to be addressed in the field of wave energy control and optimization, while also to gather novel and cutting-edge techniques and methods to advance the state-of-the-art. The scientific collection presented in this Special Issue will be valuable for both scientists and technology developers, since each paper within is moved by a bundle of theoretical and pragmatic spirits, with the objective of providing advanced and effective solutions to problems that are currently holding back the development of wave energy technologies.

From a critical analysis of the eleven different contributions of this special issue, it is possible to highlight four major connecting threads:

1. Conjunction of both technical and economic considerations to drive decision-making [1,2,8];
2. Inclusion of non-ideal power take-off (PTO) and nonlinear phenomena for the effectiveness of control strategies [4,7,10];
3. Real-time capabilities as a mandatory condition for applicability of estimation, detection, and control algorithms [3–6,10];
4. Various control strategies including all of the above [3,6,7,9–11].

Tan et al. [8] investigate the influence of the size of a heaving point absorber wave energy converter, considering the techno-economic impact of the resulting power take-off. An optimization method is proposed to reduce the Levelized Cost Of Energy (LCOE). The performance of the system is evaluated through a frequency domain model of the device considering three representative sea states for productivity assessment. In order to represent the effect of the PTO size on power production, PTO force constraints have been included in the model. A preliminary economic model is implemented to estimate costs and LCOE at an early development stage. A control-informed optimization is carried out, evaluating the influence of buoy geometry, PTO size, wave resource and control logic for LCOE reduction. The results show that, for this application, the main driver of LCOE is cost rather than productivity. In fact, smaller PTOs have lower productivity, but still achieve an LCOE reduction of 24% to 31%. It is also interesting to note that wave resource and PTO size do not influence buoy size since the main techno-economic driver is cost.

Giassi et al. [2] perform an experimental campaign to evaluate the performance of different wave energy farm configurations, considering a bottom-tethered heaving point that can move in 6 DoF. The objective of this work is to evaluate the performance of the array considering different layouts and to compare the experimental results with a numerical frequency domain model. Array configurations were obtained by optimisation with genetic algorithms. The performance of the different configurations was evaluated experimentally under regular and long-crested waves. An important result is the calculation of the interaction factor (q-factor), which lies in the range of 0.77 up to 1.06, while the optimal configuration is the staggered layout. A frequency domain model of the array that predicts the heave motion of each device has been compared with the experimental tests. The numerical results are in agreement with the experimental ones when the test conditions do not involve strongly non-linear phenomena such as parametric resonance, slack line and wave breaking.

Sirigu et al. [1] present a holistic techno-economic optimization of the Pendulum Wave Energy Converter (PeWEC), using an evolutionary-based global optimization genetic algorithm. A strong and validated statement is provided and defended about the need for the inclusion of economic functions already during the first preliminary design. A genetic algorithm is implemented in order to optimize 13 different parameters, comprising shape, dimensions, mass properties and ballast, power take-off control torque and constraints, number and characteristics of the pendula and other subcomponents. Economic estimations are included, based on the mass of the hull and the pendula, as well as the size of the PTOs. Multiple optimization objectives are considered, Capture Width Ratio (CWR) and Capital expenditure over Productivity (CoP), demonstrating that CWR and CoP may be adverse objectives: the most effective device in absorbing and converting the incoming wave energy is not, in general, economically convenient, and vice versa.

Davidson and Kalmár-Nagy [4] propose a real-time detection system to identify when parametric resonance appears in wave energy converters. Parametric resonance is a dynamic instability due to the internal transfer of energy between degrees of freedom, which is known to cause large unstable pitch and/or roll motions, usually with detrimental effects on the power extraction performance and may increase loading on the WEC structure and mooring system. To remedy such negative effects, control systems can be designed to mitigate the onset of parametric resonance. Since real-time detection is key to enabling corrective actions, this paper presents the first application of a real-time detection system for the onset of parametric resonance in WECs. The proposed detection system achieved 95% accuracy across nearly 7000 sea states, producing 0.4% false negatives and 4.6% false positives.

Bonfanti et al. [5] are concerned with the real-time estimation of wave excitation forces, considering the case study of the Inertial Sea Wave Energy Converter (ISWEC). Energy-maximizing control strategies normally require the knowledge of the incoming wave force, which cannot be measured and should be estimated; moreover, since the input (PTO) control action must be provided in real-time, also the estimation should compute faster than real time. This paper investigates the wave excitation force estimation for a non-linear WEC, using both a model-based and a model-free approach. Firstly, a Kalman Filter is implemented considering the WEC linear model with the excitation force modelled as an unknown state to be estimated. Secondly, a feed-forward Neural Network is applied to map the WEC dynamics to the excitation force by training the network through a supervised learning algorithm. Sensitivity and robustness analyses are performed to investigate the estimation error in presence of un-modelled phenomena, model errors and measurement noise.

Garcia-Violiniet al. [3] present a critical comparison of a set of five simple controllers with the common ability to compute in real-time. In fact, it is argued that the computational cost of some complex control algorithms make them inapplicable to real devices; on the other hand, having the objective of actual implementation, a number of energy-maximising wave energy controllers have been recently developed based on relatively simple strategies, stemming from the fundamentals behind impedance-matching. This paper documents this set of five controllers, which have been developed over the period 2010–2020: (i) Suboptimal causal reactive controller; (ii) Simple and effective real-time controller; (iii) Multi resonant feedback controller; (iv) Feedback resonating controller; (v) LiTe-Con. The comparison, carried out both analytically and numerically, encompass their characteristics, in terms of energy-maximising performance, the handling of physical constraints, and computational complexity. In particular, a scoring system is set, explicitly evaluating the following metrics: computational simplicity, stability, constraint handling, and resulting performance.

Mérigaud and Tona [7] propose an energy-maximisation spectral control that is able to include non-ideal PTO systems in the underlying model used to compute the optimal control law. The discontinuous PTO efficiency characteristic is included via a smooth function approximation to ensure computational efficiency. However, the cost function becomes non-quadratic, hence requires a slight generalisation of the derivative-based spectral control approach, initially introduced for quadratic cost functions. This generalisation is derived in the presented paper, providing details on its practical interest. Two application cases are considered, namely a single-body and a two-body heaving point absorbers inspired by real devices. In both cases, the spectral approach calculates WEC trajectory and control force solutions, for which the mean electrical power is shown to lie within a few percent of the true optimal electrical power. Regarding the effect of a non-ideal PTO efficiency upon achievable power production, the power achieved lies within 80–95% of that obtained by simply applying the efficiency factor to the optimal power with ideal PTO. This is a significantly less pessimistic result than the others found in the literature.

Anderlini et al. [6] implement a real-time reinforcement learning control for wave energy converters, to cope with the potential inaccuracies and uncertainties of the underlying mathematical description of model-based controllers. In particular, such uncertainties may be due to both initial limitations of the model (e.g., linear and nonlinear assumptions) and to variations of some parameters during the operative life of the device (e.g., ageing and wear). In this paper, an alternative solution is introduced to address such challenges, applying deep reinforcement learning (DRL) to the control of WECs. A DRL agent is initialised from data collected in multiple sea states under linear model predictive control in a linear simulation environment. The agent outperforms model predictive control for high wave heights and periods, but suffers close to the resonant period of the WEC. The computational cost at the deployment time of DRL is also much lower by diverting the computational effort from deployment time to training. In addition, model-free reinforcement learning can autonomously adapt to changes in the system dynamics, enabling fault-tolerant control.

Previsic et al. [9] tackle the comparison of model predictive control (MPC) and optimal causal control, applied to a heaving point absorber. In recent years, efforts by various researchers have been invested in the design of simple causal control laws, thanks to their simplicity of implementation in a real system. However, it is important to have a fair comparison, under representative conditions, with more complex non-causal controllers, in order to appropriately evaluate the trade-off between power yield and complexity, also including constraint handling ability. In this paper, a linear MPC is compared to a casual controller that incorporates constraint handling. The analysis demonstrates that the MPC provides significant performance advantages compared to an optimized causal controller, particularly if significant constraints on device motion and/or forces are imposed. It is further demonstrated that distinct control performance regions can be established that correlate well with classical point absorber and volumetric limits of the wave energy conversion device.

Haider et al. [10] propose a nonlinear model predictive controller for a wave energy converter with multiple degrees of freedom. The proposed control is computed in real-time and includes non-ideal power take-off and model non linearities. The inclusion of non-linearities in the model leads to a non-quadratic and non-standard cost function, which is challenging to solve in a computationally effective manner. The considered device is the CENTIPOD, simulated in WEC-Sim, while the extracted power is re-written in pseudo-quadratic form and polynomial decomposition. Comparing linear to nonlinear MPC, despite a computational load that is only slightly higher (+35%), an appreciable improvement in power capture is achieved (up to +10.6%).

Demonte Gonzalez et al. [11] consider the application of sliding mode control for a floating heaving wave energy converter, including nonlinear hydrodynamic effects. In fact, the effectiveness of a control strategy is tightly linked to the representativeness of the underlying model, usually related to nonlinearities. Maximising energy extraction normally implies exaggerating the motion of the floater, inducing hydrodynamic nonlinearities: the most remarkable and often impactful are nonlinear static and dynamic Froude–Krylov forces, which are herein included. A sliding mode controller is proposed, which tracks a reference velocity that matches the phase of the excitation force to ensure higher energy absorption. The control algorithm is tested in regular linear waves and is compared to a complex-conjugate control and a nonlinear variation of the complex-conjugate control. Results show that the sliding mode control successfully tracks the reference and keeps the device displacement bounded while absorbing more energy than the other, although simple, control strategies. Furthermore, due to the robustness of the control law, it can also accommodate disturbances and uncertainties in the dynamic model of the wave energy converter.

**Conflicts of Interest:** The authors declare no conflict of interest.

## References


1. Sirigu, S.A.; Foglietta, L.; Giorgi, G.; Bonfanti, M.; Cervelli, G.; Bracco, G.; Mattiazzo, G. Techno-Economic Optimisation for a Wave Energy Converter via Genetic Algorithm. *J. Mar. Sci. Eng.* **2020**, *8*, 482. [CrossRef]
2. Giassi, M.; Engström, J.; Isberg, J.; Göteman, M. Comparison of Wave Energy Park Layouts by Experimental and Numerical Methods. *J. Mar. Sci. Eng.* **2020**, *8*, 750. [CrossRef]
3. García-Violini, D.; Faedo, N.; Jaramillo-Lopez, F.; Ringwood, J.V. Simple controllers for wave energy devices compared. *J. Mar. Sci. Eng.* **2020**, *8*, 793. [CrossRef]
4. Davidson, J.; Kalmár-Nagy, T. A Real-Time Detection System for the Onset of Parametric Resonance in Wave Energy Converters. *J. Mar. Sci. Eng.* **2020**, *8*, 819. [CrossRef]
5. Bonfanti, M.; Hillis, A.; Sirigu, S.A.; Dafnakis, P.; Bracco, G.; Mattiazzo, G.; Plummer, A. Real-time wave excitation forces estimation: An application on the ISWEC device. *J. Mar. Sci. Eng.* **2020**, *8*, 825. [CrossRef]
6. Anderlini, E.; Husain, S.; Parker, G.G.; Abusara, M.; Thomas, G. Towards Real-Time Reinforcement Learning Control of a Wave Energy Converter. *J. Mar. Sci. Eng.* **2020**, *8*, 845. [CrossRef]
7. Mérigaud, A.; Tona, P. Spectral Control of Wave Energy Converters with Non-Ideal Power Take-off Systems. *J. Mar. Sci. Eng.* **2020**, *8*, 851. [CrossRef]
8. Tan, J.; Polinder, H.; Laguna, A.J.; Wellens, P.; Miedema, S.A. The Influence of Sizing of Wave Energy Converters on the Techno-Economic Performance. *J. Mar. Sci. Eng.* **2021**, *9*, 52. [CrossRef]
9. Previsic, M.; Karthikeyan, A.; Scruggs, J.; Giorgi, G.; Sirigu, S.A. A Comparative Study of Model Predictive Control and Optimal Causal Control for Heaving Point Absorbers. *J. Mar. Sci. Eng.* **2021**, *9*, 805. [CrossRef]
10. Haider, A.S.; Brekken, T.K.; McCall, A. Real-Time Nonlinear Model Predictive Controller for Multiple Degrees of Freedom Wave Energy Converters with Non-Ideal Power Take-Off. *J. Mar. Sci. Eng.* **2021**, *9*, 890. [CrossRef]
11. Gonzalez, T.D.; Parker, G.G.; Anderlini, E.; Weaver, W.W. Sliding Mode Control of a Nonlinear Wave Energy Converter Model. *J. Mar. Sci. Eng.* **2021**, *9*, 951. [CrossRef]

Journal of
*Marine Science and Engineering*

*Article*

# Techno-Economic Optimisation for a Wave Energy Converter via Genetic Algorithm

Sergej Antonello Sirigu *, Ludovico Foglietta, Giuseppe Giorgi, Mauro Bonfanti , Giulia Cervelli, Giovanni Bracco and Giuliana Mattiazzo

Department of Mechanical and Aerospace Engineering, Politecnico di Torino, 10129 Turin, Italy;
ludovico.foglietta@studenti.polito.it (L.F.); giuseppe.giorgi@polito.it (G.G.); mauro.bonfanti@polito.it (M.B.);
giulia.cervelli@polito.it (G.C.); giovanni.bracco@polito.it (G.B.); giuliana.mattiazzo@polito.it (G.M.)
* Correspondence: sergej.sirigu@polito.it

Received: 9 June 2020; Accepted: 24 June 2020; Published: 30 June 2020

**Abstract:** Although sea and ocean waves have been widely acknowledged to have the potential of providing sustainable and renewable energy, the emergence of a self-sufficient and mature industry is still lacking. An essential condition for reaching economic viability is to minimise the cost of electricity, as opposed to simply maximising the converted energy at the early design stages. One of the tools empowering developers to follow such a virtuous design pathway is the techno-economic optimisation. The purpose of this paper is to perform a holistic optimisation of the PeWEC (pendulum wave energy converter), which is a pitching platform converting energy from the oscillation of a pendulum contained in a sealed hull. Optimised parameters comprise shape; dimensions; mass properties and ballast; power take-off control torque and constraints; number and characteristics of the pendulum; and other subcomponents. Cost functions are included and the objective function is the ratio between the delivered power and the capital expenditure. Due to its ability to effectively deal with a large multi-dimensional design space, a genetic algorithm is implemented, with a specific modification to handle unfeasible design candidate and improve convergence. Results show that the device minimising the cost of energy and the one maximising the capture width ratio are substantially different, so the economically-oriented metric should be preferred.

**Keywords:** wave energy; wave energy converter; PeWEC; techno-economic optimisation; genetic algorithm; cost of energy; CaPex; CWR

## 1. Introduction

Based on the gradual depletion of conventional fossil fuels, and most importantly, the established awareness of their destructive impact on the environment, recent years have witnessed a compelling momentum towards the energy transition and decarbonisation, shared by industry, governmental bodies and policy-makers. A condition for effective and reliable energy procurement is diversification, especially concerning renewable sources, which are usually more variable and volatile. Ocean energy, wave energy in particular, can play a major role in the overall energy mix, thanks to its high power density and availability [1]. Despite its potential, wave energy still remains substantially untapped since several challenges, hindering commercialisation at industrial scale, remain to be overcome. Wave energy conversion is yet to reach economic viability and is not competitive with other renewable energy technologies, due to an excessive LCOE (levelised cost Of energy) at the current stage of development.

From a high-level perspective, one of the main reasons for the unsuccessful development of effective technologies is that typical development pathways prefer to rapidly increase readiness levels before an acceptable performance is achieved [2]. Considering a matrix composed of technology readiness level (TRL) and the technology performance level (TPL), deficient development pathways tend to increase TRL first, at low TPL; conversely, optimal trajectories should increase TPL first, at low TRL, and then increase TRL towards commercialisation [3]. Although is it challenging to perform an accurate estimation of some performance aspects at low TRL, due to inherent limitations of small-scale devices and subsystems, techno-economic aspects should be taken into account. Changing such a development paradigm involves adopting better design practices, since common approaches are inherently suboptimal. In particular, often, design is performed in sequential fashion, considering different parts of the power conversion chain independently. A representative example is the optimisation of the hydrodynamics and power absorption of an uncontrolled device, followed by a design of an optimal controller, followed by the inclusion of technical constraints and efficiency, finally reaching the stage of cost evaluation and reduction. Although such an incremental approach is justified by the complexity of the system, neglecting mutual interaction typically leads to a suboptimal solution.

An effective design requires one to follow a holistic approach, based on representative wave-to-wire mathematical models [4], including some energy-maximising control strategy [5,6] considering technological constraints [7,8] and costs [9], since all of such aspects are tightly interrelated. Representative hydrodynamic models are advisable, since a controlled device may be affected by strong nonlinearities with major negative effects on power production [10]. Furthermore, unconstrained linear or partially-linear models may predict unrealistic behaviour under the action of an optimal controller [11], which is also sensitive to modelling errors [12,13]. Note that it is essential to include an energy-maximising control strategy, since it has the ability to significantly increase power-production performance over the operational sea states. Moreover, the control strategy has a great influence on the dynamic characteristic of the response of the device [14], making the power-wise optimal solutions for uncontrolled and controlled conditions significantly different [15]. However, the ability of the controller to optimise power absorption is highly dependent on physical constraints, mainly on displacement, speed and force/torque [16]. Such constraints depend on the electrical and mechanical subcomponents characteristics, and ultimately, cost. On the other hand, the hydrodynamic behaviour of the device mainly depends on its shape and dimensions, which entail volume and mass, which are main drivers of capital cost.

A practical implementation of an all-encompassing design optimisation is challenging, mainly due to the vast multi-dimensional search space which cannot be tackled with conventional global optimisation approaches due to computational limitations. Moreover, often there is no analytical representation of the relationship between input and output variables, due to hydrodynamic curves defined by points through panel-based simulations (BEM) and implicit nonlinearities requiring simulation via time-advancing schemes [17], or in some conditions, harmonic balance [18]. Therefore, both cost function and constraints usually behave as grey or black boxes. For such problems, genetic algorithms are among the most common choice, since they can handle design spaces of high dimensionality and complex cost functions and constraints, without relying on the knowledge of their mathematical structure [19]. Genetic algorithms are metaheuristic optimisation codes based on the evolutionary theory, relying on the survival-of-the-fittest principle. Thanks to their stochastic approach, they favour the identification of the global optimum, reducing the risk of convergence to a local minimum. Furthermore, they impose no specific requirement on the system to be optimised, since it is treated as a black box. Genetic algorithms are used in this paper based on previous experience of the authors, although other metaheuristic approaches exist, able to handle a large parameter space, such as covariance matrix adaptation, differential evolution and particle swarm algorithms [20].

Following their use in the general marine engineering field [21], genetic algorithms recently gained popularity in the wave energy field, but are mainly used for the optimisation of the array layout of several wave energy converters (WECs) [22,23]. Although scarce, some single WEC design optimisations via genetic algorithm are found in the literature. Reference [24] optimised the radius and draft of an archetypal cylinder, maximising the absorbed power and bandwidth. The design space is bi-dimensional; the geometry is idealised; and the power take-off (PTO) is a simple linear damper with a frequency-independent damping coefficient. Since power is not an effective single-objective function for practical design, Ref. [25] introduces proxies for costs, structural loads and PTO rating loads into the objective function. Costs are assumed to be proportional to the submerged surface and the maximum reaction force. While [25] considered fixed-axis devices of conventional shape, Ref. [26] attempted to generalise the geometry as much as possible, using B-splines to describe wetted surfaces of arbitrary shapes. Moreover, frequency-dependent energy-maximising control is introduced, also including constraints to limit unrealistically large amplitudes of motion that follow. As a proxy for costs, length and displaced volumes are considered in the cost functions. Finally, Ref. [27] proposes a thorough optimisation of the SEAREV device, attempting to maximise the produced power while minimising the device mass (as a crude proxy for costs), including various constraints and latching-declutching control.

This paper proposes a comprehensive techno-economic optimisation of the PeWEC (pendulum wave energy converter), which is a self-reacting WEC extracting energy from the oscillation of a pendulum sealed in a floating hull [28]. Several novelties, with respect to other design optimisation of WECs, are included. First and foremost, similar only to [27], a high level of detail of a realistic device is considered, as opposed to archetypal or generic geometries. However, a wide design space with high dimensionality is investigated, counting up to 13 design variables, comprising shape and dimensions of the floater; ballast and inertial properties; pendulum properties and numerosity; PTO rated torque; velocity and power; and gearbox speed conversion ratio. Moreover, realistic geometrical, technical and structural constraints are implemented, based on subcomponents characteristics. No idealised proxy is used for estimating costs, which are directly computed based on the cost of different material used for construction and different PTOs from a catalogue. Finally, a simple but realistic energy-maximising control strategy is implemented, including explicit constraint handling, in order to perform a control-informed optimisation. Due to the high-dimensionality and mutual dependence of design variables, a stochastic optimisation approach, treating different variables as independent and random, encounters a high level of unfeasible combinations, which may potentially hinder the success of the optimisation. In order to deal with unfeasible solutions, a tournament-based genetic algorithm is implemented, reducing the impact of mortality rate on the convergence rate. The objective function, which the genetic algorithm tries to minimise, is the ratio between capital cost and annual energy production (AEP). However, in order to highlight the impact of the performance index and to generate a Pareto front, the capture width ratio is also considered as an additional objective. It is shown that hydrodynamic efficiency and cost of energy are contrasting objectives; therefore, this paper provides evidence that challenges the common practice of optimising the hydrodynamic performance during early-stage development, suggesting companies should pursue economic performance instead.

The remainder of the paper is organised as follows: Section 2 describes the technology while Section 3 presents its parametrisation for the optimisation algorithm, also focusing on constraints, cost functions and different objective functions. Finally, Section 4 provides main significant results with a critical discussion. Some final remarks are given in Section 5. Appendix A provides full general details about the genetic algorithm that effectively deals with discrete multi-variate problems with high mortality rate.

## 2. Technology

The system under consideration in this optimisation study is the PeWEC, an acronym of pendulum wave energy converter. PeWEC is a self-referenced inertial-based floating WEC, composed of a sealed hull enclosing a pendulum and the power take-off (PTO). There are notional similarities between PeWEC and the SEAREV device [29], both based on an inertial-activated body contained in a floating hull; however, while SEAREV is designed for ocean applications, PeWEC is initially aiming at lower energetic closed seas with lower-period waves (e.g., the Mediterranean Sea). The device is anchored to the seabed by means of a 3-line-3-segment mooring system, with each line composed of a jumper (riser) and a clump-weight [30]. When the device is moved by the waves, a relative motion between the hull and the internal pendulum is obtained, mostly due to the pitching motion. The kinetic energy related to this movement is converted into electrical energy by the PTO connected to the pendulum's hinge. A control law is implemented in the PTO driver to adjust the pendulum dynamics to the instantaneous sea conditions, with the purpose of maximising the output energy. A schematic of the PeWEC system is shown in Figure 1.

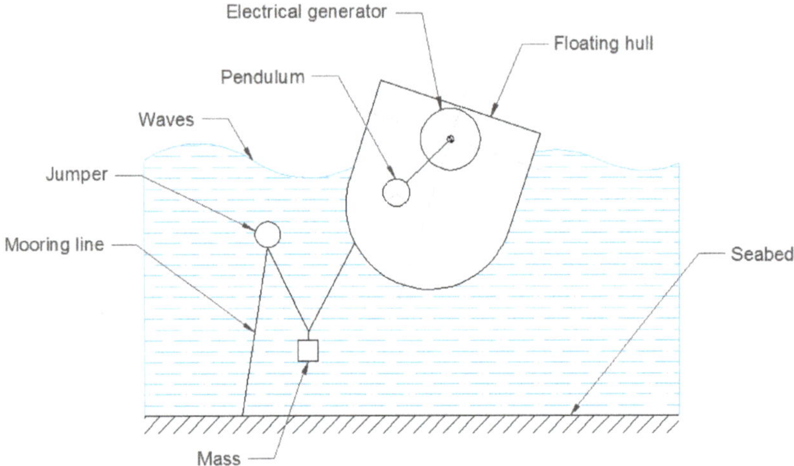


**Figure 1.** Schematic of the PeWEC working principle. Incoming waves excite the floating hull in pitch, transferring energy to the oscillation of an inner pendulum, whose motion is damped by a PTO. Three 3-segment mooring system, composed of a jumper and a clump-weight, keeps the system in place.

The hull is a sealed steel structure composed of a curved keel, two side walls and a flat topping; three internal sand ballasts (on the keel, stern and bow) ensure the mass distribution necessary to guarantee the required inertial properties, while a trellis structure is used to support the pendulum and the electronic equipment. The pendulum is composed of a cylindrical steel oscillating mass and a shaft, which is connected to the PTO system through a gearbox that ensures an appropriate coupling between the pendulum oscillation velocity and the PTO nominal speed. In order to extract energy from the pendulum movement, the PTO generates a reaction torque, which is adjusted by the driver's control system. Such constructive and technical features, based on early-development of the device, are here optimised via genetic algorithm. The technology has already been proven to be effective and promising, mainly thanks to wave tank tests at 1:45 and 1:12 scales [31]. Figure 2 shows a 3D digital model of the 1:45 prototype while Figure 3 shows a photo of the 1:12 prototype in the testing wave tank.

The concept of PeWEC germinated from the ISWEC (inertial sea wave energy converter), which generates the inertial coupling between PTO and the hull by means of the gyroscopic effect of a spinning flywheel [32]. While ISWEC requires power to keep the flywheel in rotation and create the self-reaction force [33], PeWEC is entirely passive, hence reducing explicit power loss sources. Further technological improvements are being explored for both the PeWEC and the ISWEC in order to increase the overall power conversion efficiency, such as resonant U-tanks [34], advanced control techniques [35], estimation and forecasting [36]. Moreover, the mooring system is under development, since it can affect the device performance [37]. However, this paper deals with a more fundamental early holistic optimisation of the PeWEC device shape, geometry and subcomponents, as further discussed in Section 3. The mathematical model, used to predict motion and loads of a given device, is presented in Section 2.1. The strategy to parametrically define multiple devices that the genetic algorithm can automatically optimise is presented in Section 3.

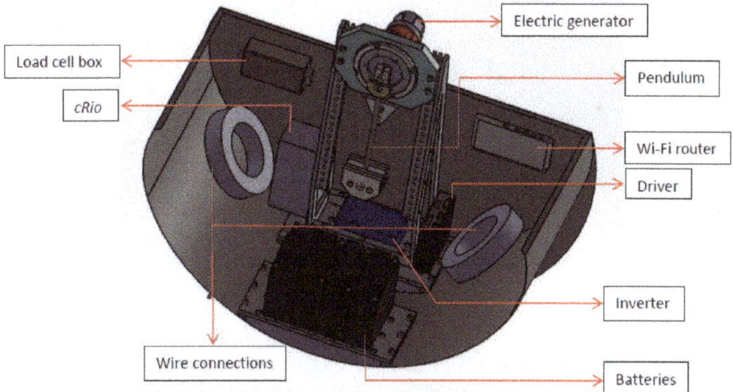


**Figure 2.** Digital representation of the 1:45 PeWEC prototype and its subcomponents.

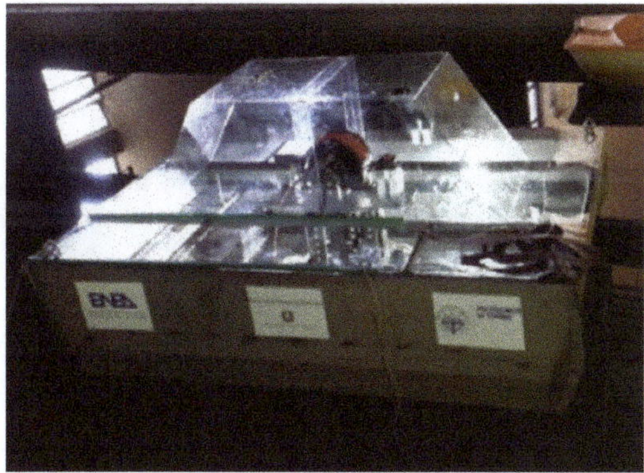

**Figure 3.** Photo of the 1:12 PeWEC prototype tested a wave tank [31].

*2.1. Mathematical Model*

The optimisation software relies on a mathematical model to evaluate the objective function which, in this paper, depends on the response of the device to incoming waves. An essential requirement for such a mathematical model is low computational speed, since a high number of individuals (order of tens of thousands) have to be assessed in a reasonable time. Therefore, a fully-linear model based on linear potential flow is implemented in frequency domain, thanks to its computational convenience. However, in order to avoid unrealistic motion that a linear model would predict under controlled conditions, various constraints are included, as further discussed in Section 3.2. Potential-flow-based models are conceptually divided into two phases: Firstly, the linear hydrodynamic curves, namely, excitation coefficients ($\mathbf{F}_w$), added mass ($\mathbf{A}$) and radiation damping ($\mathbf{B}$), are evaluated at a representative set of frequencies by means of a boundary element method (BEM) software, such as NEMOH [38] (this must be automatised in the genetic algorithm). Secondly, the dynamic response of the device is evaluated for a comprehensive set of sea states, characteristic of the designed installation site, taking care of determining appropriate energy-maximising control parameters. Although the run time of BEM codes depends on the number of simulated frequencies and on the discretised geometry, it is of the order of magnitude of $10^2$ s, on average. Since shape and dimensions of the hull are among the optimisation parameters, a detailed description is provided in Section 3.1. On the other hand, the structure of the dynamic response simulator is invariant:

$$[\mathbf{M} + \mathbf{A}(\omega)]\ddot{\mathbf{X}} + \mathbf{B}(\omega)\dot{\mathbf{X}} + (\mathbf{K}_h + \mathbf{K}_p)\mathbf{X} = A_w(\omega)\mathbf{F}_w(\omega) + \mathbf{T}_{PTO} \tag{1}$$

where $\mathbf{M}$ is the mass matrix, $\mathbf{K}_h$ is the hydrostatic stiffness, $\mathbf{K}_p$ the restoring force of the pendulum, $A_w$ is the wave amplitude, $\mathbf{T}_{PTO}$ is the PTO action and $\mathbf{X}$ is the state vector. Under the assumption of mono-directional waves aligned with the longitudinal axis of the hull, a planar motion of the hull can be assumed, namely, in surge ($x$), heave ($z$) and pitch ($\delta$). Including the pendulum oscillation ($\varepsilon$), the state vector has four dimensions:

$$\mathbf{X} = \begin{bmatrix} x \\ z \\ \delta \\ \varepsilon \end{bmatrix} \tag{2}$$

The PTO acts on the rotational degree of freedom of the pendulum, applying active and reactive power, proportional to displacement and velocity with constant coefficients, $k_{PTO}$ and $c_{PTO}$, respectively:

$$\mathbf{T}_{PTO} = \begin{bmatrix} 0 \\ 0 \\ 0 \\ -k_{PTO}\varepsilon - c_{PTO}\dot{\varepsilon} \end{bmatrix} \tag{3}$$

Coupling between the pendulum and the hull is provided by the restoring force of the pendulum and inertial reaction forces, which are expressed in $\mathbf{K}_p$ and $\mathbf{M}$, respectively. After linearisation [39]:

$$\mathbf{K}_p = \begin{bmatrix} 0 & 0 & 0 & 0 \\ 0 & 0 & 0 & 0 \\ 0 & 0 & -gm_p(d-l) & gm_pl \\ 0 & 0 & gm_pl & gm_pl \end{bmatrix} \tag{4}$$

$$\mathbf{M} = \begin{bmatrix} (M+m_p+m_b) & 0 & m_p(d-l) & -m_pl \\ 0 & (M+m_p+m_b) & 0 & 0 \\ m_p(d-l) & 0 & I_y+I_p+I_b+m_p(d-l)^2 & I_p+m_pl^2-m_pdl \\ -m_pl & 0 & I_p+m_pl^2-m_pdl & I_p+m_pl^2 \end{bmatrix} \tag{5}$$

where $M$ is the mass of the hull; $I_y$ is the hull moment of inertia; $g$ is the acceleration of gravity; $d$ is the distance between the device centre of gravity (CoG) and the pendulum fulcrum; $l$ is the pendulum length; $m_p$ and $m_b$ are the masses of the pendulum and the bar holding the pendulum, respectively, and $I_p$ and $I_b$ are their moments of inertia, respectively.

In order to maintain the model linear, mooring effects are neglected, as are mean drift forces and second order effects. Moreover, viscous effects are neglected, since they would require specific tuning for each device [40]. However, the linearised model has been validated through an experimental campaign [39], ensuring appropriate accuracy and representativeness for global design optimisation problems [41]. Nevertheless, it is worth noticing that these assumptions (planar waves and motion, and absence of a mooring system) are simplifications, since real waves may be short-crested, inducing complex motion in 6 degrees of freedom, which is also possible due to parametric instabilities and nonlinear coupling [42], and the mooring system may significantly affect the dynamics of the system [37].

The fully-linear model can be efficiently solved in the frequency domain with a very low computational time. For each wave, the optimal $k_{PTO}$ and $c_{PTO}$ are sought by means of a multi-variate simplex algorithm [43], respecting global constraints on loads and displacements, as discussed in Section 3.2.

It is worth remarking that the optimal control (performed by the simplex algorithm for each device) is completely independent of the design optimisation (performed by the genetic algorithm modifying geometrical and technical characteristics of the device). For each individual, the response to each irregular wave (modelled according to a Jonswap spectrum), comprising the simplex optimisation, requires about one second of computation. The productivity is computed over a scatter diagram densely described by 128 waves, accounting for about 100–150 seconds of computation. The targeted installation site is near Pantelleria island, in the Mediterranean Sea. The set of representative waves is chosen in order to adequately cover areas of the scatter diagram with non-negligible occurrence, as shown in Figure 4.

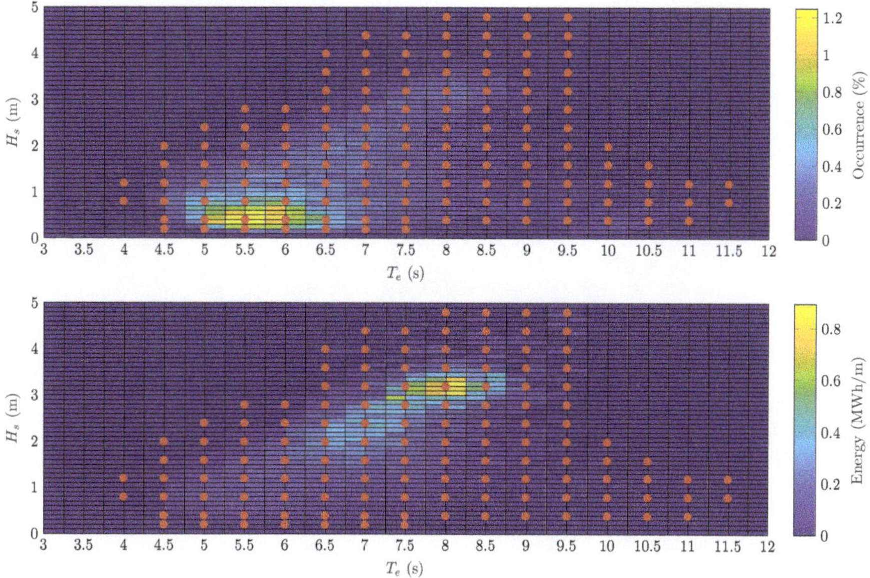

**Figure 4.** Occurrences and energy scatter diagrams for the installation site close to Pantelleria island, Italy; 128 representative waves were selected for the productivity estimation, shown by the red markers, where the occurrence percentage is not negligible.

## 3. Optimisation Software

The implementation of the optimisation software, as presented in Figure A1 and discussed in Appendix A, requires the definition of the design variables, the objective function and the tuning parameters of the genetic algorithm. The system, presented in Section 2, has virtually an infinite number of design variables that define its hydrodynamic behaviour, inertial properties and power absorption/conversion characteristics. Therefore, making some assumptions on the geometry and subcomponents is mandatory. Moreover, a parametric definition is required, allowing the software to fully define the system without direct human intervention. Section 3.1 details such a parametric representation, univocally identifying a device by its shape, dimensions, mass, inertia, ballast's distribution, pendulum and PTO characteristics. Starting from the shape of the hull, a BEM code automatically computes hydrodynamic curves, used to predict the device response to waves. However, techno-economic considerations should be included in the optimisation process, as extensively discussed in Section 1. Section 3.2 provides all functions that are used to estimate the cost of material and components. Moreover, constraints are imposed on displacements, velocities and loads, in accordance to the characteristics (and costs) of subcomponents. Finally, Section 3.3 presents tuning parameters of the genetic algorithm and the definitions of different performance indexes that the software aims at optimising.

### 3.1. Parametric Definition

In this work, 13 parameters are chosen to describe the device, summarised in Table 1 and hereafter discussed in detail: six for the hull, five for the pendulum and two for the PTO.

The shape of the hull is based on previous experience matured during the development of the PeWEC devices. For both manufacturability and hydrodynamic performance, the hull profile is assumed to be composed of a bottom circumference, tangential to two circumferences in the bow/stern sections, as shown in Figure 5, while the transversal section is assumed to be constant. The floater is symmetric with respect to the $y$-$z$ and $x$-$z$ planes and it results from the extrusion of the floater profile shape along the $y$-axis. The $x$-axis coincides with the floater deck and the $z$-axis goes from the deck of the floater upwards. The smooth surface with large radius of curvature is easier and cheaper to manufacture [44]. Moreover, the absence of sharp edges in the hull profile guarantee a low impact of vortex shedding and viscous drag losses [41]. Finally, the absence of re-entrant angles significantly reduces the risk of slamming events [27].

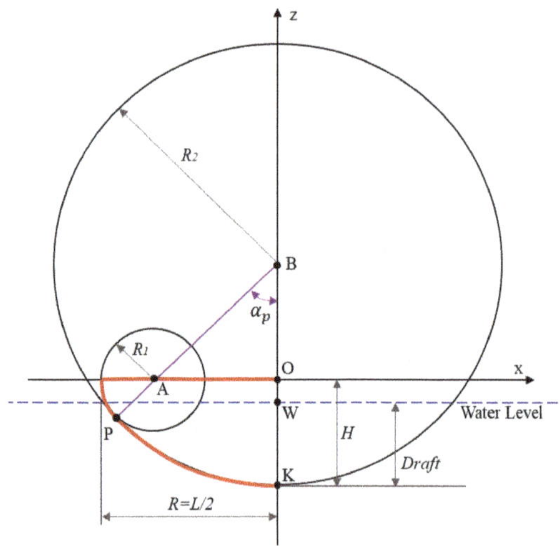


**Figure 5.** Parametric definition of the cross sectional of the floater.

The following relevant parameters, with respect to Figure 5, characterise the hull's profile:

- $R$: semi-length of the floater;
- $H$: overall height of the hull—i.e., the distance between the keel and the deck;
- $D$: draft of the hull;
- $R_1$: radius of circumference $C_1$;
- $(x_A, z_A)$: $x$- and $z$-coordinates of the centre of $C_1$, respectively, with $z_A = 0$;
- $R_2$: radius of circumference $C_2$;
- $(x_B, z_B)$: $x$- and $z$-coordinates of the centre of $C_2$, respectively, with $x_B = 0$;
- $\alpha_P = \angle PBO$.

A subset of six independent geometrical and inertial parameters is defined, which are used as design parameters in the genetic algorithm:

- $L$: length of the floater;
- $W$: width of the floater;
- $h = x_A / R$: bow/stern circumference ratio;

- $k = H/R$: height ratio;
- $j = D/H$: draft ratio;
- BFR: ballast filling ratio, defined as the ratio of ballast located in aft/fore ballast tanks over the total ballast (BFR = 1: all the ballast is stored in aft/fore ballast tanks; BFR = 0: all the ballast is stored in bottom ballast tank).

Figure 6 shows four examples of the effects of the geometric parameters $h$ and $k$. Figure 7 shows examples of draft and ballast allocation; namely, $j$ and BFR. Note that the amount of ballast required ($M_{bal}$) is univocally defined by the mass of the displaced volume of water ($M_{tot}$), the mass of the hull structure ($M_h$) and the mass of the pendulum-PTO units. The allocation of $M_{bal}$ between bow/stern and keel compartments determines the total moment of inertia of the system, which is an important parameter to conveniently change the resonant period in pitch. The ballast material is assumed to be sand, with density ($\rho_{bal}$) of 1400 kg m$^{-3}$, which is considered to be an appropriate compromise between cost and specific weight. Filling of the ballast compartments is assumed from the extremities inwards, providing higher inertia and being easier from a practical point of view.

The hull is made out of standard naval carpentry steel, with a density ($\rho_h$) of 7800 kg m$^{-3}$. Based on previous experience with prototyping of the PeWEC device, $M_h$ is assumed to be 90 times the total volume of the floater. Consequently, assuming the walls of the floater as thin plates, an equivalent thickness is computed from the lateral surface of the floater and $M_h$, making the computation of inertial properties of the hull possible.

On the hull's centreline, multiple pendulums can be placed, and their number ($N_p$) is one of the design parameters of the genetic algorithm. Each pendulum is composed of an oscillating cylindrical steel-made mass and a steel shaft. A trellis structure envelopes and supports the pendulum, a gearbox and the PTO; the ensemble of these components is referred to as unit, and its mass, inertia and volume are parametrically estimated starting from the pendulum's properties. The mass and inertia of the entire unit are 20% and 30% higher than the mass and inertia of the pendulum, respectively. The extension of the unit in the $y$-direction is 2 m longer than the pendulum dimension, in order to account for the required space for machinery and technical operators during assembly and maintenance. The overall mass of the device depends on the submerged volume, defined by $j$, $k$ and $L$. Since the mass of the hull is proportional to its total volume, the remaining part has to be composed of the ballast and units. A design parameter ($\beta_U$) is defined so that the mass of each unit is:

$$M_{unit} = \frac{\beta_U (M_{tot} - M_h)}{N_p} \tag{6}$$

Consequently, the ballast mass is the complementary fraction of $\beta_U$.

Each pendulum is composed of a cylinder with axis parallel to the $y$-direction, swinging around a fulcrum. The volume of the cylinder is computed from its mass and density ($\rho_p = 7800$ kg m$^{-3}$). Known volume, radius and height are defined by a shape factor ($\sigma_p$), which is a free design parameter deciding if the pendulum is large and short ($\sigma_p = 0$) or small and long ($\sigma_p = 10$), with respect to the available space. Once the geometry is defined, the length of the swinging arm of the pendulum is defined by a design ratio $\gamma_p$, which takes the available space in the hull into account. Finally, the height of the fulcrum is defined by a further design parameter $\lambda_p$. Note that the relative geometry of the hull and the pendulum allow for full rotation of the swinging mass around the fulcrum.

Note that not all possible combinations of geometric and pendulum property parameters are feasible due to volume and weight physical restrictions. Consequently, during the optimisation, unfeasible individuals are discarded and the death penalty function is computed, contributing to increase the mortality rate of the genetic algorithm.

Finally, the actual conversion stage is performed by a permanent magnet syncronous motor (PMSM), one for each pendulum, combined with a planetary gearbox that is well suited for high torque and low speed applications. The gearbox ratio $r_g$ is a further design parameter. Finally, different PTO systems are used ($ID_{PTO}$), with different combinations of nominal speed ($n_{PTO}$) and nominal torques ($T_{PTO}$), hence the costs, as shown in Figure 8.

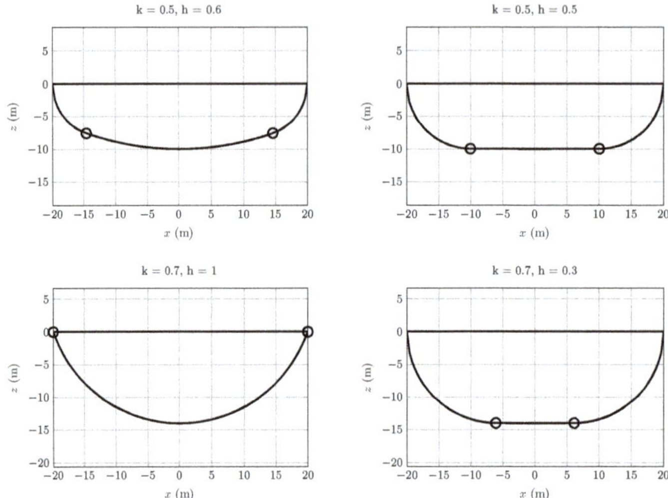

**Figure 6.** Examples of the effects of the *h* and *k* geometric design parameters.

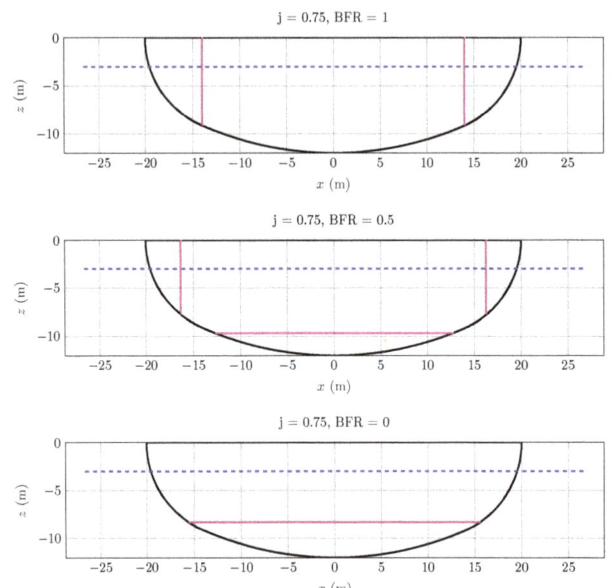

**Figure 7.** Examples of the effects of the *j* and ballast filling ratio (BFR) design parameters.

A summarising table of the 13 design parameters is provided in Table 1, including lower and upper bounds, and number of equally spaced steps. Bounds and step sizes are based on the authors' experience in developing the technology and running the model, ensuring that the minimum allowed variations of the design parameters produce small but relevant variations in the device performance. In this way, the algorithm is forced to consider only substantially different individuals, preventing the risk of simulating almost-equal devices. Although this objective could be achieved with a continuous parameter space by applying appropriate tolerances and checks on parameter variations, the authors believe that a gridded approach is more convenient, since it is easier to inspect and implement.

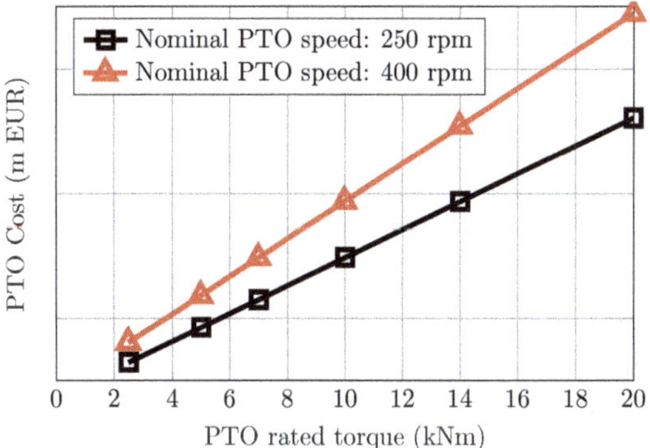


**Figure 8.** Cost of each power take-off, linearly proportional to the rated torque, with slope depending on the nominal speed.

**Table 1.** Thirteen design parameters used by the genetic algorithm to univocally define an individual, as discussed in Section 3.1. Discrete values with constant spacing are used, with lower and upper bounds.

| Design Parameter | Symbol | Units | Lower Bound | Upper Bound | # Discretisation |
|---|---|---|---|---|---|
| Hull length | $L$ | (m) | 8 | 30 | 23 |
| Hull width | $W$ | (m) | 5 | 30 | 26 |
| Bow/stern circumference ratio | $h$ | (-) | 0.5 | 1 | 6 |
| Height ratio | $k$ | (-) | 0.5 | 1 | 6 |
| Draft ratio | $j$ | (-) | 0.65 | 0.8 | 4 |
| Ballast filling ratio | BFR | (-) | 0.5 | 1 | 6 |
| Number of pendulum/PTO | $N_p$ | (-) | 1 | 15 | 15 |
| Unit mass ratio | $\beta_u$ | (-) | 0.05 | 0.95 | 18 |
| Pendulum shape factor | $\sigma_p$ | (-) | 0.05 | 0.95 | 10 |
| Pendulum arm factor | $\gamma_p$ | (-) | 0.1 | 1 | 10 |
| Pendulum fulcrum factor | $\lambda_p$ | (-) | 0.1 | 1 | 10 |
| Gearbox ratio | $r_g$ | (-) | 10 | 30 | 3 |
| PTO ID | $\text{ID}_{\text{PTO}}$ | (-) | 1 | 11 | 11 |

### 3.2. Techno-Economic Definitions and Constraints

The definition of the fitness index beholds primary importance in the identification of an optimal device: the genetic algorithm, similarly to other optimisation codes, iteratively tests several devices to find the one that minimises the objective function. In this work, three different fitness functions are used in three independent optimisation cycles, in order to have a deeper understanding of the phenomenon by building the Pareto front. In fact, multi-objective algorithms are usually performed as several single-objective optimisations where the performance index is computed as a weighted average of the multiple objectives [45].

The first objective is to minimise the capital expenditure (CapEx) over productivity (CoP) ratio, computed as the ratio between the overall capital expenditure (CapEx) and the produced energy over the lifetime of the plant ($N_y$, assumed 25 years for PeWEC):

$$\text{CoP} = \frac{\text{CapEx}}{N_y\,\text{AEP}} \tag{7}$$

where the annual energy production (AEP) is computed considering net power, affected by efficiencies and baseload: the gearbox is assumed to have a 0.95 mechanical efficiency; the PTO conversion system is considered to have a conversion efficiency of 0.95; the baseload considers all the electrical machinery input energy necessary for their correct operation: 500 W for each pendulum. Power losses from the bearings have been neglected, since the pendulum rotates at low speed [39].

The overall cost considers the device's capital expenditures to be classified in three main terms: hull's materials and construction, units' materials and construction, PTO. The mooring system is not considered, since its layout and cost are device-specific and no clear automatic parametrisation could be tackled in such an early design stage. Similarly, operational and maintenance expenditure (OpEx) is not considered in this work: on the one hand, its assessment requires extensive experience and a mature technology; on the other hand, in sealed WECs such as PeWEC but also ISWEC and SEAREV, OpEx is usually negligible compared to CapEx, since there are no moving parts subject to hostile marine environment (salty water, dust and debris). In fact, for common WECs presenting moving parts in hostile environment, it is estimated that the CapEx accounts for up to 35% only [46]. However, for the SEAREV device (with all moving parts enclosed in a sealed hull, as the PeWEC) it is estimated that OpEx is about 2% of the overall expenses [27]. Moreover, in such conditions, the levelised cost of energy (LCOE) is shown to be mainly sensitive to AEP and CaPex, while Opex, inflation rate, lifetime and expected internal rate of return are less impactful [27].

Several PTO models are considered in this analysis; each of them is characterised by its own maximum and nominal values in torque and speed. PTO costs are assumed to be significantly proportional to the nominal torque, with slope proportional to the nominal velocity, as shown in Figure 8. The PTO costs comprise those of the generator, the gearbox and various power electronic ancillary systems. However, the explicit costs of PTOs have been hidden for confidentiality. Note that a larger PTO provides higher freedom to the controller and is able to convert more energy. However, such a flexibility requires higher investments. The optimisation algorithm attempts to find a suitable compromise, also taking into account technical constraints on speed, torque and power of the PTO.

Material and manufacturing costs of the hull ($C_h$) and the pendulums ($C_p$) are estimated to be proportional to their respective mass, with the specific cost of the pendulums being almost twice the specific cost of the hull. Although simplified, all cost functions (PTO, hull construction and pendulum construction) are notionally based on the experience of the authors in the development and deployment of actual small and full-scale devices.

As discussed in Section 1, it is advisable to optimise the overall CoP, since it takes into account the compromise between techno-economic performance and hydrodynamic performance. However, a popular approach in the literature is to optimise the AEP or the hydrodynamic absorption abilities, regardless of the cost and/or technical characteristics and the limitations of the subcomponents required to achieve such a high conversion efficiency. In order to highlight substantial differences in convergence direction and the obtained optimal device, a second objective is considered: the capture width ratio (CWR), which is a common measure of energy conversion ability, defined as ratio between AEP and annual energy available in the resource ($E_w$):

$$CWR = \frac{AEP}{E_w} \tag{8}$$

Note that the correspondent performance index (to be minimised) is defined as -CWR, since CWR should be maximised. For each sea state, the incoming wave power on the device is defined as:

$$P_w = \frac{\rho g^2}{64\pi} H_s^2 T_e W \tag{9}$$

where $\rho$ is the water density, $H_s$ the significant wave height and $T_e$ the energy period. Equation (9) shows how the incoming available energy is proportional to the width, which is one of the main design parameters in the optimisation. In practice, while maximising AEP will likely maximise ($W$) in order to have access to larger power, maximising CWR seeks to increase the portion of the converted energy from the wave to the grid, regardless of price of conversion. Finally, maximising CoP focuses on the cost of the conversion, regardless of its efficiency.

Since there are two competing objectives (minimising CoP and maximising CWR), a multi-objective optimisation approach is adopted. By means of a scalarisation technique [47], the two different objectives are aggregated into a unique objective by means of a weighted average, using a pair of weights ($w_1$, $w_2$). Through the choice of different weights, it is possible to build the Pareto-optimal set. Three pairs of weights have been considered, as shown in Table 2. Effectively, sets (1) and (3) are single-objective optimisations, while set (2) is a proper multi-objective one. However, by aggregating the results of the three optimisations, a Pareto front can be constructed.

**Table 2.** Multi-objective weights.

| Weight Set ID | Explanation | CoP-Weight ($w_1$) | CWR-Weight ($w_2$) |
|---|---|---|---|
| (1) | CoP-driven | 1 | 0 |
| (2) | Weight-driven | 0.5 | 0.5 |
| (3) | CWR-driven | 0 | 1 |

Technological constraints are applied to all three optimisations. For each sea state, the optimal control parameters are chosen in order fulfil constraints tabulated in Table 3. Note that velocity and torque constraints depend on the particular PTO and the gearbox installed, and hence on the design parameters $ID_{PTO}$ and $r_g$, respectively.

**Table 3.** Technological constraints, fulfilled by tuning the control parameters.

| Constrained Variable | Units | Constrained Value |
|---|---|---|
| $\text{rms}(\delta)$ | (°) | 15 |
| $\text{rms}(\varepsilon)$ | (°) | 40 |
| $\max(\dot{\varepsilon})$ | (rpm) | $n_{PTO}^{\max}/r_g$ |
| $\max(T_{PTO})$ | (kNm) | $T_{PTO}^{\max}r_g$ |

*3.3. Genetic Algorithm*

In Appendix A, a detailed description of the genetic algorithm is provided. Considering the specific problem, presented in Section 3.1, the tuning factors of the algorithm in Table 4 were chosen. The MATLAB built-in implementation of the genetic algorithm has been used, along with its default tuning parameters, which are considered to be suitable. Although beyond of the scope of this paper, a throughout sensitivity analysis can potentially improve the convergence rate of the algorithm.

**Table 4.** Tuning parameters of the genetic algorithm, described in Appendixes A.1–A.6 and shown in red in the flow chart in Figure A1.

| Name | Symbol | Value |
|---|---|---|
| Population size | $N$ | 150 |
| Maximum generation count | $M$ | 200 |
| Maximum stall generation | $M_\Delta$ | 50 |
| Convergence threshold | $\Delta$ | $1.00 \times 10^{-6}$ |
| Elitism percentage | $\epsilon$ | 5% |
| Tournament size | $k$ | 4 |

## 4. Results

The MI-LXPM genetic algorithm has been chosen since it can effectively handle integer optimisation problems with potentially high number of unfeasible solutions. Figure 9 shows the development of the mortality rate through generations for the example of the CoP-driven optimisation. The first population has a significant number of unfeasible solutions, since it results from a purely stochastic process. However, the mortality rapidly drops to an average value (excluding the first two generations) of 25.5% with a standard deviation of 4.3%. Furthermore, note that there are 118 generations, meaning that the algorithm met convergence criteria before the maximum generation count of 200.

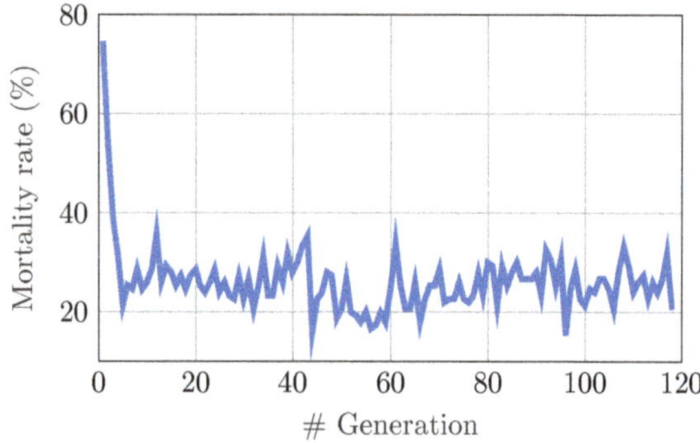

**Figure 9.** Mortality rate for the capital expenditure (CapEx) over productivity (CoP)-driven optimisation. Excluding the first two generations, the average is 25.5% and the standard deviation is 4.3%.

Note that two different seeds for the randomisation of the initial population of each optimisation have been used, reaching consistent results, hereafter discussed. The overall development of the algorithm is shown in Figure 10 for the example of the CoP-driven optimisation. The generic objective function $f^*$ to be minimised is on the vertical axis. For each generation, individuals are sorted according to $f^*$, so that the $N$th individual has the highest $f^*$. Unfeasible individuals are missing markers at lowest individual count. The colour code is in accordance with the generation count, showing the evolution of the population. The final population count is 118, as also shown in Figure 9. Therefore, since the maximum stall generation criterion is 50 (see Table 4), improvements of the overall-optimum are found until generation 68. The cloud of points is also projected onto the vertical planes, highlighting relevant trends. The projection on the vertical right plane has the purpose of showing the potentially significant change of theslope as generations advance; i.e., if the ratio of fit individuals over the whole population changes. The fact that such a projection preserves about the same shape shows that the overall distribution of fit individuals remains about constant, apart from early generations. This is also shown in the projection on the vertical left plane: the minimum of $f^*$ decreases as generations progress, as expected and desired; however, the improvement is abrupt in early generations (until generation 40) and slows down afterwards (until generation 68). Furthermore, the number of devices at a low $f^*$ is relatively scarce, with a rarefied gap between the fittest individuals and the rest of the population. This suggests that the degradation of performance is quite sensitive to the choice and combination of design parameters. Moreover, the presence of about 20% of unfeasible devices may have an impact on the uniform improvement of the whole population. An alternative, smoother definition of the performance index, a finer parameter grid and different cross-over/mutation strength may smooth the convergence and increase the convergence rate. Moreover, the definition of the design parameters may be appropriately revised in order to structurally reduce the likelihood of unfeasible individuals. However, although such an extensive sensitivity analysis is an interesting topic for future investigation, it is believed that the current setup provides appropriate convergence and satisfactory results to discuss techno-economic aspects of the PeWEC device optimisation.

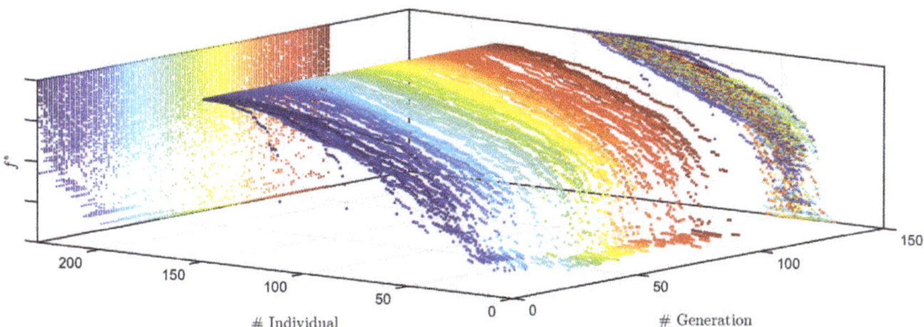

**Figure 10.** CoP-driven optimisation results: for each generation, individuals are sorted according to the objective function ($f^*$) to be minimised. The cloud of points is projected onto the vertical planes. The colour code corresponds to the generation count.

Three different optimisations have been performed, pursuing different objective functions in order to generate a Pareto front, as discussed in Section 3.2. Figure 11 shows the ensemble of all individuals generated from the three separate optimisations aggregated, plotting on the productivity versus the device cost, with the colour code being proportional to the CoP (left) and CWR (right). Elliptical annotations on the graphs indicated the most convenient region according to the respective metric (low CoP and high CWR, respectively). As expected, the optimal device changes with the optimality metric. In fact the highest AEP, the highest CWR and the lowest CoP are achieved by three significantly different individuals. In particular, it is evident that the increase in AEP is slower than the required increase in cost, so that the lowest CoP is found in the low cost region, in spite of a lower AEP. Moreover, low-CoP devices have also low CWRs, demonstrating the fact that the main driver is the device cost. Conversely, achieving high CWR and/or high AEP typically requires such an increase in device cost that the overall conversion becomes economically unfavourable.

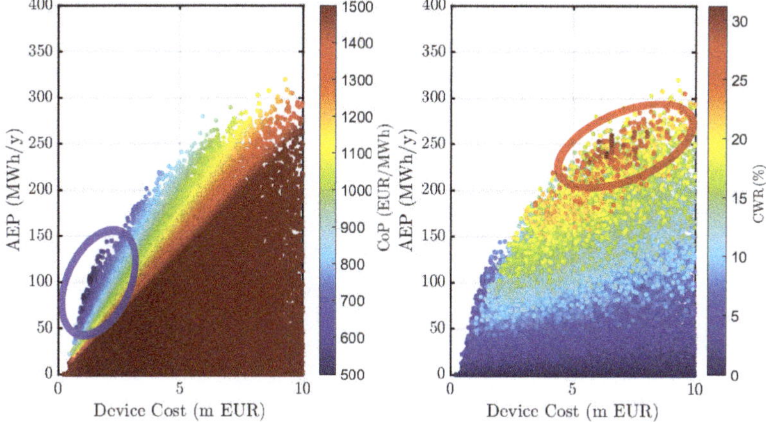

**Figure 11.** Productivity versus cost of the ensemble of the three optimisations aggregated; with each marker (individual), colour is proportional to the CoP (left) and capture width ratio (CWR) (right). The elliptical annotation highlight the most convenient region according to the evaluation metric (low CoP and high CWR).

It is interesting to analyse the direction that each optimisation tries to follow in order to improve performance index. Figure 12 shows a map of CWR versus COE, with three shaded areas encompassing individuals generated according to the three optimisation objectives. Individuals are also represented by markers of the same colour of the respective shaded area. As expected, the COE-driven area is attracted towards the bottom CoP-axis and achieves the overall minimum CoP. Similarly the CWR-driven optimisation extends the furthest along the CWR-axis, while obtaining higher CoP. In between, the weighted optimisation reaches good compromises of the two metrics.

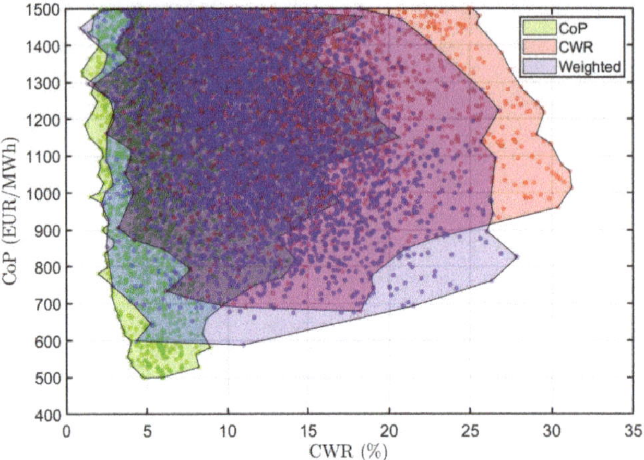

**Figure 12.** Ensemble of the three optimisations (CoP-driven, CWR-driven and weighted), with shaded areas encompassing each one of the three populations.

Overall, it is important to remark that the CWR-driven optimum has a significantly high CoP, twice the overall minimum CoP. Similarly, also the device with lowest CoP from the CWR-driven optimisation has a high CoP, compared to the entire populations. Conversely, the lowest CoP is obtained by a device with a CWR 6 times smaller than the highest-overall CWR.

From Figure 12 it is possible to extract the Pareto front, shown in Figure 13. Moreover, it is interesting to highlight the evolution of the Pareto front as the algorithm advances through generations. Five different stages of progression are shown, equally spaced from 0% to 100% progression. It is evident that an abrupt improvement is achieved in early stages of the optimisation. Afterwards, although the improvement slows down, the Pareto front is gradually improved, moving downwards and rightwards, namely, to lower CoP and higher CWR.

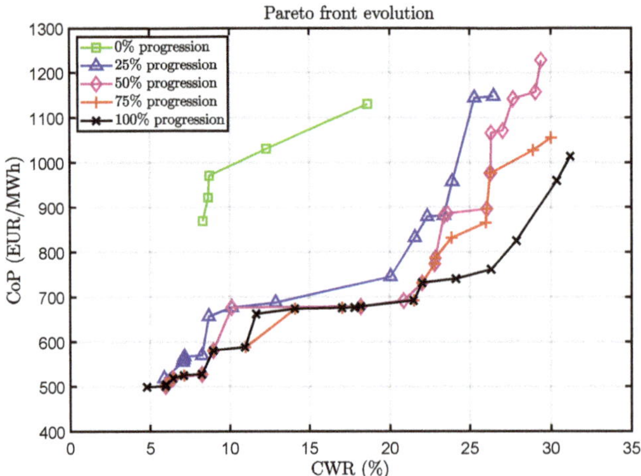

**Figure 13.** Evolution of the Pareto front, from 0% to 100% progression.

### 4.1. CoP and CWR Optima

The characteristics of the optimum devices according to different objective functions are hereafter discussed. Figures 14 and 15 show the cross-section and the top-view, respectively, of the optimal devices according to the three optimisations; namely, the CoP-driven optimum, the weight-driven optimum and the CWR-driven optimum. The same axis scale and dimension are used in order to highlight differences in size. Moreover, position, size and number of pendulums are shown.

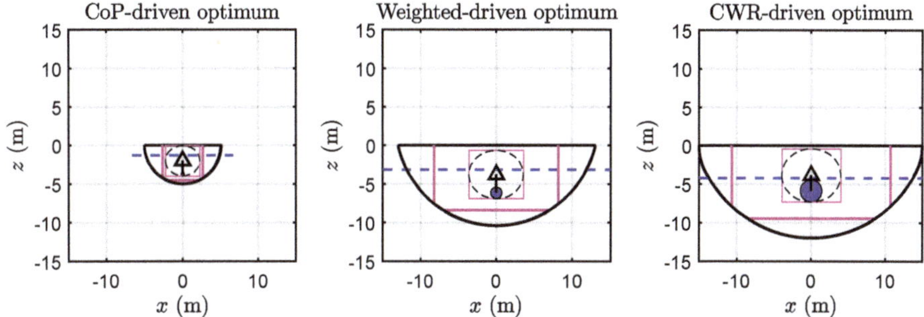

**Figure 14.** Cross-section of the CoP-driven, the weight-driven and the CWR-driven optima, with common scale and axis dimensions. The pink rectangles represents the ballast compartments and the bulk of the unit. The blue circle is the cross-section of the pendulum mass. The triangular marker represents the fulcrum. The blue dashed line is the still water level.

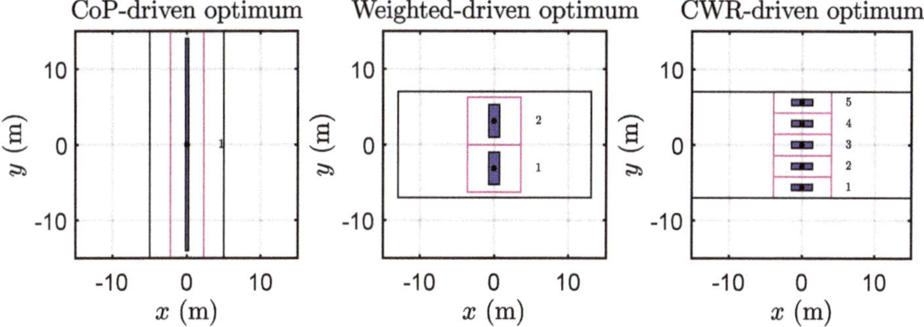

**Figure 15.** Top view of the CoP-driven, the weight-driven and the CWR-driven optima, with common scale and axis dimensions. The pink rectangles represents the bulk of the unit. The blue rectangle is the pendulum mass.

Clearly, dimensions in the $x - z$ plane increase from left (CoP-driven) to right (CWR-driven), suggesting that a long and high hull is more hydrodynamically efficient for the considered installation site. However, the CoP-optimum is wider than the CWR-optimum, intercepting more wave-crest, hence the high amount of incoming energy. This suggests that increasing the width requires a lower increment in cost than the consequent increase in AEP, hence lowering the resulting CoP. Moreover, the number of pendulums (and PTOs) increases from left to right, hinting that the cost of the PTO is a limiting constraint in the CoP-driven optimisation. Conversely, higher CWR requires more pendulums with larger cross sectional areas and smaller radius/length ratios.

Figure 16 shows the response amplitude operator (RAO) of the CoP-optimum (left) and CWR-optimum (right), for the pitch ($RAO_{55}$, top) and pendulum oscillation ($RAO_{77}$, bottom) degrees of freedom. The PTO-free RAO is shown in solid lines and square markers, showing that the natural attitudes of the devices differ according to the pursued objective. In particular, maximising the CWR leads to a higher natural period (about 7 s) than optimising the CoP (about 5.4 s), so that it can be inferred that converting lower energetic waves (lower period) is more economically convenient. Furthermore, Figure 16 also shows RAOs, including optimal control parameters, shown in Figure 17, which significantly modify the dynamics of the system in order to maximise the performance, while remaining compliant with technical constraints. Since the PTO acts directly on the pendulum (under-actuated system), $RAO_{77}$ is more affected by the control, with significant increases in amplitude and bandwidth. Conversely, the $RAO_{55}$ is damped out, due to the inertial coupling between the hull and the pendulum. Overall, the control increases the energy flux from the hull to the pendulum, and ultimately, the PTO.

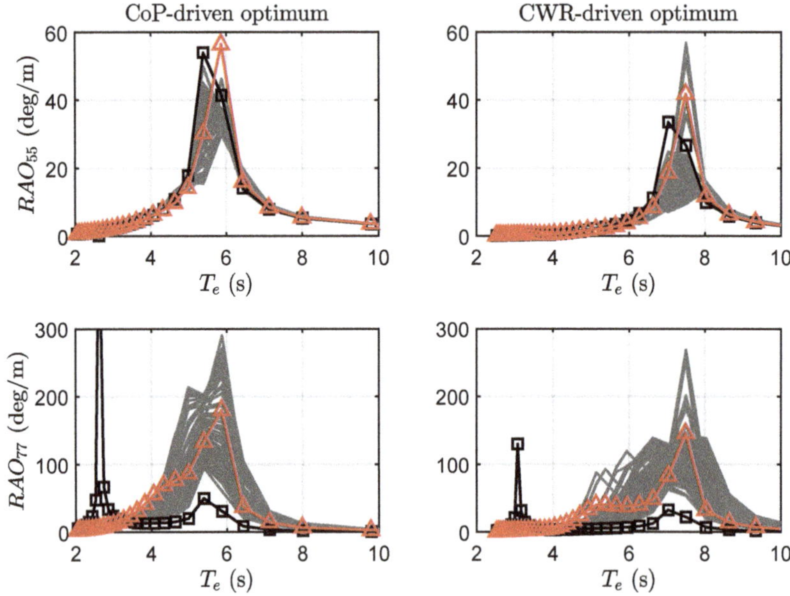

**Figure 16.** Response amplitude operator of the CoP-optimum (left) and CWR-optimum (right), for the pitch ($RAO_{55}$, top) and pendulum oscillation ($RAO_{77}$, bottom) degrees of freedom. The black solid line with square markers is for the PTO-free condition. The set of grey RAO lines is for the various controlled conditions, shown in Figure 17, among which a representative example is the red solid line with triangular markers, referring to $c_{PTO}$ and $k_{PTO}$ weighted over the occurrences matrix.

Overall, Figure 16 highlights the importance of a controlled-informed optimisation. In fact, the algorithm selects a device that is structurally apt to perform well when controlled, as opposed to select a performing uncontrolled device, and then applies a control.

Figure 17, in accordance with Figure 16, shows that the peak of conversion efficiency, i.e., CWR, is around lower wave periods for the CoP-optimum than the CWR-optimum. The control parameters maps show that $c_{PTO}$ is higher for the CWR-driven optimum, since it has to apply a higher damping torque to convert power in higher-energetic sea states. The reactive term of the PTO control, i.e., $k_{PTO}$, is negative almost across the whole scatter in order to obtain the resonant period of the pendulum $T_{res}$, shown in the last row of Figure 17. The natural period of the pendulum can be affected by the properties of the pendulum (mass, inertia and length) and $k_{pto}$. However, the inertial properties of the pendulum also affect the coupling between hull and pendulum, as shown in the off-diagonal terms of (5) and (4), while $k_{pto}$ only affects the diagonal term of the pendulum. Therefore, it appears that the control-informed optimisation algorithm favours lighter pendulums with negative $k_{PTO}$ rather than heavier pendulums and higher $k_{PTO}$.

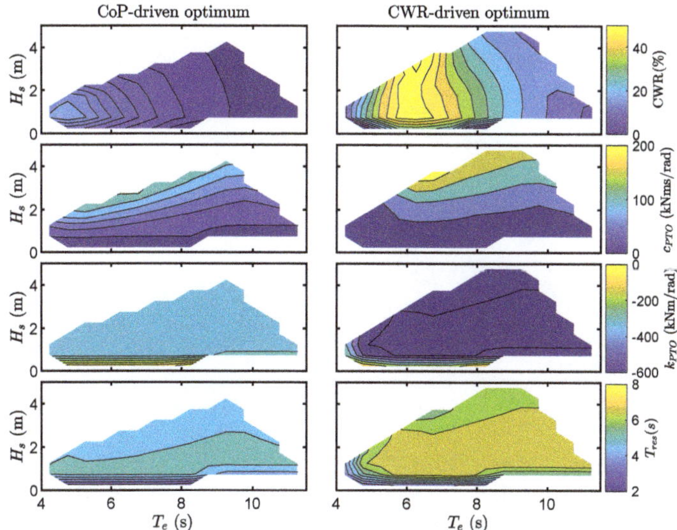

**Figure 17.** Capture width ratio, reactive control parameters and resonant period of the pendulum for the CoP-driven and CRW-driven optima. The same colour bar is used for each row, in order to favour comparison.

### 4.2. Techno-Economic Trends and Considerations

It is important to remark that Figures 14 and 15, although representative, are just three examples. Moreover, comments so far have no general or absolute validity, since many interrelated parameters comes into play. In order to investigate significant and more informative trends, Figures 18–21 show relevant results and design parameters for a selection of 28 devices, chosen in order to be representative of the fittest individuals in the whole population, according to the two evaluation metrics considered (CWR and CoP). In practice, the range of CWR from 0% to 30% is equally divided into 14 bins, and the individual with highest CWR of each bin is selected. Likewise, the range of CoP from 500 EUR/MWh to 1000 EUR/MWh is equally divided into 14 bins, and the individual with lowest CoP of each bin is selected. In order to highlight the most convenient individual, according to each metric, bars colour is proportional to CWR and 1/CoP, for the left and right graphs, respectively, so that best devices are represented by blue bars, while worst devices are in red.

The first row of Figure 18 highlights that high CWRs consistently result in high CoPs and vice versa, remarking that conversion efficiency and economic ability can be contrasting objective. It follows that the best devices in the CWR sense (blue bars in the left column) have high CoP. Conversely, the best devices in the CoP sense (blue bars in the right column) are economically performant in spite of their low CWR. In fact, the second and third rows of Figure 18 show that high AEPs come at high device costs and their ratio is more convenient at low AEP.

Figure 19 explores variations in the hull dimensions, which are the main drivers of hydrodynamic performance, and consequent mass, which is proportional to the hull cost. On the one hand, while CWR consistently increases with the hull length, longer hulls generate high CoP. Conversely, the device width is about constant for all CWR and CoP bins, with strong variations just around the second and third bins of the CWR and CoP scales, respectively. Finally, mainly as a consequence of length and width trends, the hull mass increases for increasing CWR and CoP. In particular, low CoP devices require lighter and cheaper hulls.

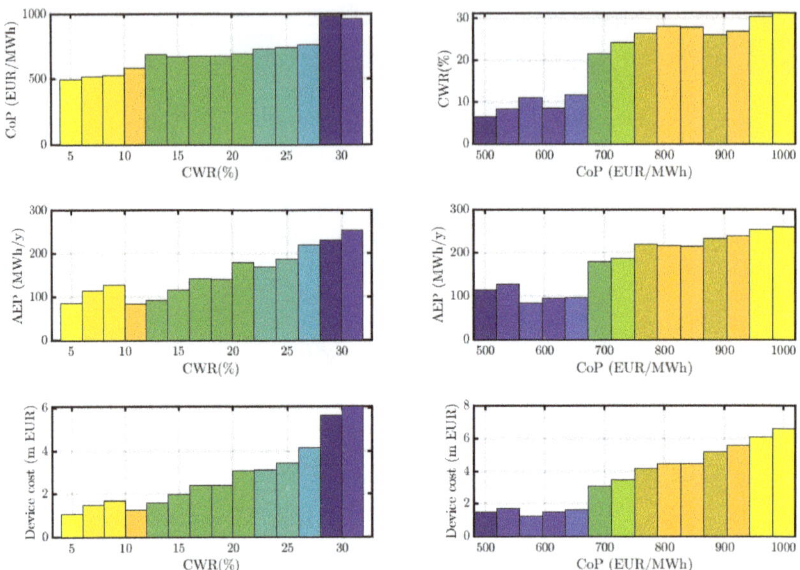

**Figure 18.** On the left: 14 individuals with the highest CWRs for each CWR-bin, with bars coloured proportionally to CWR. On the right: 14 individuals with the lowest CoPs for each CoP-bin, with bars coloured proportionally to 1/CoP.

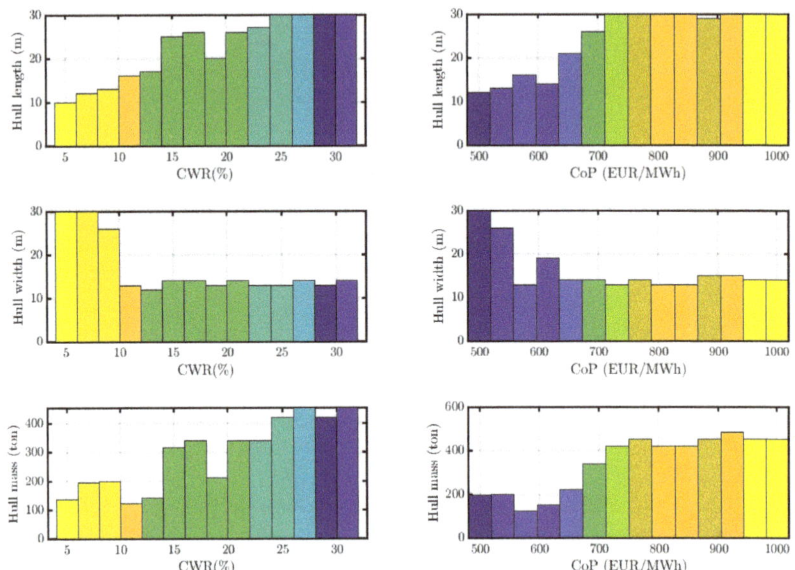

**Figure 19.** On the left: 14 individuals with highest CWR for each CWR-bin, with bars coloured proportionally to CWR. On the right: 14 individuals with lowest CoP for each CoP-bin, with bars coloured proportionally to 1/CoP.

Figure 20 shows the total pendulum mass, the geometry ratio and the number of pendulums. The cost of the unit is proportional to the total masses of pendulums and their numbers, since there is one PTO attached to each pendulum. Figure 20 shows that more and overall heavier pendulums increase the CWR but also the CoP, in a similar trend found for the hull dimension and mass, confirming that the incremental investment required to produce more energy is not cost-effective. However, since the pendulums mass and number of pendulums follow the same trend, the single pendulum has about a constant mass in all bins. The geometry ratio $\sigma_p$ defines how slender the pendulum is, considering the available space. Overall, although no strong correlation is found, it seems that once the masses and number of pendulums are defined, the algorithm leads to cylinders as long and slim as possible (high $\sigma_p$).

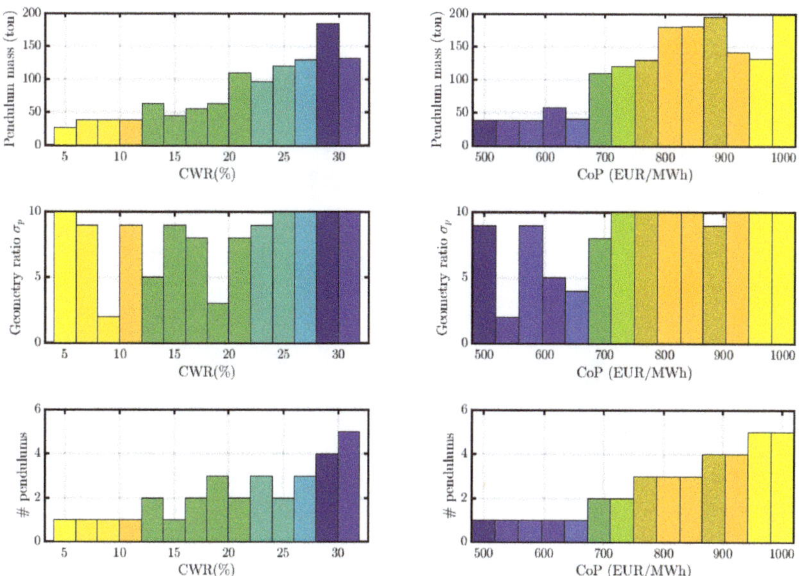

**Figure 20.** On the left: 14 individuals with the highest CWRs for each CWR-bin, with bars coloured proportionally to CWR. On the right: 14 individuals with the lowest CoPs for each CoP-bin, with bars coloured proportionally to 1/CoP.

Figure 21 shows that CWR and CoP are not particularly sensitive to the single-PTO parameters. In fact, the gearbox ratio is almost always the highest available, while the highest rated torque is often selected, despite the absence of a clear trend, especially in the CWR bins. Similarly, the rated torque is about equally split across both CWR and CoP bins, without any significant pattern. Consequently, the PTOs affect the performance mainly by means of their number, since there is an equal number of PTOs and pendulums: as the optimisation seeks more torque to increase the power conversion, since the largest PTO is already used, the number of pendulum units is increased. Likewise, it seems more cost effective to have fewer pendulums with the largest PTOs than more pendulums with small-medium PTOs.

Finally, Figure 22 shows the total and percentage costs, divided in the hull, pendulum and PTO subcomponents. On the one hand, the absolute cost of each subcomponent tends to increase with both CWR and CoP. On the other hand, the rates of increase seem to be different, since the hull tends to become less relevant in favour of the pendulum/PTO unit. Although the mass (and cost) of the hull increases (from 300 tons to 450 tons), as shown in Figure 19, the number (and cost) of pendulums/PTOs increases at a higher rate (from 1 to 5), as shown in Figure 20. Therefore, Figure 22 shows that the proportional increase of pendulum units has a higher impact than the increase of hull dimension and mass. Alternatively, it can be inferred that the optimisation algorithm reduces the number of pendulum units in order to reduce the CoP. Conversely, in order to increase the CWR, regardless of subcomponents cost, there must be a higher number of conversion units.

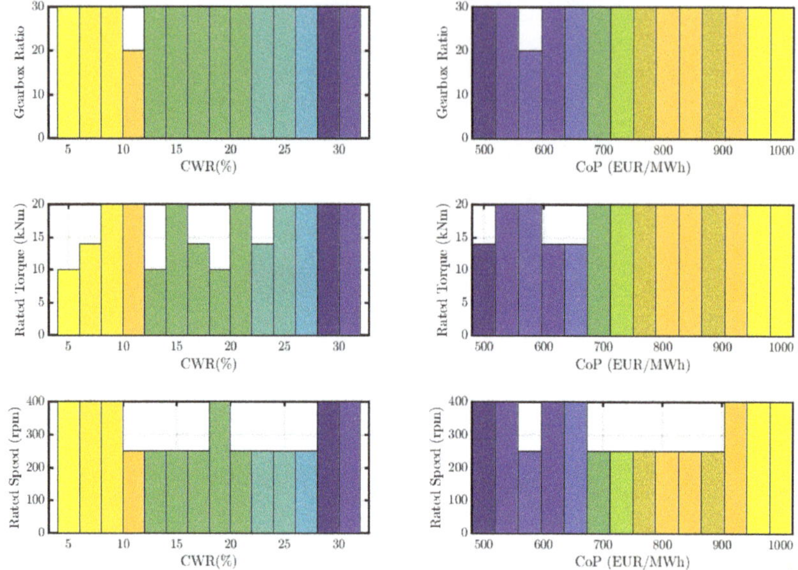

**Figure 21.** On the left: 14 individuals with the highest CWRs for each CWR-bin, with bars coloured proportionally to CWR. On the right: 14 individuals with lowest the CoPs for each CoP-bin, with bars coloured proportionally to 1/CoP.

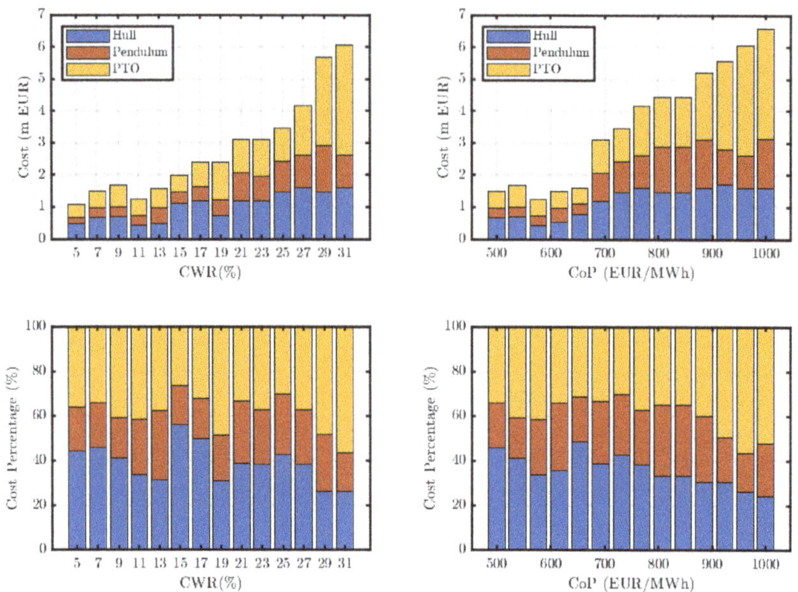

**Figure 22.** Hull, pendulum and PTO absolute and percentage cost. On the left: 14 individuals with the highest CWRs for each CWR-bin. On the right: 14 individuals with the lowest CoPs for each CoP-bin.

The results discussed so far highlight a limitation of the current concept, which is also an opportunity for substantial improvement. In particular, CWR and CoP are partially competing objectives, due to intrinsic limitations and assumptions. In all configurations, while the largest PTO and gearbox ratio possible are selected, the numbers of pendulum/PTO units are mainly selected in order to either maximise CWR or minimise CoP. Therefore, if the total cost of the PTO system could be decreased while maintaining the same conversion ability, both CWR and CoP would benefit. This can be achieved by connecting each PTO to more than one pendulum. However, since this solution implies requiring an even higher PTO torque, electrical generators become ill-suited. Conversely, hydraulic PTOs work better with high loads at low speed, so the gearbox could also be removed. Moreover, virtually, just one hydraulic PTO could be connected to all pendulums by means of an appropriate hydraulic system. Although the hydraulic PTO system introduces more power losses, the overall cost benefits could lead to a significantly lower CoP.

## 5. Conclusions

The present paper tackles a comprehensive techno-economic optimisation of a floating wave energy converter via genetic algorithm, considering a wide multi-variate design space, including shape and dimensions of the floater, and subcomponent configuration and characteristics. It is crucial to consider both power conversion ability and total cost of the device, taking physical constraints and control into account for a realistic assessment of the overall performance. Results herein presented are preliminary and not meant to have general validity since they are the consequence of specific techno-economic assumptions and hypotheses on the working principle and configuration. However, there is general validity in stating that the optimisation objective has a great influence on the optimised design, so it should be carefully chosen to be as close as possible to the actual intended target. In this paper it is shown that *efficient* conversion does not necessary imply *economical* conversion. On the contrary, for the specific case analysed, it resulted that increasing conversion efficiency required an excessive increase in costs, so that the most convenient solution in the economic sense had a relatively low conversion efficiency. Moreover, specific and detailed techno-economic considerations at an early design optimisation stage can lead to substantial conceptual modifications of the system, impossible at a mature development stage, which may be key for the success of such technology.

**Author Contributions:** Conceptualisation, L.F., S.A.S., G.G., M.B., G.C., G.B. and G.M.; methodology, L.F., S.A.S., G.G., M.B., G.C., G.B. and G.M.; software, L.F., S.A.S., G.G., M.B.; validation, L.F., S.A.S., G.G., M.B.; formal analysis, L.F., S.A.S., G.G., M.B.; investigation, L.F., S.A.S., G.G., M.B., and G.C.; resources, G.G., G.B. and G.M.; data curation, L.F., S.A.S.; writing—original draft preparation, L.F., S.A.S. and G.G.; writing—review and editing, L.F., S.A.S., G.G. and G.C.; visualisation, L.F., S.A.S., G.G. and G.C.; supervision, G.B. and G.M.; project administration, G.B. and G.M.; funding acquisition, G.G., G.B. and G.M. All authors have read and agreed to the published version of the manuscript.

**Funding:** This research has received funding from the Italian National Agency for New Technologies, Energy and Sustainable Economic Development (ENEA), under the project PTR-PTR_19_21_ENEA_PRG_7. Furthermore, this research was also partially funded by the European Research Executive Agency (REA) under the European Union's Horizon 2020 research and innovation programme under grant agreement No 832140.

**Acknowledgments:** Computational resources provided by hpc@polito (http://hpc.polito.it).



## Appendix A. Genetic Algorithm

Genetic algorithms find their best applications in complex and multi-dimensional optimisation problems, especially those with no analytical formulation of cost and constraint functions. The usual way to handle constraints is via penalty functions, worsening the fitness value in order to avoid unfeasible individuals to proceed in the optimisation. However, in approaches implementing multi-level [48] or dynamic [49] penalty functions, deciding on the penalty parameter is challenging. Moreover, since there is no direct analytical formulation of the constraints based on the input variables, specific constraint handling based on the knowledge of the mathematical structure (such as the cutting plane method, reduced gradient method or gradient projection method) is not applicable. On the other hand, since genetic algorithms are population based, pair-wise comparison tournament selection is possible [19]. In a tournament, no penalty parameter is required, but only a one-to-one comparison of the severity of the constraint violation. The design optimisation problem presents real variables, which can be handled more efficiently than binary variables, also avoiding the Hamming-Cliff effect [50]. Moreover, in order to improve the convergence rate, variables are defined in a discrete way through integers. The size of each step for each variable is decided by the engineer, based on common sense and previous experience. In this way, the number of possible combinations is reduced, and only significant variations of control parameters are allowed.

Based on previous consideration, the MI-LXPM genetic algorithm has been chosen; it is schematically shown in Figure A1, where:

- **MI**: mixed-integer [51];
- **LX**: Laplace crossover [52];
- **PM**: power mutation [53].

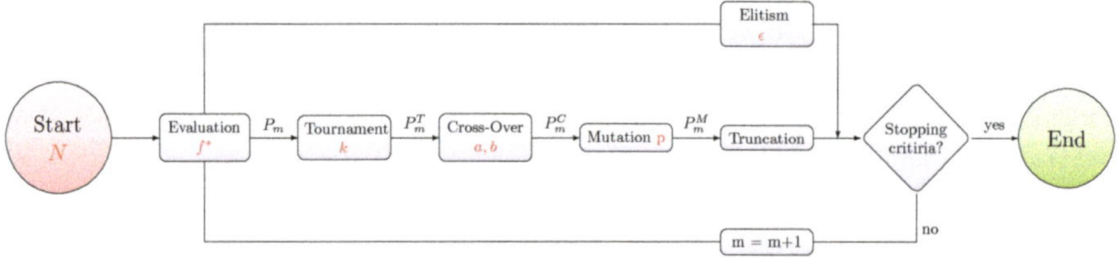


**Figure A1.** Flowchart of the MI-LXPM genetic algorithm. All phases are described in detail in Appendixes A.1–A.6. Major tuning parameters of the genetic algorithm are shown in red in the flow chart, and summarised in Table 4.

*Appendix A.1. Optimisation Problem*

The optimisation problem under study can be framed as the minimisation of a generic scalar function $f(x)$ of a vector $x$, subject to various scalar constraints $g_i(x)$. Note that functions $f$ and $g_i$ present a notional dependence on $x$, which does not necessary require an explicit mathematical formulation, since they can be evaluated via numerical simulation. Furthermore, each element $x(j)$ of the input vector $x$ is bounded, with $x^L(j)$ being the lower bound and $x^L(j)$ being the upper bound. Finally, note that all elements of $x$ have discrete variations, defined as integer times appropriate steps. Therefore, the optimisation problem can be presented as follows:

$$\min_{x} \quad f(x)$$
$$\text{subject to} \quad g_i(x) \geq 0 \qquad \forall i, \tag{A1}$$
$$x^L(j) \leq x(j) \leq x^U(j) \quad \forall j.$$

Constraints are enforced by applying a parameter-free penalty to the objective function when there is a constraint violation. As a condition, constraint functions $g_i(x)$ should also express the severity of the constraint violation, so that ranking between unfeasible position is possible. Therefore, the following fitness index $(f^*)$ is defined:

$$f^*(x) = \begin{cases} f(x) & \text{if } g_i(x) \geq 0 \ \forall i, \\ f_{\text{death}} + \frac{\Delta_C}{C} & \text{otherwise} \end{cases} \tag{A2}$$

where $f_{\text{death}}$ is a death penalty and $\frac{\Delta_c}{C}$ represents the distance from of the unfeasible point to the violated constraint, normalised by the constraint itself, as shown in Figure A2. Note that, if $f$ is bounded, $f_{\text{death}}$ should be higher than the upper boundary. If $f$ has no upper-boundary, then $f_{\text{death}}$ can be iteratively updated with the maximum $f$ until the current generation. Such a formulation, advantageous in handling feasibility and high mortality rates, requires and is made possible by a tournament-type selection of the fittest individuals, as explained in Appendix A.3.

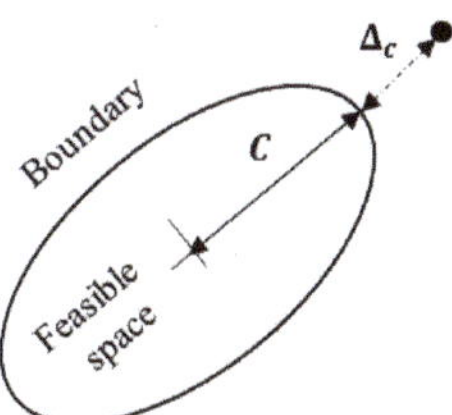


**Figure A2.** Schematic representation of the constraint violation handling, with reference to (A2).

*Appendix A.2. Initial Population and Stopping Criteria*

The algorithm starts from an initial population of $N$ individuals, randomly chosen in the design space. The success and computational requirements of the optimisation depend, among other things, on $N$, which should be large enough to ensure diversification of the genetic code, but not excessively large, which would demand longer simulation time tp progress through generations.

In each iteration $m$ of the algorithm, with $m = 1$ at the start, the population $P_m$ goes through a tournament stage, a crossover stage, a mutation stage, and a truncation stage, described in Appendixes A.3–A.6, respectively. In this process, the population is appropriately modified in order to eventually reach a global optimum. However, according to the elitism principle, at each iteration there is a small percentage ($\epsilon$) of the current population which by-passes the mixing-scheme without modifying its genetic information. Such an elite group is formed by the best individuals, whose genetic information should be passed down to the next generation.

The algorithm iterates until one of the two following stopping criteria is met:

(a)   The relative variation of the best fitness index ($f^*_{\min}$) over the previous $M_\Delta$ generations is less than a threshold tolerance $\Delta$;

(b)   The generation count $m$ reaches the allowed limit ($M$).

Satisfying criterion (a) represents a successful convergence to a minimum. If the population is large enough and with appropriate genetic algorithm parameters (see following sections), such a minimum can be confidently assumed to be also the absolute minimum within given boundaries. On the contrary, if the exit condition is (b), either $\Delta$ is overly strict, or the algorithm did not converge, so the tuning parameters of the algorithm should be adjusted.

*Appendix A.3. Tournament*

The tournament stage implements the survival-of-the-fittest principle by creating an appropriate mating pool $P_m^T$, where $P_m^T$ stands for the m-population $P_m$ after the tournament phase ($T$). In each tournament, there are $k$ individuals competing, where $k$ is the tournament size. The winner of each tournament is copied in the mating pool and can continue competing. Every individual takes part to exactly $k$ tournaments so that $(1 - \epsilon)N$ tournaments are played, and in turn, the mating pool is finally populated by $(1 - \epsilon)N$ individuals. It is worth noting that the best-overall individual wins all matches, hence sending $k$ copies of itself into the mating pool and ensuring higher chances to pass their genes over to the next generation. Conversely, the worst-overall individual loses all matches and disappear from the population. In each tournament, the $k$ individuals are compared in pairs until the one that wins all comparisons is found.. If both individuals are feasible, the one with lowest fitness value wins. If only one is feasible, it survives. If both individuals are unfeasible, the one with lowest constraint violation wins.

*Appendix A.4. Laplace Crossover*

The crossover phase aims at appropriately mixing genetic information of two parents, randomly taken within the mating pool $P_m^T$, hence generating a set of off-springs, two from each parents, in order to preserve the total number of individuals. The population of off-springs is called $P_m^C$, which refers to the m-population $P_m$ after crossover (C). The mixing of information should be a compromise between the need to preserve optimal characteristics of the parents and introducing variability in the genetic code.

Let us consider two arbitrary individuals $n_1$ and $n_2$ from the m-population mating pool, namely $(x_{n_1}^T, x_{n_2}^T) \in P_m^T$. Two off-springs are generated, namely $(x_{n_1}^C, x_{n_2}^C) \in P_m^C$, in the following manner:

$$x_n^C(j) = x_n^T(j) + \beta_j \left| x_{n_1}^T(j) - x_{n_2}^T(j) \right| \ , \ \forall j \ \text{and} \ n \in [n_1, n_2] \tag{A3}$$

where $\beta_j$ satisfies the Laplace distribution thanks to the following formulation:

$$\beta_j = \begin{cases} a - b \log(u_j), & \text{if } r_j \leq \frac{1}{2} \\ a + b \log(u_j), & \text{if } r_j > \frac{1}{2} \end{cases} \tag{A4}$$

with $u_j$ and $r_j$ uniform random numbers between 0 and 1, $a$ the location parameter (usually set to 0) and $b > 0$ the scaling parameter. From (A3) and (A4), it is clear that the scaling parameter determines how far the off-springs fall from the parents, hence affecting the convergence rate and success.

*Appendix A.5. Power Mutation*

Once off-springs are generated, it is necessary to add random variations in the population in order to appropriately investigate the whole design space and avoid the risk of convergence to a local but not global minimum. This is the objective of the mutation phase, generating the $P_m^M$ population, referring to the m-population $P_m$ after mutation (M). A power-distribution mutation scheme is adopted, creating a new individual $x_n^M$ in the proximity of $x_n^C$ as follows:

$$x_n^M(j) = \begin{cases} x_n^C(j) - s_j^p \left( x_n^C(j) - x^L(j) \right), & \text{if } t_j < r_j \\ x_n^C(j) + s_j^p \left( x^U(j) - x_n^C(j) \right), & \text{if } t_j \geq r_j \end{cases} \ , \ \forall j \tag{A5}$$

with

$$t_j = \frac{x_n^C(j) - x^L(j)}{x^U(j) - x_n^C(j)} \tag{A6}$$

and $s_j$ and $r_j$ uniform random numbers between 0 and 1, and $p$ is the index of mutation, determining the strength of the perturbation induced by the power mutation. It follows that, if $p < 1$, $s_j^p$ is statistically closer to 1, amplifying the mutation; conversely, if $p > 1$, $s_j^p$ is statistically closer to 0, attenuating the mutation [51].

*Appendix A.6. Truncation*

In general, after crossover and mutation, design variables are real values different from the discrete set in the problem definition, shown in (A1). Therefore, for each individual, each element is randomly approximated to either the discretised input variable right below or above, with equal probability. Such an approach ensures to keep the optimisation problem discrete, while adding further randomness in the process.

## References


1. Gunn, K.; Stock-Williams, C. Quantifying the global wave power resource. *Renew. Energy* **2012**, *44*, 296–304. [CrossRef]
2. Weber, J. WEC Technology Readiness and Performance Matrix—Finding the best research technology development trajectory. In Proceedings of the 4th International Conference on Ocean Energy ( ICOE), Dublin, Ireland, 17–19 October 2012; pp. 1–10.
3. Weber, J.; Roberts, J. Cost, time and risk assessment of different wave energy converter technology development trajectories. In Proceedings of the Twelfth European Wave and Tidal Energy Conference, Cork, Ireland, 27 August–1 September 2017.
4. Penalba, M.; Ringwood, J.V. A high-fidelity wave-to-wire model for wave energy converters. *Renew. Energy* **2019**, *134*, 367–378. [CrossRef]
5. Genuardi, L.; Bracco, G.; Sirigu, S.A.; Bonfanti, M.; Paduano, B.; Dafnakis, P.; Mattiazzo, G. An application of model predictive control logic to inertial sea wave energy converter. In *Mechanisms and Machine Science*; Springer Nature: Heidelberg, Germany, 2019; Volume 73, pp. 3561–3571. [CrossRef]
6. Bracco, G.; Casassa, M.; Giorcelli, E.; Giorgi, G.; Martini, M.; Mattiazzo, G.; Passione, B.; Raffero, M.; Vissio, G. Application of sub-optimal control techniques to a gyroscopic Wave Energy Converter. *Renew. Energ. Offshore* **2014**, *1*, 265–269.
7. Sirigu, A.S.; Gallizio, F.; Giorgi, G.; Bonfanti, M.; Bracco, G.; Mattiazzo, G. Numerical and Experimental Identification of the Aerodynamic Power Losses of the ISWEC. *J. Mar. Sci. Eng.* **2020**, *8*, 49. [CrossRef]
8. de Andres, A.; Maillet, J.; Todalshaug, J.H.; Möller, P.; Bould, D.; Jeffrey, H. Techno-economic related metrics for a wave energy converters feasibility assessment. *Sustainability* **2016**, *8*, 1109. [CrossRef]
9. Farrell, N.; Donoghue, C.; Morrissey, K. Quantifying the uncertainty of wave energy conversion device cost for policy appraisal: An Irish case study. *Energy Policy* **2015**, *78*, 62–77. [CrossRef]
10. Giorgi, G.; Ringwood, J.V. Analytical formulation of nonlinear Froude-Krylov forces for surging-heaving-pitching point absorbers. In Proceedings of the ASME 2018 37th International Conference on Ocean, Offshore and Arctic Engineering, Madrid, Spain, 17–22 June 2018.
11. Giorgi, G.; Ringwood, J.V. Articulating parametric nonlinearities in computationally efficient hydrodynamic models. In Proceedings of the 11th IFAC Conference on Control Applications in Marine Systems, Robotics, and Vehicles, Opatija, Croatia, 10–12 September 2018.
12. Ringwood, J.V.; Merigaud, A.; Faedo, N.; Fusco, F. Wave Energy Control Systems: Robustness Issues. In Proceedings of the IFAC Conference on Control Aplications in Marine Systems, Robotics, and Vehicles, Opatija, Croatia, 10–12 September 2018.
13. Ringwood, J.V.; Merigaud, A.; Faedo, N.; Fusco, F. An Analytical and Numerical Sensitivity and Robustness Analysis of Wave Energy Control Systems. *IEEE Trans. Control Syst. Technol.* **2019**, *28*, 1337–1348. [CrossRef]
14. Giorgi, G.; Ringwood, J.V. Parametric motion detection for an oscillating water column spar buoy. In Proceedings of the 3rd International Conference on Renewable Energies Offshore RENEW, Lisbon, Portugal, 8–10 October 2018.
15. Gilloteaux, J.C.; Ringwood, J.V. Control-informed geometric optimisation of wave energy converters. *IFAC Proc. Vol. (IFAC-PapersOnline)* **2010**, *43*, 366–371. [CrossRef]
16. Garcia-Rosa, P.; Bacelli, G.; Ringwood, J. Control-Informed Geometric Optimization of Wave Energy Converters: The Impact of Device Motion and Force Constraints. *Energies* **2015**, *8*, 13672–13687. [CrossRef]
17. Giorgi, G.; Ringwood, J.V. Articulating parametric resonance for an OWC spar buoy in regular and irregular waves. *J. Ocean Eng. Mar. Energy* **2018**, *4*, 311–322. [CrossRef]
18. Novo, R.; Bracco, G.; Sirigu, S.; Mattiazzo, G.; Merigaud, A.; Ringwood, J. Non-linear simulation of a wave energy converter with multiple degrees of freedom using a harmonic balance method. In Proceedings of the International Conference on Offshore Mechanics and Arctic Engineering (OMAE), Madrid, Spain, 17–22 June 2018; Volume 10. [CrossRef]
19. Deb, K. An efficient constraint handling method for genetic algorithms. *Comput. Methods Appl. Mech. Eng.* **2000**, *186*, 311–338. [CrossRef]
20. Göteman, M.; Giassi, M.; Engström, J.; Isberg, J. Advances and Challenges in Wave Energy Park Optimization— A Review. *Front. Energy Res.* **2020**, *8*, 26. [CrossRef]
21. Birk, L. Application of constrained multi-objective optimization to the design of offshore structure hulls. *J. Offshore Mech. Arct. Eng.* **2009**, *131*, 011301. [CrossRef]

22. Sharp, C.; DuPont, B. Wave energy converter array optimization: A genetic algorithm approach and minimum separation distance study. *Ocean Eng.* **2018**, *163*, 148–156. [CrossRef]

23. Giassi, M.; Göteman, M. Layout design of wave energy parks by a genetic algorithm. *Ocean Eng.* **2018**, *154*, 252–261. [CrossRef]

24. Shadman, M.; Estefen, S.F.; Rodriguez, C.A.; Nogueira, I.C. A geometrical optimization method applied to a heaving point absorber wave energy converter. *Renew. Energy* **2018**, *115*, 533–546. [CrossRef]

25. Kurniawan, A.; Moan, T. Optimal geometries for wave absorbers oscillating about a fixed axis. *IEEE J. Ocean. Eng.* **2013**, *38*, 117–130. [CrossRef]

26. McCabe, A.P. Constrained optimization of the shape of a wave energy collector by genetic algorithm. *Renew. Energy* **2013**, *51*, 274–284. [CrossRef]

27. Cordonnier, J.; Gorintin, F.; De Cagny, A.; Clément, A.H.; Babarit, A. SEAREV: Case study of the development of a wave energy converter. *Renew. Energy* **2015**, *80*, 40–52. [CrossRef]

28. Pozzi, N.; Bonetto, A.; Bonfanti, M.; Bracco, G.; Dafnakis, P.; Giorcelli, E.; Passione, B.; Sirigu, S.; Mattiazzo, G. PeWEC: Preliminary design of a full-scale plant for the mediterranean sea. In Proceedings of the NAV International Conference on Ship and Shipping Research, Trieste, Italy, 20–22 June 2018; pp. 504–514. [CrossRef]

29. Ruellan, M.; Benahmed, H.; Multon, B.; Josset, C.; Babarit, A.; Clement, A. Design methodology for a SEAREV wave energy converter. *IEEE Trans. Energy Convers.* **2010**, *25*, 760–767. [CrossRef]

30. Sirigu, S.A.; Bonfanti, M.; Begovic, E.; Bertorello, C.; Dafnakis, P.; Bracco, G.; Mattiazzo, G. Experimental Investigation of Mooring System on a Wave Energy Converter in Operating and Extreme Wave Conditions. *J. Mar. Sci. Eng.* **2020**, *8*, 180. [CrossRef]

31. Pozzi, N.; Bracco, G.; Passione, B.; Sirigu Sergej, A.; Vissio, G.; Mattiazzo, G.; Sannino, G. Wave Tank Testing of a Pendulum Wave Energy Converter 1:12 Scale Model. *Int. J. Appl. Mech.* **2017**, *9*, 1750024. [CrossRef]

32. Sirigu, S.A.; Bonfanti, M.; Passione, B.; Begovic, E.; Bertorello, C.; Dafnakis, P.; Bracco, G.; Giorcelli, E.; Mattiazzo, G. Experimental investigation of the hydrodynamic performance of the ISWEC 1:20 scaled device. In Proceedings of the NAV International Conference on Ship and Shipping Research, Trieste, Italy, 20–22 June 2018; pp. 551–560. [CrossRef]

33. Pozzi, N. Numerical Modeling and Experimental Testing of a Pendulum Wave Energy Converter (PeWEC). Ph.D. Thesis, Polytechnic of Turin, Torino, Italy, 2018. [CrossRef]

34. Sirigu, S.; Bonfanti, M.; Dafnakis, P.; Bracco, G.; Mattiazzo, G.; Brizzolara, S. Pitch Resonance Tuning Tanks: A novel technology for more efficient wave energy harvesting. In Proceedings of the OCEANS 2018 MTS/IEEE Charleston (OCEAN 2018), Charleston, SC, USA, 22–25 October 2019. [CrossRef]

35. Bonfanti, M.; Bracco, G.; Dafnakis, P.; Giorcelli, E.; Passione, B.; Pozzi, N.; Sirigu, S.; Mattiazzo, G. Application of a passive control technique to the ISWEC: Experimental tests on a 1:8 test rig. In Proceedings of the NAV International Conference on Ship and Shipping Research, Trieste, Italy, 20–22 June 2018; pp. 60–70. [CrossRef]

36. Sirigu, S.A.; Bracco, G.; Bonfanti, M.; Dafnakis, P.; Mattiazzo, G. On-board sea state estimation method validation based on measured floater motion. *IFAC-PapersOnLine* **2018**, *51*, 68–73. [CrossRef]

37. Giorgi, G.; Gomes, R.P.F.; Bracco, G.; Mattiazzo, G. The effect of mooring line parameters in inducing parametric resonance on the Spar-buoy oscillating water column wave energy converter. *J. Mar. Sci. Eng.* **2020**, *8*, 29. [CrossRef]

38. Babarit, A.; Delhommeau, G. Theoretical and numerical aspects of the open source BEM solver NEMOH. In Proceedings of the 11th European Wave and Tidal Energy Conference, Nantes, France, 6–11 September 2015; pp. 1–12.

39. Pozzi, N.; Bracco, G.; Passione, B.; Sirigu, S.A.; Mattiazzo, G. PeWEC: Experimental validation of wave to PTO numerical model. *Ocean Eng.* **2018**, *167*, 114–129. [CrossRef]

40. Fontana, M.; Casalone, P.; Sirigu, S.A.; Giorgi, G. Viscous Damping Identification for a Wave Energy Converter Using CFD-URANS Simulations. *J. Mar. Sci. Eng.* **2020**, *8*, 355. [CrossRef]

41. Penalba, M.; Giorgi, G.; Ringwood, J.V. Mathematical modelling of wave energy converters: A review of nonlinear approaches. *Renew. Sustain. Energy Rev.* **2017**, *78*, 1188–1207. [CrossRef]

42. Giorgi, G.; Gomes, R.P.F.; Bracco, G.; Mattiazzo, G. Numerical investigation of parametric resonance due to hydrodynamic coupling in a realistic wave energy converter. *Nonlinear Dyn.* **2020**. [CrossRef]

43. Lagarias, J.C.; Reeds, J.A.; Wright, M.H.; Wright, P.E. Convergence properties of the nelder–mead simplex method in low dimensions. *SIAM J. Optim.* **1998**, *9*, 112–147. [CrossRef]

44. Garcia-Teruel, A.; Forehand, D.I.; Jeffrey, H. Wave Energy Converter hull design for manufacturability and reduced LCOE. In Proceedings of the 7th International Conference on Ocean Energy (ICOE), Cherbourg, Normandy, France, 12–13 June 2018; pp. 1–9.
45. Emmerich, M.T.; Deutz, A.H. A tutorial on multiobjective optimization: Fundamentals and evolutionary methods. *Nat. Comput.* **2018**, *17*, 585–609. [CrossRef]
46. Astariz, S.; Iglesias, G. Enhancing Wave Energy Competitiveness through Co-Located Wind and Wave Energy Farms. A Review on the Shadow Effect. *Energies* **2015**, *8*, 7344–7366. [CrossRef]
47. Miettinen, K. *Nonlinear Multiobjective Optimization*, 1st ed.; Springer: Boston, MA, USA, 1998; p. 298. [CrossRef]
48. Homaifar, A.; Qi, C.X.; Lai, S.H. Constrained Optimization Via Genetic Algorithms. *Simulation* **1994**, *62*, 242–253. [CrossRef]
49. Joines, J.A.; Houck, C.R. On the use of non-stationary penalty functions to solve nonlinear constrained optimization problems with GA's. In Proceedings of the First IEEE Conference on Evolutionary Computation (IEEE World Congress on Computational Intelligence), Orlando, FL, USA, 27–29 June 1994; pp. 579–584. [CrossRef]
50. Cheung, B.K.; Langevin, A.; Delmaire, H. Coupling Genetic Algorithm with a grid search method to solve Mixed Integer Nonlinear Programming problems. *Comput. Math. Appl.* **1997**, *34*, 13–23. [CrossRef]
51. Deep, K.; Singh, K.P.; Kansal, M.L.; Mohan, C. A real coded genetic algorithm for solving integer and mixed integer optimization problems. *Appl. Math. Comput.* **2009**, *212*, 505–518. [CrossRef]
52. Deep, K.; Thakur, M. A new crossover operator for real coded genetic algorithms. *Appl. Math. Comput.* **2007**, *188*, 895–911. [CrossRef]
53. Deep, K.; Thakur, M. A new mutation operator for real coded genetic algorithms. *Appl. Math. Comput.* **2007**, *193*, 211–230. [CrossRef]

Journal of
*Marine Science
and Engineering*

*Article*

# Comparison of Wave Energy Park Layouts by Experimental and Numerical Methods

**Marianna Giassi *, Jens Engström, Jan Isberg and Malin Göteman**

Department of Electrical Engineering, Uppsala University, Ångströmlaboratoriet, Box 65,
75103 Uppsala, Sweden; jens.engstrom@angstrom.uu.se (J.E.); jan.isberg@angstrom.uu.se (J.I.);
malin.goteman@angstrom.uu.se (M.G.)
* Correspondence: marianna.giassi@angstrom.uu.se

Received: 11 September 2020; Accepted: 23 September 2020; Published: 26 September 2020

**Abstract:** An experimental campaign of arrays with direct-driven wave energy converters of point-absorbing type is presented. The arrays consist of six identical floats, moving in six degrees of freedom, and six rotating power take-off systems, based on the design developed at Uppsala University. The goals of the work were to study and compare the performances of three different array layouts under several regular and irregular long-crested waves, and in addition, to determine whether the numerical predictions of the best performing array layouts were confirmed by experimental data. The simulations were executed with a frequency domain model restricted to heave, which is a computationally fast approach that was merged into a genetic algorithm optimization routine and used to find optimal array configurations. The results show that good agreement between experiments and simulations is achieved when the test conditions do not induce phenomena of parametric resonance, slack line and wave breaking. Specific non-linear dynamics or extensive sway motion are not captured by the used model, and in such cases the simulation predictions are not accurate, but can nevertheless be used to get an estimate of the power output.

**Keywords:** wave energy; arrays; experiments; wave tank; scale test; point absorber; array optimization

## 1. Introduction

Within ocean waves, an incredibly huge amount of energy is stored which can potentially be extracted and transformed into electricity. An efficient way to produce power in the MW range and to reduce the power fluctuations and the overall costs, is to construct and deploy parks of wave energy converters (WECs). As wave energy technologies are now facing the challenge of early stage commercialization, it is crucial to have access to accurate hydrodynamic and optimization simulation tools for such arrays of converters.

However, analytical and numerical modeling is always connected with some approximations and uncertainties, and physical experiments are required to validate the numerical results. Experiments of wave energy arrays are both expensive and complex to carry out, and it is only in recent years that several large-scale physical experiments have been conducted and their results published [1]. Nevertheless, experimental data are required to validate the numerical models, allowing for an understanding of the levels of uncertainty of the tools commonly used in the design process. In particular, due to the high computational cost, optimization of WEC arrays is typically carried out using simplified models based on assumptions such as restricted degrees of freedom, linear equations that can be solved in the frequency domain, unidirectional waves, etc. There has been little published work comparing these models to the experimental data of arrays and evaluating their validity.

Only a few physical experiments on arrays of point-absorbing WECs have been carried out during the past few years. Konispoliatis and Mavrakos [2] experimentally studied three floating oscillating water column (OWC) devices and compared the results with a semi-analytical multiple scattering method. Experimental results from a three spar-buoy type OWC in scale 1:32 were presented in da Fonseca et al. [3] and compared to the case of an isolated device. The interaction factor in an array of OWC was measured by Magagna et al. [4]. Wolgamot et al. [5] studied trapped modes within an array of eight fixed truncated cylinders, and Ji et al. [6] observed near-trapping in an array of fixed cylinders in short-crested waves. da Fonseca et al. [3] studied a triangular array of three spar-buoy WECs in scale 1:32, and it was observed that the interaction factor was larger than 1 in wave climates with large energy periods. Five fixed OWCs deployed in two different layouts (chosen based on earlier layout optimization) were studied experimentally and numerically by Bosma et al. [7], both in regular and irregular waves: a maximal power increase of 12% was found between the non-optimal and the optimal layout. In Mackay et al. [8], an array consisting of four Wavebob models at scale 1:19 was studied experimentally and compared to a numerical time-domain model. Mercadé Ruiz et al. [9] presented results from numerical simulations and experiments for five WaveStar devices at scale 1:20. Mackay et al. [8] and Mercadé Ruiz et al. [9] studied one array layout, typically a staggered park geometry or WECs in line. On the other hand, Troch et al. [10], Stratigaki et al. [11,12] studied several array layouts with up to 25 heaving point-absorber buoys connected to friction-damping power take-off systems (PTO). In these studies it was found that the wave height was reduced by up to 18% due to park effects. Similarly, Child and Weywada [13] presented results from an analogous experimental set-up with different array layouts of 24 heaving buoys; it was found that the park could experience up to 26% energy loss due to park interactions. Experimental measurements of the responses of an array of five heaving floats in regular waves were compared to numerical predictions in Thomas et al. [14], and in irregular waves by Weller et al. [15]. Nader et al. [16] and Thomas et al. [17] performed array experiments of up to four point-absorber WECs moving in six degrees of freedom. The former experiment, performed at the Australian Maritime College, measured the interaction factor for 1 or 2 floats moving in heave and surge, by tracking the surface elevation with videogrammic measurements. The latter consisted of testing both a linear damping and an advanced control algorithm based on machine learning and artificial neural networks [18]. The energy absorption of three array layouts of WECs with linear damping was compared for six devices in Giassi et al. [19].

Uppsala University (UU) has developed a wave energy converter concept consisting of a buoy and a linear generator (LG) [20]. The buoy transmits the movement of the waves to the translator of the LG on the sea bed and is free to move in six degrees of freedom (DoF). Permanent magnets are mounted on the translator and, during the vertical movement of the translator, electricity is induced in the coils located in the stator, according to Faraday's law of induction.

This paper presents results from an experimental campaign of wave energy converter arrays. The work was carried out in the COAST Lab at Plymouth University, UK: three different arrays of six devices have been tested under several regular and irregular wave conditions. The model represents a 1:10 scaled prototype of an array of point absorbers, based on the WEC concept developed at Uppsala University. The experimental set-up consisted of six identical ellipsoidal floats free to move in six degrees of freedom and connected via ropes and pulleys to six power take-off systems (PTO) located on a gantry above the water. The PTO was performed by passive damping using a rotatory system appositely designed for the wave tank experiment, described in detail in [21]. To the knowledge of the authors, this experiment represents the first one on point absorbing WEC arrays moving in six degrees of freedom published so far.

The goals of the work are multiple: to provide a set of data for validation of the array model, to study the performances of different array layouts under several wave conditions and to understand whether the genetic algorithm's optimization tool developed for the specific UU WEC technology [22] can predict optimal layouts in realistic situations. In fact, if the objective function of the optimization is the maximization of the power output, different possible solutions (i.e., array layouts) must be ranked according to their hydrodynamic performance. This experiment intended to clarify whether the predictions of the simulations of the best performing array layouts would be confirmed by experimental data.

In [23] we published the results obtained with a single device in isolation. In [19] the physical set-up was described and some preliminary results in terms of measurements of the wave elevation in the tank and energy absorption for three different array configurations are presented. Here, we present results in terms of array power absorption for different sea states, and for each buoy within the arrays. Through the calculation of the interaction factor ($q$-factor), the results have been compared to the prediction of the model.

This paper is organized as follows: the numerical model used for the simulation is described in Section 2. Section 3 describes the physical experiments, including the set-up, the array layouts and the tests that were carried out. In Section 4 we describe the parameters used in the array simulations. In Section 5, results of the modeling and experiment are presented and discussed. Further conclusions are drawn in Section 6.

## 2. Theory and Numerical Model

*Mathematical Model*

A semi-analytical frequency domain model based on linear potential flow theory and implemented in MATLAB has been used to model the array of converters. We consider a park of $n_b$ WECs labelled by $i = 1, \ldots, n_b$. Each WEC consists of a float, which is a truncated cylinder with individual radius $R^i$ and draft $d^i$, and the generator is characterized by individual power take-off constants $\gamma^i$. The model's equation of motion is restricted to heave motion and unlimited stroke length is assumed. The float and the generator are assumed to be connected by a stiff line, so that one single coupled equation of motion can be written for the float-translator system

$$m\ddot{z}^i(t) = F_{\text{exc}}^i(t) + F_{\text{rad}}^i(t) + F_{\text{PTO}}^i(t) + F_{\text{stat}}^i(t) \tag{1}$$

where $m = m_b + m_t$ is the total mass of the float and the translator, $F_{\text{exc}}^i$ and $F_{\text{rad}}^i$ are the hydrodynamic excitation and radiation forces, respectively, $F_{\text{PTO}}^i(t) = -\gamma^i \dot{z}^i(t)$ is the power take-off force and $F_{\text{stat}}^i(t) = \rho g \pi R^{i2}(d^i - z^i(t))$ is the hydrostatic restoring force.

By Fourier transformation, the problem can be considered in the frequency domain, in which the equation of motion (1) takes the form

$$\left[ -\omega^2(m^i + m_{\text{add}}^i(\omega)) - i\omega(B^i(\omega) + \gamma^i) - \rho g \pi R^{i2} \right] z^i(\omega) = f_{\text{exc}}^i(\omega)\eta^i(\omega) \tag{2}$$

where the radiation force has been divided into added mass ($m_{add}$) and radiation damping ($B$) as $F_{\text{rad}}^i(\omega) = [\omega^2 m_{\text{add}}^i(\omega) + i\omega B^i(\omega)]z^i$ and $f_{\text{exc}}^i(\omega)$ and $\eta^i(\omega)$ are the frequency domain amplitudes of the excitation force and of the incident waves, respectively. The hydrodynamic forces are computed as surface integrals over the wetted surfaces of the buoys, $\bar{F} = i\omega\rho \iint_S \phi \, d\bar{S}$, where $\phi$ represents the fluid velocity potential that satisfies the Laplace equation in the fluid domain and which is a superposition of incident, scattered and radiated waves, $\phi = \phi_I + \phi_S + \phi_R$. To evaluate the velocity potential in the fluid domain, a semi-analytical hydrodynamic model based on the multiple scattering method has been used, as previously presented in [24,25].

After computation of the forces, the dynamics of each WEC are determined in the time domain by inverse Fourier transform of the solution to the equation of motion (2), and the instantaneous power absorption computed as

$$P_{\text{SIM}}^i(t) = \gamma \cdot [\dot{z}^i(t)]^2. \tag{3}$$

The power output of the park will be the sum of all $n_b$ converters:

$$P_{\text{SIM}}(t) = \sum_{i=1}^{n_b} P_{\text{SIM}}^i(t). \tag{4}$$

The frequency domain model has been coupled with an optimization model based on a genetic algorithm in previous works [22,26]. One of the goals of the present study was to verify the feasibility of optimizing parks of the Uppsala University WEC with such tools. The optimization routine aims to find the best array layout (i.e., set of coordinates of the WECs) which maximizes the power absorption. The genetic algorithm optimization was inspired by Charles Darwin's theory of evolution and iteratively follows these evolutionary steps: *natural selection*, *pairing*, *mating* or *crossover* and *mutation*. Each chromosome in the evolution represents a wave power park, and contains a number of genes equal to the number of WECs in the park. Each gene contains a couple of coordinates for each WEC. The optimization algorithm is described in detail in [22] and it was validated against a parameter sweep optimization for a single WEC, with excellent agreement.

## 3. Physical Experiments

### 3.1. Wave Tank

The experimental campaign was carried out in the COAST Lab at the University of Plymouth, UK. The Ocean basin is 35 m long and 15.5 m wide, with a moveable floor set to operate at a depth of 2.5 m. The waves are produced by 24 flap wave makers of 2.0 m hinge depth. At the end of the wave basin a parabolic beach provides energy dissipation and minimizes reflected waves.

A set of ten twin-wire resistance wave gauges have been used to measure the wave elevation at different locations around the basin. To avoid interference with the optical system for motion capture, all the probes are located on the right side of the tank (following the direction of the waves). Their position within the wave tank is shown in Figure 1a (G1 to G10), and was decided in order to enable measurement of the wave elevation at each line of the arrays, and the incoming and outgoing waves. The results of these data have been studied in [19]. On the basin floor floor, connection points for the WECs were installed in a 2 × 2 m grid (Figure 1b).

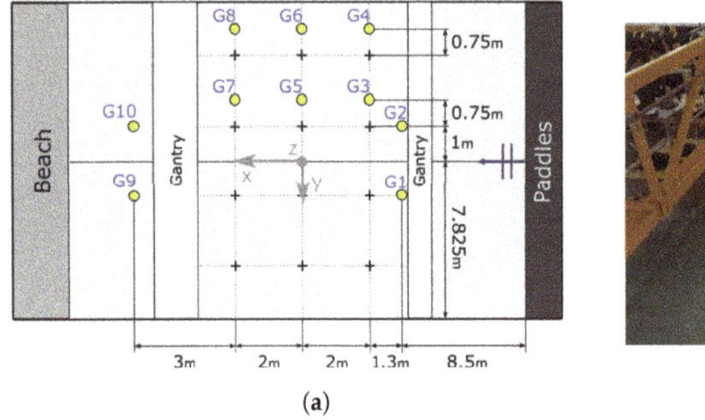

(a)

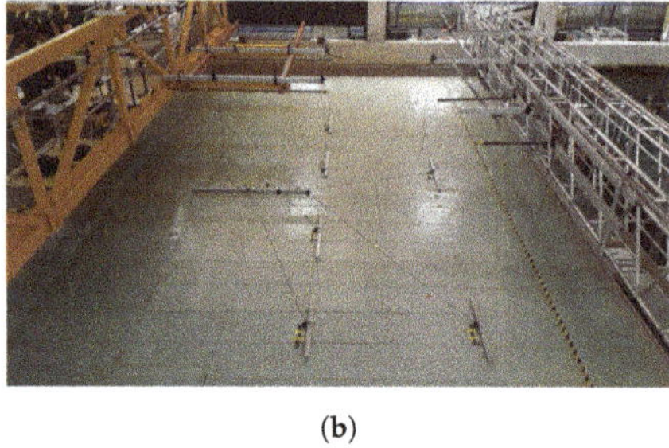
(b)

**Figure 1.** (**a**) Schematic top view of the COAST wave basin and location of the the attachment grid (+) with respect to the paddles and the beach. G1–G10 show the locations of the ten wave gauges installed during the experiment. (**b**) Picture of the attachment bars on the basin floor.

### 3.2. Experimental Set-Up

The set-up used for this experiment is based on the one introduced in [21,23]. The wave energy converter model tested is a 1:10 scaled prototype based on the point-absorber wave energy concept developed at Uppsala University [20]. In the full-scale device, a buoy is connected via a rope to a linear generator positioned on the sea floor; the motion of the waves induces the motion of the buoy and of the sub-surface generator, inducing electricity in the coils of the stator.

Differently from the work carried out in [23], six devices have been tested simultaneously, forming a park of wave energy converters. Six identical ellipsoidal floating buoys made of polyethylene (0.488 m in diameter and 0.280 m in height; total mass 4378 g) have been used, connected via a system of highly stiff ropes (100% Dyneema, 3 mm) and pulleys (Harken, 57 mm) to the power take-off systems located on the main gantry (Figure 2). Each float was equipped with five reflective markers for motion capture, which were tracked by a set of eight Qualisys six DoF cameras.

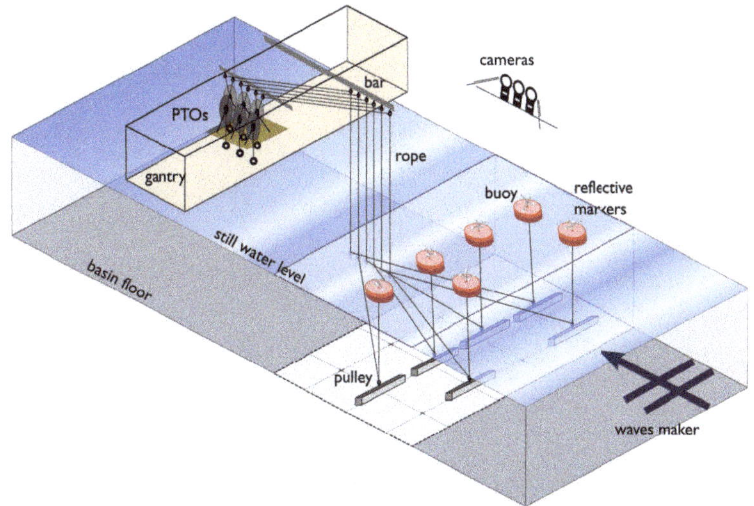


**Figure 2.** Sketch of the set-up's components.

The PTO is a rotatory system consisting of a table securely holding a bicycle truing stand (Tacx T3175) with a bicycle front hub (Shimano XT HB-M756) and a vice (Clarke CSV4E 100mm). On one external side of the bicycle hub, which has a low friction ball bearing, an aluminum disc (2 mm thickness × 40 mm diameter) was mounted, and on the internal part two ropes were wound in opposite directions. One line went towards the upper part through the pulley system to the float; the other line went downwards and held a 5 kg mass. On the vice's moveable clamps a set of magnets (5 × Neodymium N42 20 × 20 × 10 mm) were positioned. The damping $\gamma$ of the PTO was changed by changing the air gap between the magnets and the aluminum disc, which changed the magnetic flux inside the disc. An inertia measurement unit (IMU, Intel Curie SoC) positioned on the disc recorded its rotational speed $\omega_d^i$ and accelerations in the disc plane. The experimental power output of each converter was then obtained as

$$P_{\mathrm{PTO}}^i(t) = \gamma \cdot [r\omega_d^i(t)]^2, \tag{5}$$

where $\omega_d^i(t)$ is the angular velocity of the i-*th* disc and $r$ is the distance between the IMU and the center of the disc. Different damping values were achieved by varying the magnetic flux inside the disc, i.e., by manually changing the distance between the magnets and the aluminum disc. The damping was evaluated by measuring the achieved speed at steady-state with a known mass. In these experiments, a constant damping value of 306 Ns/m for all the converters was used. A more detailed discussion on the PTO and its damping can be found in [21].

### 3.3. Tests

Table 1 reports the characteristics of the sea states used in the experiment, together with the duration of each test (100 wave periods for regular waves tests and 20 min full scale for irregular wave tests). Waves were long-crested (unidirectional) and propagating along the positive $x$-axis direction in Figure 1a.

Five regular and five irregular sea states were studied for each layout, defined such that they had the same energy per ocean area and energy transport per meter wave crest. The significant wave height was $H_s = 0.175$ m (scaled value), and the wave (energy) periods spanned the range 1.11–2.37 s. The time series of the irregular waves were obtained from the Bretschneider spectrum.

**Table 1.** Sea states tested in the wave tank. Top: regular waves, bottom: irregular waves. Scaled values (1:10) (left) and full scale values (right).

| **Regular Waves** | | | | | | |
|---|---|---|---|---|---|---|
| | Scale 1:10 | | | Full scale | | |
| | $H_{rms}$ | $T$ | Run time | $H_{rms}$ | $T$ | Run time |
| Wave ID | [m] | [s] | [min] | [m] | [s] | [min] |
| RW1 | 0.124 | 1.11 | 1.8 | 1.24 | 3.5 | 5.8 |
| RW2 | 0.124 | 1.42 | 2.4 | 1.24 | 4.5 | 7.5 |
| RW3 | 0.124 | 1.74 | 2.9 | 1.24 | 5.5 | 9.2 |
| RW4 | 0.124 | 2.06 | 3.4 | 1.24 | 6.5 | 10.8 |
| RW5 | 0.124 | 2.37 | 4.0 | 1.24 | 7.5 | 12.5 |

| **Irregular Waves** | | | | | | |
|---|---|---|---|---|---|---|
| | Scale 1:10 | | | Full scale | | |
| | $H_s$ | $T_e$ | Run time | $H_s$ | $T_e$ | Run time |
| Wave ID | [m] | [s] | [min] | [m] | [s] | [min] |
| IW1 | 0.175 | 1.11 | 6 | 1.75 | 3.5 | 20 |
| IW2 | 0.175 | 1.42 | 6 | 1.75 | 4.5 | 20 |
| IW3 | 0.175 | 1.74 | 6 | 1.75 | 5.5 | 20 |
| IW4 | 0.175 | 2.06 | 6 | 1.75 | 6.5 | 20 |
| IW5 | 0.175 | 2.37 | 6 | 1.75 | 7.5 | 20 |

*3.4. Array Layouts*

The basin floor was prepared to test three different array layouts, under the ten wave conditions specified in Table 1. All the array configurations were obtained by positioning the floats over a fixed grid of $2 \times 2$ m wide cells, over a total area of $4 \times 6$ m$^2$. The locations of the buoys within the three layouts are shown in Figures 3–5, and referred to as array 1 (A1), array 2 (A2) and array 3 (A3), respectively. Each buoy, and corresponding PTO, is identified by a letter (A, B, C, D, E, F).

**Figure 3.** Layout A1.

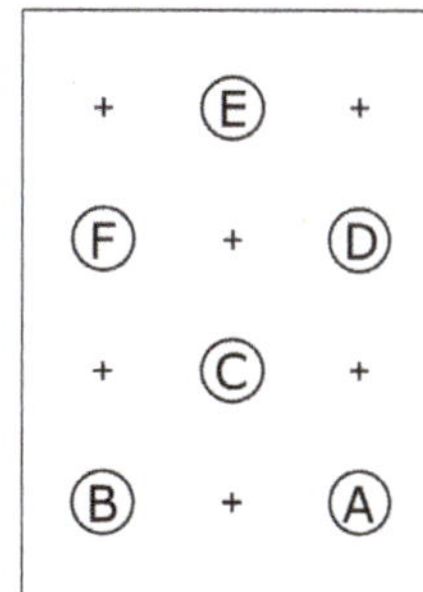



**Figure 4.** Layout A2.

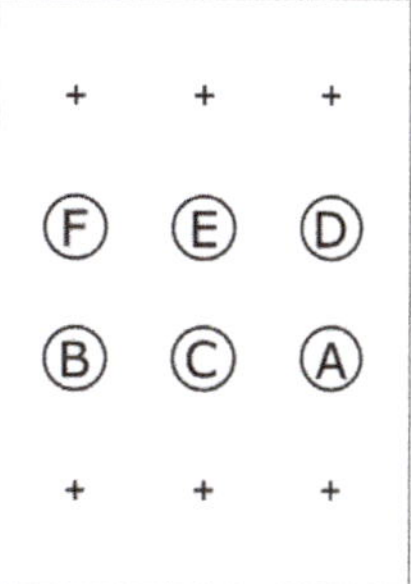



**Figure 5.** Layout A3.

A1 consisted of a two-line array with a complete line on the back and a line filled every second spot in the front; A2 consisted of a classical three-line staggered configuration; and A3 was a three-line aligned array. Each array's upstream line was located 9.8 m from the wave makers. Layout A1 has been tested twice on two different days, while layout A2 and A3 have been tested once.

## 4. Simulations

As in [23], the float's geometry has been simulated in two ways: as identical equivalent cylinders of 0.244 m radius and 0.0501 m draft, which correspond to the same submerged volume and radius as the physical elliptical buoy, and as identical equivalent cylinders of 0.178 m radius and 0.0948 m draft, which correspond to the same submerged volume and draft of the physical elliptical buoys. The results corresponding to these two geometries are referred to as frequency domain simulations R1 and R2, respectively. This is due to a restriction of the semi-analytical hydrodynamical model, which can only handle cylinder-shaped floats or cylinders with moon-pool. In the equation of motion an extra rotational inertia of the PTO of 48 kg was added, together with an additional damping of 63.2 Ns/m to account for the friction of the ropes and pulley movement inside the water [21,23].

## 5. Results and Discussion

In order to analyze the motion data acquired with the Qualisys optical system and understand the results shown in the upcoming section, the authors would like to make the following clarification. Let us consider a cylindrical buoy $i$ moving in three translational degrees of freedom (neglecting the rotational motion for simplicity), $\left[x^i(t), y^i(t), z^i(t)\right]$ (surge, sway and heave, respectively) and the translator moving only vertically, $\zeta^i(t)$. From the geometry of the problem (Figure 6), the vertical position of the translator is given in terms of the buoy position as

$$\zeta^i \approx \Delta L^i = \sqrt{\left(x^{i2} + y^{i2} + (L + z^{i2})\right)} - L \approx z^i + \frac{1}{2L}(x^{i2} + y^{i2}) \tag{6}$$

where $L$ is the length of the line connecting the translator and buoy; $\Delta L^i$ is the vertical PTO position variation; and the approximation holds up to second order in the small parameter $\varepsilon = \sqrt{x^{i2} + y^{i2}}/L$. Here, the case of a slack line is neglected. This approach has been used to estimate the translator position from the float's motion data.

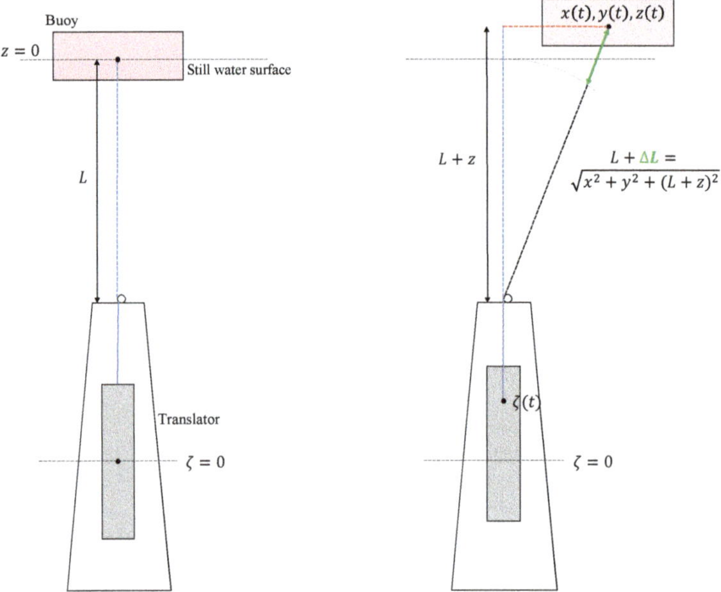


**Figure 6.** The wave energy converter (WEC) developed at Uppsala University at equilibrium and with waves, respectively. The translator is situated in a sealed generator hull. The distance L is measured from the top of the generator hull to the water's surface at equilibrium. In the experimental tests, the distance L was measured from the water's surface at equilibrium to the pulley on the seabed beneath the buoy. Reproduced from [23].

In the results section the power computed numerically is referred to as $P_{SIM}$, computed from the buoy displacement $z(t)$, which is assumed to be equal to the translator displacement $\zeta(t)$ (Equations (3) and (4)). From the experimental results, we display the power absorption measured from the power take-off system $P_{PTO}$ (Equation (5)) and the power absorption computed from the buoy motion $P_{\Delta L}$, computed analogously to Equation (3) where $z$ is replaced with $\Delta L$. In other words, $P_{\Delta L}$ is calculated considering all the three translational DoFs of the floats. This latter is called virtual power of the buoy, as it does not represent the actual motion of the PTO, from which power is absorbed. In fact, the buoy motion along the connection line $\Delta L$ would correspond to the position of the translator $\zeta$ only assuming a stiff connection between them, which is assumed in the numerical model, but is not realistic in the physical experiment.

## 5.1. Power Output

Figure 7 shows the results in terms of average power output over the time window tested, including both experimental and simulation results, for all the three layouts. Simulations are shown for the two buoy geometries mentioned in Section 4.

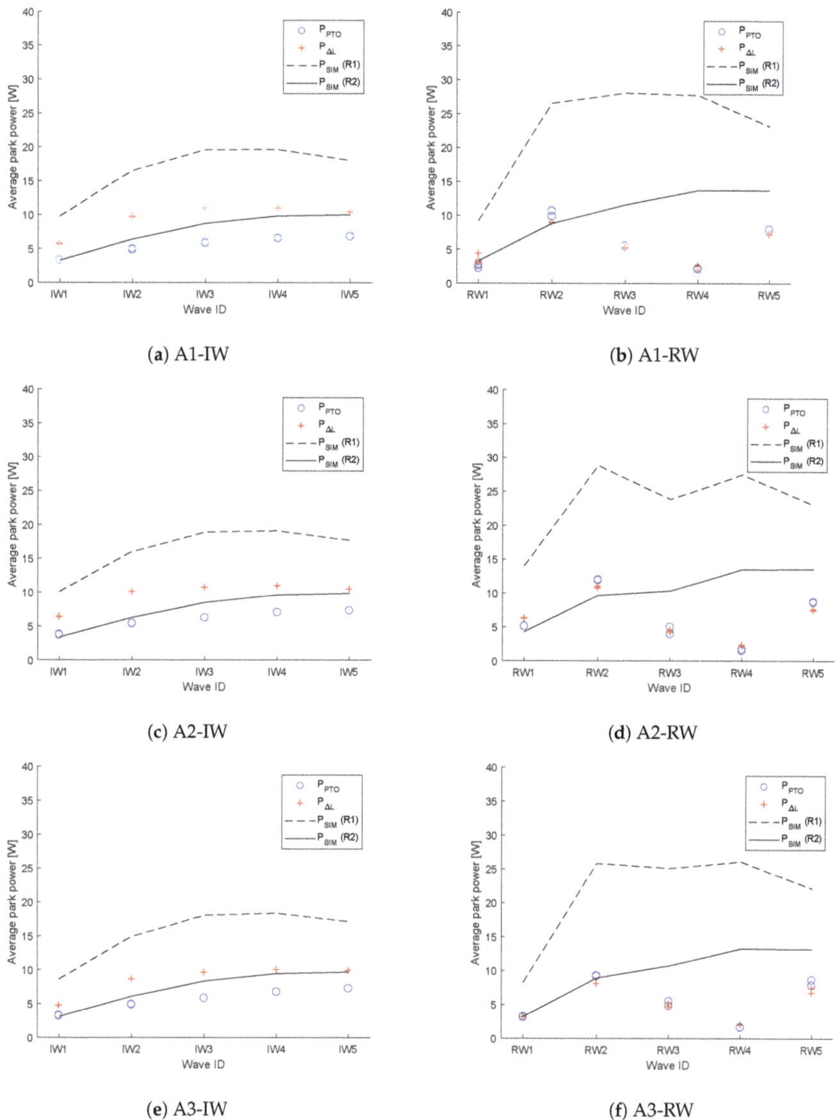

**Figure 7.** Average power absorbed by the park. (**a,b**) Layout A1. (**c,d**) Layout A2. (**e,f**) Layout A3. Results from irregular waves are shown in (**a,c,e**) and regular waves in (**b,d,f**).

In irregular waves (Figure 7a,c,e), the simulations reproduced the power behavior quite well, even though the error between simulations and experiments increased with increasing energy period. In the case of float geometry R1 the power was overestimated, while in the case of R2 the one-DoF-restricted, simplified simulation fit the experimental data quite well. Within the goals of the present work, it was important to reproduce the qualitative behavior of the power output; i.e., the results obtained with R1 are still useful if the goal is the park optimization by relative comparison between layouts.

The wave height is constant for all wave conditions and the (energy) period increases linearly. The energy flux incident on the buoy thus increases linearly with the wave conditions. As can be seen from Figure 7, the absorbed power increases with the period, but not linearly. The increase is largest for the lowest periods, after which the increase is only small. This can be explained from the fact that the natural period is $T_{\mathrm{nat}} = 1.36$ s, i.e., it lies just below wave condition 2. Hence, the WECs will absorb most power relative to the incident energy in the wave conditions with period around wave condition 2.

From the experimental data, it can be observed that the power computed from the power take-off system ($P_{\mathrm{PTO}}$) differs from the power computed from the optical system of the float ($P_{\Delta L}$). This means that the connection line does not behave as a stiff connection, as is the case here. In fact, the intermittent nature of the irregular waves make the float be exposed to sudden and quick changes of its motion, which is not synchronously followed by the rotating PTO, due also to the pulley system and PTO inner inertia. This does not happen in the regular waves, where the motion of the PTO follows the motion of the buoy more easily, at least for the PTO damping used. As we showed in [23], for other damping values, the buoy and PTO motion could differ also in regular waves.

However, in the regular wave results, the experimentally obtained power undergoes a significant reduction in RW3 and RW4, while a lesser reduction is observed in RW5. This can be explained by referring to the results obtained in [23]; a single WEC in isolation was studied in terms of the Mathieu-type instability. Results have shown that in RW3 and RW4 the buoy is affected by a high transverse instability, due to the coupling between heave and sway, resulting in a high sway motion. RW5 displayed only little sway instability. As for the irregular waves, the simulations with geometry R1 overestimated the power (including stable waves); however, in this case we did not expect agreement between the heave model and the experiments that displayed Mathieu instabilities.

An example of results of the buoys' power output within array A1 is shown in Figure 8, while for array A2 and A3 the data are reported in the Appendices (Figures A1 and A2). The power output of the buoys within the same layout is strongly dependent on the wave condition (regular, irregular) and the wave period. In the regular tests the maximum power is achieved by different buoys according to the wave period; in irregular waves the relative level of power output is more or less kept even over different waves periods; an example for layout A2 is shown in Figure 9. As noted before, the two-body dynamics of the UU WEC are noticeable, especially in irregular waves, from the difference between $P_{\mathrm{PTO}}$ and $P_{\Delta L}$ measured for each buoy.

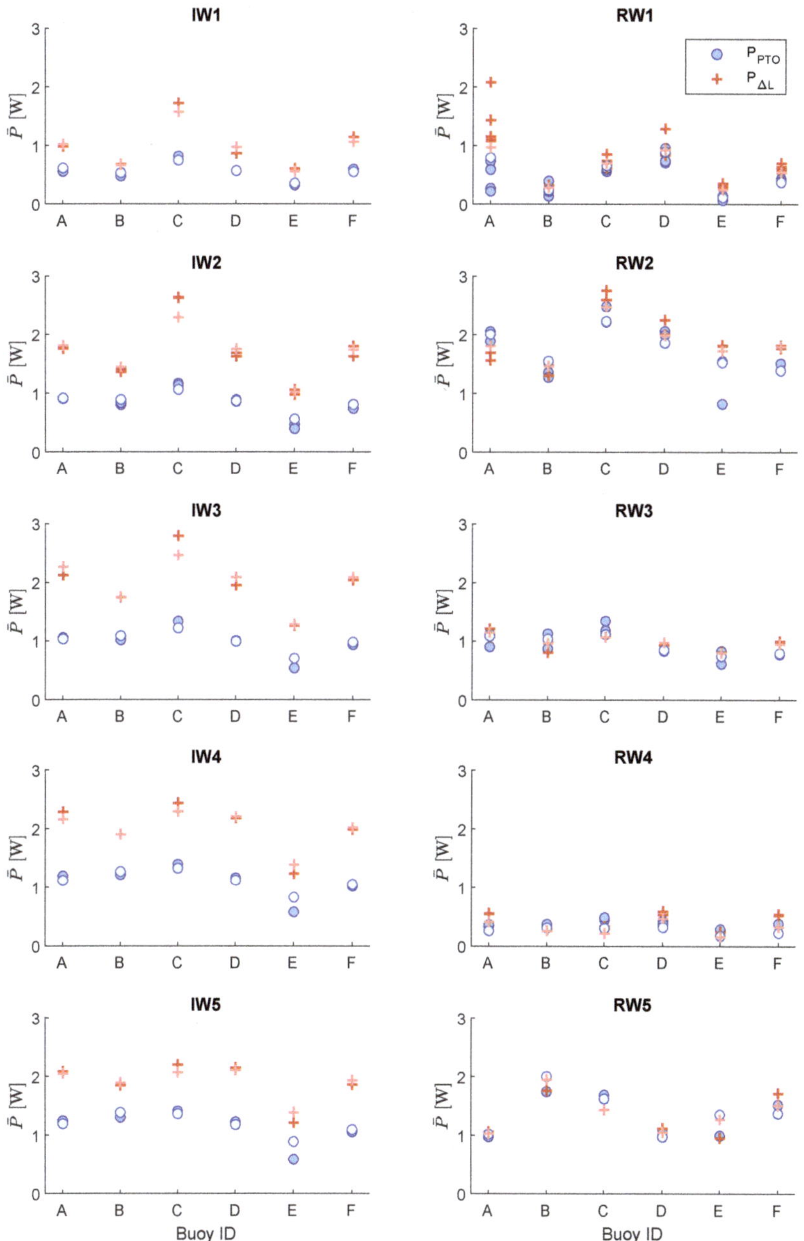

**Figure 8.** Average power of each buoy for layout A1 in all the tested sea states. Comparison between $P^i_{\Delta L}$ obtained from the Qualisys data (+) and $P^i_{\mathrm{PTO}}$ obtained from the PTO data (o). Multiple markers refer to repeated tests. Light blue and light red markers refer to layout A1 tested for the second time. The left column shows results in irregular waves, the right column shows results in regular waves.

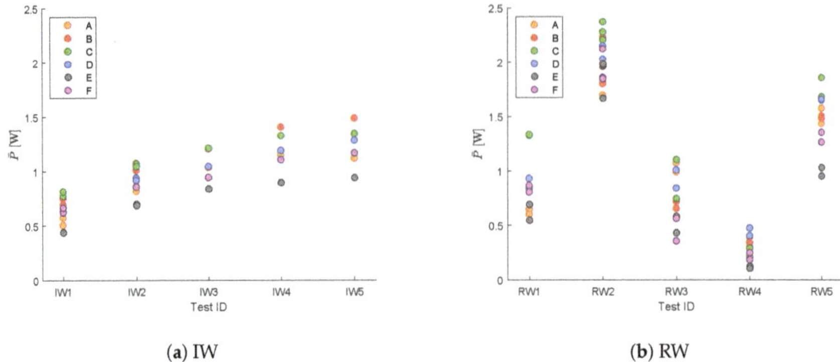

(**a**) IW          (**b**) RW

**Figure 9.** Average power output ($P^i_{PTO}$) of the buoys in layout A2. (**a**) irregular waves; (**b**) regular waves.

The experimental testing of layout A1 performed at two different occasions has shown consistent results (Figures 7a,b and 8).

Layouts A1 and A2 are not symmetric with respect to the center of the tank, whereas A3 is symmetric with respect to the center of the tank and the lateral walls. However, due to disturbances in the wave tank and high sway motion, there was no symmetry in the experimental results (Figures 8, A1 and A2). The buoys moved in heave, surge and sway in the experiment (if we neglect rotations), and therefore the shadowing effect was minimized. We can notice that on several occasions, the WECs that should have been shadowed produced higher power output than the non-shadowed ones.

*5.2. q-Factor*

Different layouts have been compared in terms of the interaction factor (or *q*-factor), defined as:

$$q = \frac{P}{n_b \cdot P^i_{IS}}. \tag{7}$$

$P$ represents the total power of the array and $n_b$ is the number of WECs in the park. $P^i_{IS}$ is the power of the device in isolation, obtained from the previous work for the isolated system in [23]. The experimental *q*-factor has been compared to the simulated one in Figure 10, for the three layouts. The *q*-factor values of the WECs within the three layouts are reported in the Appendices (Figures A3–A5).

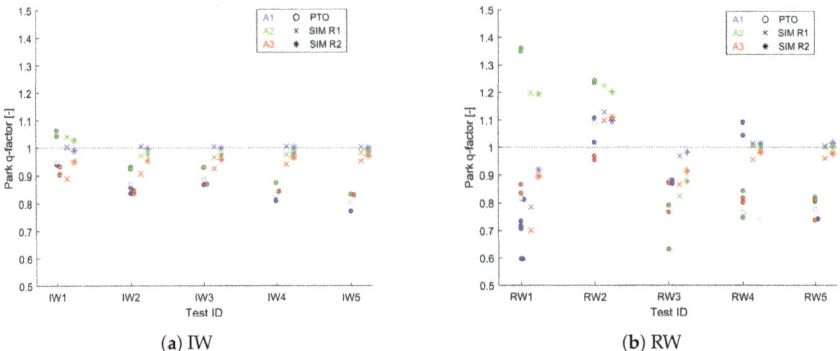

(**a**) IW          (**b**) RW

**Figure 10.** Comparison of the park *q*-factor values of layouts A1, A2 and A3, in all the tested sea states, calculated from PTO experimental data (o), from simulations with R1 (x) and from simulations with R2 (*). (**a**) Irregular waves; (**b**) regular waves.

The experimental results of Figure 10 show that the best layout was A2 for all the irregular wave conditions and for regular wave conditions RW1, RW2 and RW5. In RW3 and RW4 the best layout was A1.

The best performing layout according to the simulations was A2 in IW1, RW1, RW2 and A1 in IW2, IW3, IW4, IW5, RW3, RW4 and RW5; however, the difference between A1 and A2 was very small.

The shadowed layout, i.e., A3, should have been the worst performing layout for all the wave conditions according to the simulations, but this was not always the case in the experiment. This could be explained by the fact that the experimental WECs were moving in sway, interfering with the expected hydrodynamical interaction and the shadowing effects. In fact, rear devices did experience complete shadowing in the simulations, but not in the wave tank experiment. On the other hand, in the simulations, perfectly heaving devices influenced the hydrodynamic interaction in a different way.

For realistic optimization, we should take into account WECs with at least three DoFs, or better, six DoFs, and waves coming from different directions (multi-directional waves), which would lead to less influence in the shadowing of WECs during the simulations.

The experimental conditions are more controlled in IW1 and IW2, where waves are less energetic and we have less phenomena of slack line and breaking. In these cases we see better agreement of the arrays' $q$-factor values with the experimental results. The same is valid for RW1 and RW2, in which the devices do not present parametric resonance. However, there is still some ambiguity between the simulations performed with R1 vs R2.

Both in experiments and simulations, the difference between the arrays in terms of $q$-factor is small in most of the cases, in particular for all the irregular waves. When we have non-linear dynamics, the discussed uncertainties in the experimental data imply that the modeling is not so accurate as to capture such small variations among layouts. Nevertheless, although the WEC dynamics and the hydrodynamic model is limited by constrained heave motion, the stiff connection between buoys and PTOs and linear dynamics, the $q$-factor estimation is rather good, and the results are close to the experimental region (gray zone in Figure 10).

In general, we expect the hydrodynamic interaction to be larger in larger parks (i.e., parks with more units) [24], resulting in a larger difference between best and worst layouts. In such cases, the model can be expected to better capture such discrepancies. The present study was based on a trade-off between a feasible experimental campaign and simulation work. A larger layout would have given more accurate information, but would also have been difficult to test.

It is worth mentioning that the sources of uncertainty in an experimental campaign were multiple: reflection of waves against the walls of the tank, possibly unbalanced buoys due to the optical markers, uncertainty in the damping and friction estimation, varying damping due to unbalanced discs, elasticity of the connection line and friction in the pulley system. Whereas the optical markers are not expected to influence the dynamics to a relevant degree, the other uncertainties might affect the results. In [21], the friction and line elasticity for the system were evaluated, and it was found that adding a slight elasticity better reproduced the experimental results, and the friction was measured qualitatively. In this paper, friction has been included in the simulations as an additional damping term. Other potential error sources in wave tank testing, highlighted in [27], are: the spatial variation of the wave-field within the wave basin; the temporal variation of the wave-field from one repeat to another; the repeatability of a model's response for any single individual WEC; the reproducibility of a model's response for various identical WECs. In [19] the spatial variation of the wave field around the basin was studied and the results showed that, in irregular waves, the expected significant wave height was always under-reproduced. Moreover, it was found that the discrepancies are larger in relation to shorter periods. To partially asses these problems, repeated tests were executed, when possible, and measurements of the wave field in the empty tank were conducted and used as inputs series in the simulations for the irregular waves. The results from the repeated tests show good agreement, which increases the trustworthiness of the results.

## 6. Conclusions

The work presented an experimental campaign conducted with six point-absorber WECs moving in six DoFs. The goals were to (a) study the performances of different array layouts under several wave conditions and to (b) understand whether the simulation optimization predictions of best performing array layouts would be confirmed by experimental data. The experimental campaign represents a unique example of array testing with a point-absorber moving in six degrees of freedom. The simulations have been carried out by a frequency domain model restricted to heave, with unlimited stroke length and stiff connection between the float and the generator, so that each WEC is considered as a single body.

### 6.1. Performances of Different Array Layouts under Several Wave Conditions

Although there is a vast literature on numerical simulations on wave energy parks, there is very little published material on experimental data. The paper is the first publication of experimental data of several array layouts of WECs moving in six DoFs, subjected to both regular and irregular waves.

In both the simulations and the experimental data, arrays A1 and A2 were the best performing layouts in most wave conditions. However, although the numerical simulations predicted that the array A3 should be the worst performing layout in all wave conditions due to the shadowing effect, in the experiments this did not always happen. This can be understood from the fact that in reality, the floats moved also in sway, reducing the shadowing effect. It is worth mentioning that the differences between the best and worst layouts were quite small in irregular waves, both in the simulations and experiments (within 14% in the experiments and within 15.5% and 8% in simulations with R1 and R2, respectively). For larger parks, the shadowing effect would be larger and be present also in realistic conditions, and we would expect both a bigger difference between the layouts and a closer agreement between the simulations and the experimental results. In addition, short-crested waves can decrease the shadowing effect, and thus experiments and simulations using short-crested waves would probably agree to a larger extent.

The power absorption of the array in irregular waves increases with increased energy flux up to an optimum energy period, after which the power is constant or decreases slightly. This behavior is shown both in the simulations and experimental data. Since the natural period of the device is $T_{nat} = 1.36$ s, i.e., it lies between just below wave condition 2, the WECs will absorb most power relative to the incident energy in the wave conditions with a period around wave condition 2.

The highest $q$-factor was obtained for the staggered array A2 in the irregular waves with the shortest energy period, in both the experiments ($q = 1.04$, $q = 1.06$) and simulations ($q = 1.03, 1.04$). In the simulations, the $q$-factor was close to 1 in all irregular sea states, whereas it decreased with increasing energy period in the experiments, down to a minimum of $q = 0.77$. In the regular waves, the range of the $q$-factor obtained in different wave conditions and for different arrays was much larger, both in the simulations and experiments. Additionally, in the regular waves, the highest $q$-factor was obtained for the staggered array A2 in the waves with the shortest energy period, both in the simulations and in the experiments.

### 6.2. Comparison of Simulation Optimization Predictions with Experimental Data

The paper gives several important insights into common assumptions and models used for wave energy park optimization. In most of the modeling works of wave energy parks, in particular in array optimization studies, a stiff connection between the float and the PTO is assumed, and the numerical model of the float is restricted to heave, although the physical float is free to move in six DoFs. Do these two assumptions hold in realistic conditions?

- The experimental data show that in regular waves and for the PTO damping used, the power obtained from the float and the PTO motion is the same, so the assumption of a stiff line is valid. In the irregular waves, however, the WEC shows clear two-body dynamics with different displacement of the float and the PTO due to slack and elasticity in the connection line. As a result, the power obtained from the float and the PTO can differ significantly, and the assumption no longer holds. The virtual power obtained from the float's motion is in this case higher than the actual power obtained from the PTO; in other words, if a stiff connection between the float and the PTO were to be assumed, the predicted power would be higher than the one that will be obtained in reality. This is confirmed by comparing the power computed in the simulations (which assume a stiff line) with the experimental data: the simulations generally overestimate the absorbed power by the PTO. This conclusion holds both for the total power of the park (Figure 7) and for the power absorbed by the individual buoys (Figure 8).

- The results from the simple one-DoF simulation model are able to reproduce the qualitative behavior of the absorbed power at least to an approximate degree (Figures 7 and 10), as long as non-linear dynamics such as Mathieu instabilities are not present. In the experiments, Mathieu instabilities mostly occurred in some of the regular wave cases; hence, in realistic, irregular waves, the simulations can give an estimation of the power output of the UU WEC. In summary, in the wave conditions wherein we neither observe extensive sway nor non-linear behavior, the results of the optimization are reliable. However, when non-linear dynamics or motions in multiple degrees of freedom are present, then the optimization routine does not provide accurate results. In other words, the discrepancy between the experimental and numerical results is mostly due to either slack line condition (mostly in irregular waves) or Mathieu-type instability (mostly in regular waves). In addition, losses, friction and reflections in the wave tank, and other sources of uncertainty in the experimental set-up, also play roles in the outcome of the present study.

In order to improve the output of the presented model, an array hydrodynamical model with six DoFs and multi-directional waves could be implemented in the optimization routine. To capture the non-linear effects and two-body dynamics, a non-linear time-domain model would be required. However, the computational costs of such a model would probably be too high to enable large-scale optimization.

**Author Contributions:** Conceptualization, M.G. (Marianna Giassi), J.E., J.I. and M.G. (Malin Göteman); data curation, M.G. (Marianna Giassi); formal analysis, M.G. (Marianna Giassi); funding acquisition, J.E., J.I. and M.G. (Malin Göteman); investigation, M.G. (Marianna Giassi) and M.G. (Malin Göteman); project administration, J.E., J.I. and M.G. (Malin Göteman); supervision, J.E. and M.G. (Malin Göteman); writing—original draft, M.G. (Marianna Giassi); writing—review and editing, J.E., J.I. and M.G. (Malin Göteman). All authors have read and agreed to the published version of the manuscript.

**Funding:** This research is supported by the the Swedish Research Council (VR, grant number 2015-04657), the Swedish Energy Agency (project number 40421-1), StandUp for Energy, the Lundström-Åman foundation and Miljöfonden.

**Acknowledgments:** The authors would like to thank Martyn Hann and Edward Ransley for discussions and input provided; and Kieran Monk and Tom Tosdevin for the technical support and help in the COAST Lab ocean basin.

**Conflicts of Interest:** The authors declare no conflict of interest.

## Nomenclature

Numerical model:

| | |
|---|---|
| $x^i$ | buoy position in surge |
| $y^i$ | buoy position in sway |
| $z^i$ | buoy position in heave |
| $\zeta^i$ | PTO position |
| $\gamma$ | PTO damping |
| $P^i_{\mathrm{SIM}}$ | WEC power output ($i^{\mathrm{th}}$ WEC) |
| $P_{\mathrm{SIM}}$ | park power output |

Experimental data:

| | |
|---|---|
| $X^i$ | buoy position in surge measured by the optical system |
| $Y^i$ | buoy position in sway measured by the optical system |
| $Z^i$ | buoy position in heave measured by the optical system |
| $\Delta L^i$ | buoy position along the connection line calculated from the optical measurements |
| $\gamma$ | PTO damping |
| $P^i_{\mathrm{PTO}}$ | WEC experimental PTO power output |
| $P_{\mathrm{PTO}}$ | park experimental PTO power output |
| $P^i_{\Delta L}$ | WEC experimental virtual power output computed from the buoy motion along the connection line |
| $P_{\Delta L}$ | park experimental virtual power output computed from the buoy motion along the connection line |

## Abbreviations

The following abbreviations are used in this manuscript:

| | |
|---|---|
| DoF | Degrees of freedom |
| IW | Irregular wave |
| LG | Linear generator |
| PTO | Power take-off |
| RW | Regular wave |
| UU | Uppsala University |
| WEC | Wave energy converter |

## Appendix A

*Appendix A.1. WEC's Power Output*

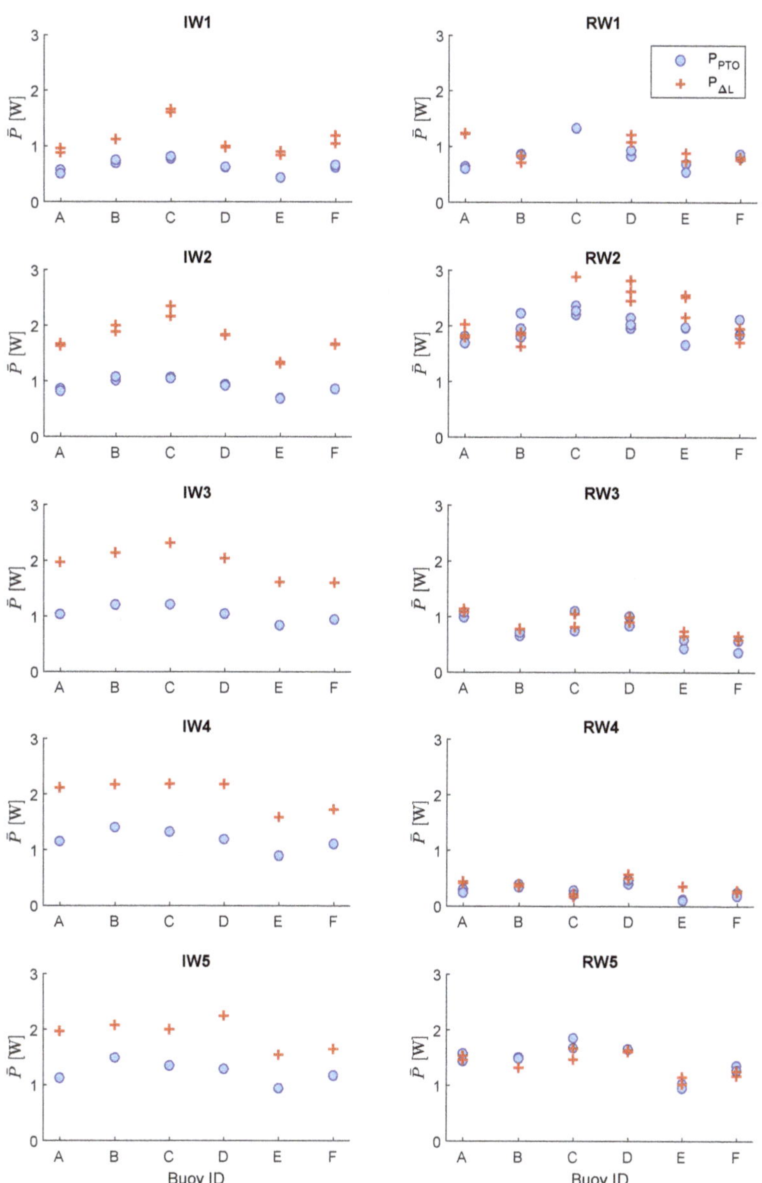

**Figure A1.** Average power of each buoy for Layout A2 in all the tested sea states. Comparison between $P_{\Delta L}^{i}$ obtained from the Qualisys data (+) and $P_{PTO}^{i}$ obtained from the PTO data (o). Multiple markers refer to repeated tests. Left column shows results in irregular waves, right column in regular waves.

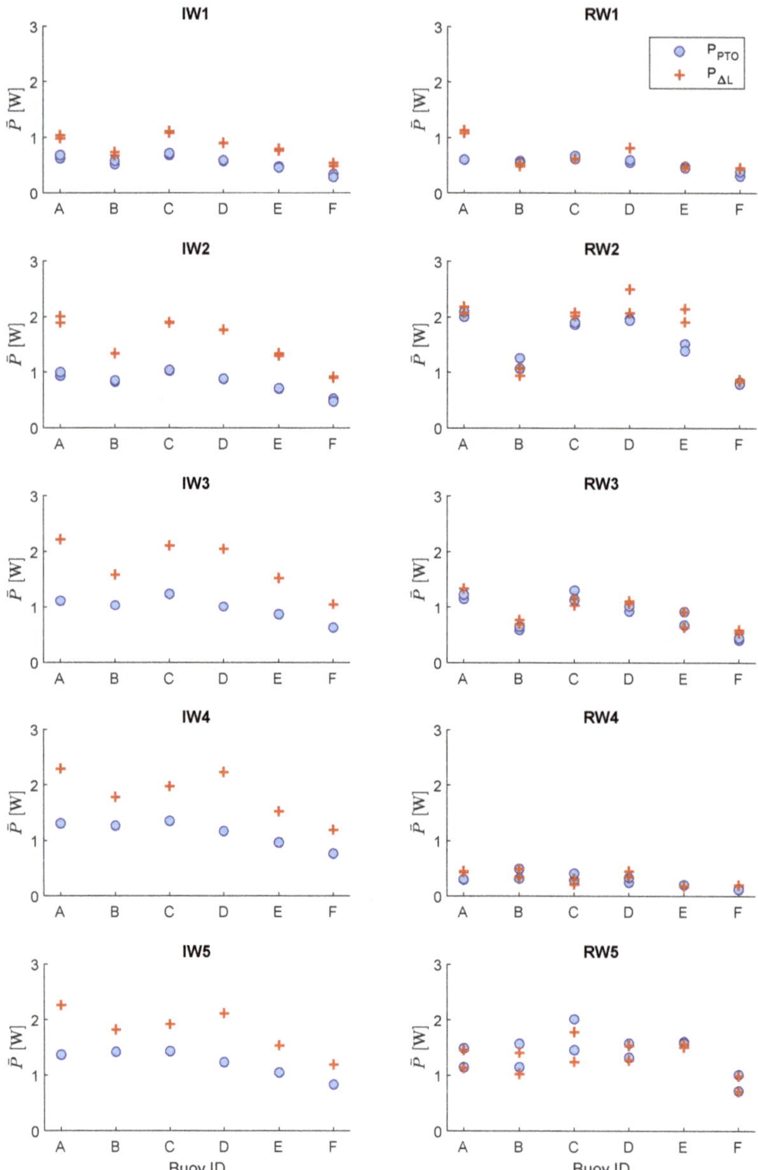

**Figure A2.** Average power of each buoy for Layout A3 in all the tested sea states. Comparison between $P^i_{\Delta L}$ obtained from the Qualisys data (+) and $P^i_{PTO}$ obtained from the PTO data (o). Multiple markers refer to repeated tests. Left column shows results in irregular waves, right column in regular waves.

*Appendix A.2. WEC's q-Factor*

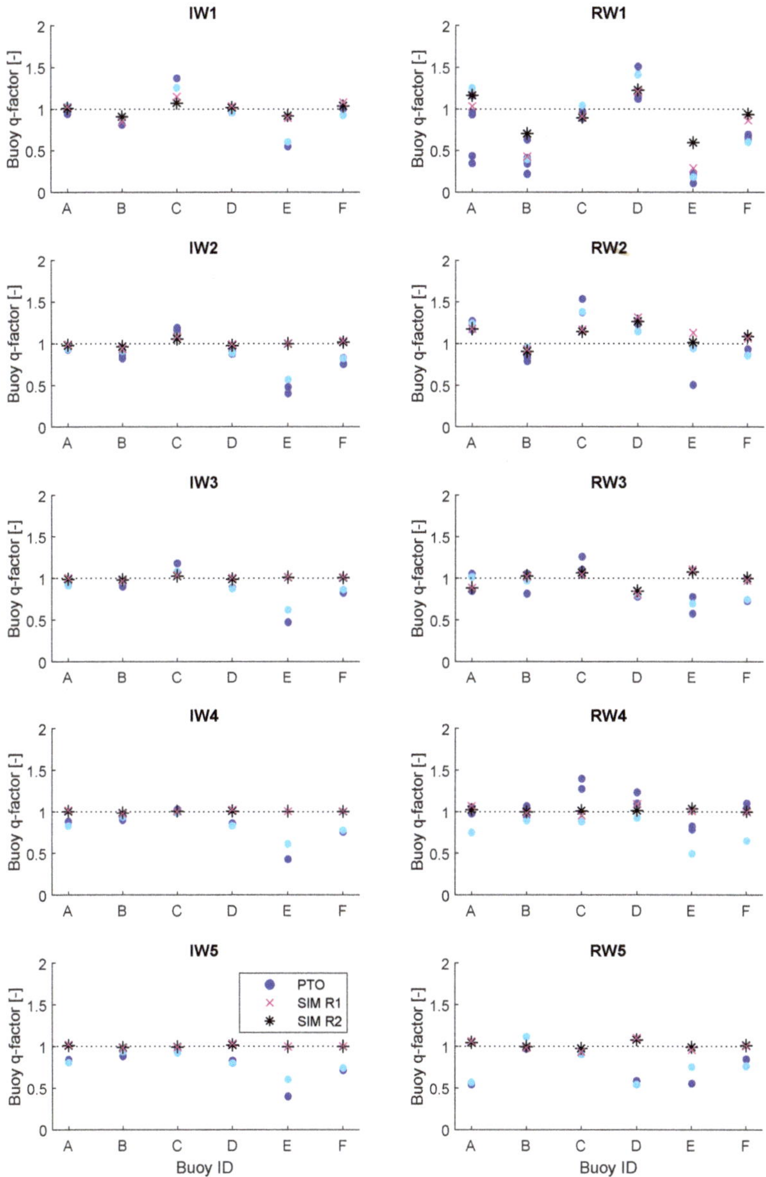

**Figure A3.** The buoys' *q*-factor for layout A1 computed from PTO experimental data and simulations. Left column shows results in irregular waves, right column in regular waves.

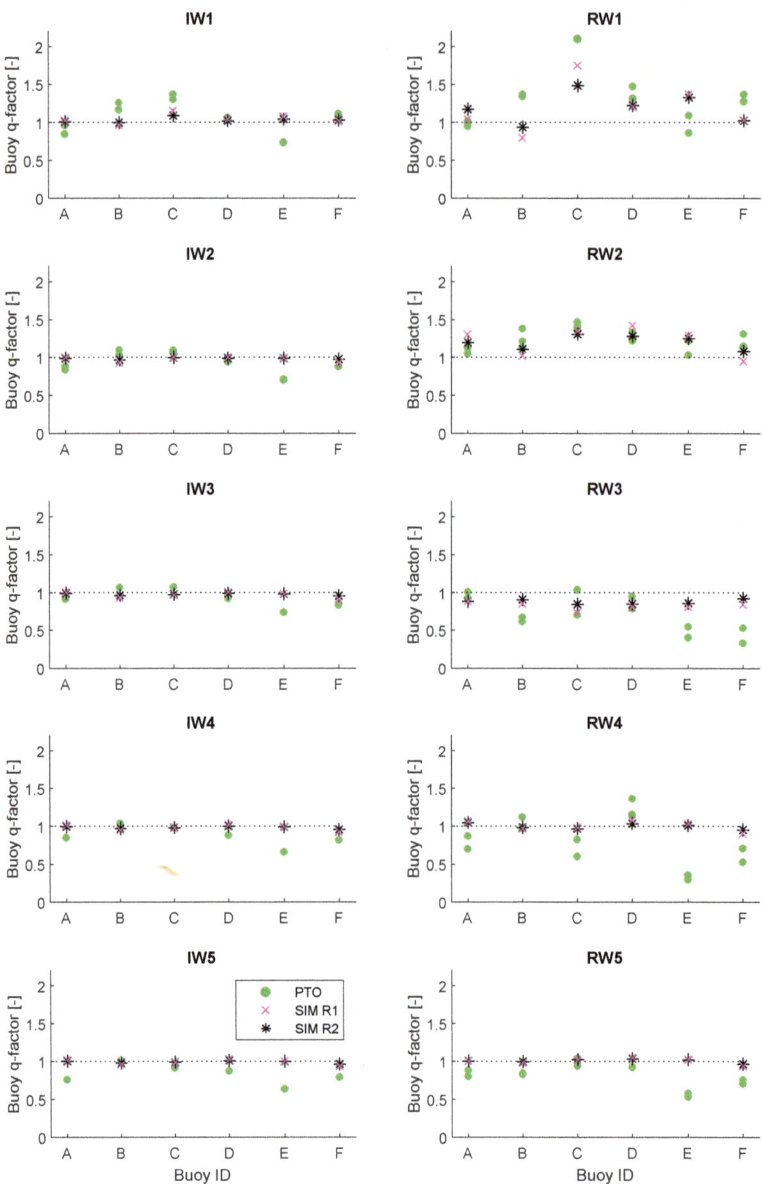



**Figure A4.** The buoys' *q*-factor for layout A2 computed from PTO experimental data and simulations. Left column shows results in irregular waves, right column in regular waves.

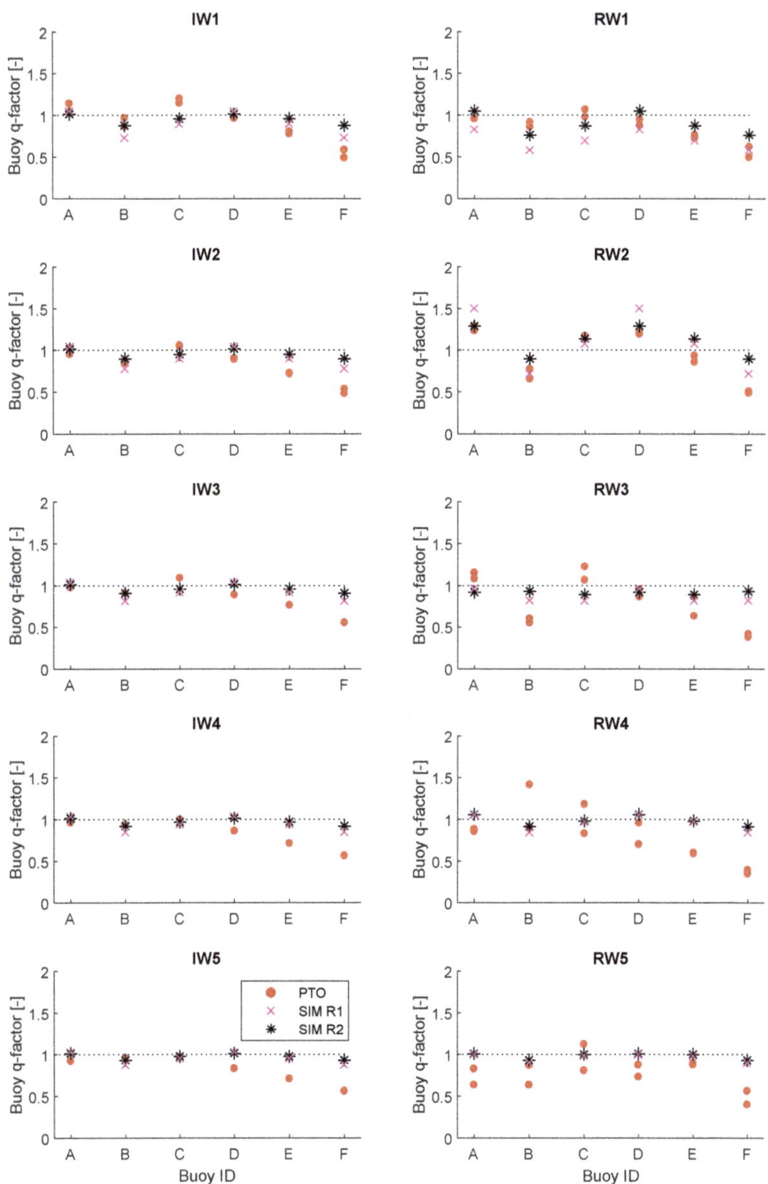



**Figure A5.** The buoys' *q*-factor for layout A3 computed from PTO experimental data and simulations. Left column shows results in irregular waves, right column in regular waves.

# References

1. Göteman, M.; Giassi, M.; Engström, J.; Isberg, J. Advances and Challenges in Wave Energy Park Optimization—A Review. *Front. Energy Res.* **2020**, *8*, 26. doi:10.3389/fenrg.2020.00026. [CrossRef]
2. Konispoliatis, D.; Mavrakos, S. Mean drift loads on arrays of free floating OWC devices consisting of concentric cylinders. In Proceedings of the 29th International Workshop on Water Waves and Floating Bodies (IWWWFB2014), Osaka, Japan, 30 March– 2 April 2014.

3. Da Fonseca, F.; Gomes, R.; Henriques, J.; Gato, L.; Falcao, A. Model testing of an oscillating water column spar-buoy wave energy converter isolated and in array: Motions and mooring forces. *Energy* **2016**, *112*, 1207–1218. [CrossRef]

4. Magagna, D.; Carr, D.; Stagonas, D.; Mcnabola, A.; Gill, L.; Muller, G. Experimental Evaluation of the Performances of an Array of Multiple Oscillating Water Columns. In Proceedings of the 9th European Wave and Tidal Energy Conference (EWTEC), Southampton, UK, 5–9 September 2011.

5. Wolgamot, H.; Taylor, P.; Eatock Taylor, R.; Van Den Bremer, T.; Raby, A.; Whittaker, C. Experimental observation of a near-motion-trapped mode: Free motion in heave with negligible radiation. *J. Fluid Mech.* **2016**, *786*. [CrossRef]

6. Ji, X.; Liu, S.; Bingham, H.; Li, J. Multi-directional random wave interaction with an array of cylinders. *Ocean. Eng.* **2015**, *110*, 62–77. [CrossRef]

7. Bosma, B.; Brekken, T.; Lomonaco, P.; DuPont, B.; Sharp, C.; Batten, B. Array modeling and testing of fixed OWC type Wave Energy Converters. In Proceedings of the 13th European Wave and Tidal Conference (EWTEC), Napoli, Italy, 1–6 September 2019

8. Mackay, E.; Cruz, J.; Livingstone, M.; Arnold, P. Validation of a Time-Domain Modelling Tool for Wave Energy Converter Arrays. In Proceedings of the 10th European Wave and Tidal Energy Conference, Alborg, Denmark, 2–5 September 2013.

9. Mercadé Ruiz, P.; Ferri, F.; Kofoed, J. Experimental validation of a wave energy converter array hydrodynamics tool. *Sustainability* **2017**, *9*, 115. [CrossRef]

10. Troch, P.; Stratigaki, V.; Stallard, T.; Forehand, D.; Folley, M.; Kofoed, J.; Benoit, M.; Babarit, A.; Sánchez, D.; De Bosscher, L.; et al. Physical modelling of an array of 25 heaving wave energy converters to quantify variation of response and wave conditions. In Proceedings of the 10th European Wave and Tidal Energy Conference Series (EWTEC), Aalborg, Denmark, 2–5 September 2013; pp. 2–5.

11. Stratigaki, V.; Troch, P.; Stallard, T.; Forehand, D.; Kofoed, J.; Folley, M.; Benoit, M.; Babarit, A.; Kirkegaard, J. Wave basin experiments with large wave energy converter arrays to study interactions between the converters and effects on other users in the sea and the coastal area. *Energies* **2014**, *7*, 701–734. [CrossRef]

12. Stratigaki, V.; Troch, P.; Stallard, T.; Forehand, D.; Folley, M.; Kofoed, J.; Benoit, M.; Babarit, A.; Vantorre, M.; Kirkegaard, J. Sea-state modification and heaving float interaction factors from physical modelling of arrays of wave energy converters. *J. Renew. Sustain. Energy* **2015**, *7*, 061705. [CrossRef]

13. Child, B.; Weywada, P. Verification and validation of a wave farm planning tool. In Proceedings of the 10th European Wave and Tidal Energy Conference (EWTEC) conference, Aalborg, Denmark, 2–5 September 2013; pp. 2–5.

14. Thomas, S.; Weller, S.; Stallard, T. Float response within an array: Numerical and experimental comparison. In Proceedings of the 2nd International Conference on Ocean Energy (ICOE), Brest, France, 15–17 October 2008; Volume 1517.

15. Weller, S.; Stallard, T.; Stansby, P. Interaction factors for a rectangular array of heaving floats in irregular waves. *IET Renew. Power Gener.* **2010**, *4*, 628–637. [CrossRef]

16. Nader, J.; Fleming, A.; Macfarlane, G.; Penesis, I.; Manasseh, R. Novel experimental modelling of the hydrodynamic interactions of arrays of wave energy converters. *Int. J. Mar. Energy* **2017**, *20*, 109–124. [CrossRef]

17. Thomas, S.; Eriksson, M.; Göteman, M.; Hann, M.; Isberg, J.; Engström, J. Experimental and numerical collaborative latching control of wave energy converter arrays. *Energies* **2018**, *11*, 3036. [CrossRef]

18. Thomas, S.; Giassi, M.; Eriksson, M.; Göteman, M.; Isberg, J.; Ransley, E.; Hann, M.; Engström, J. A model free control based on machine learning for energy converters in an array. *Big Data Cogn. Comput.* **2018**, *2*, 36. [CrossRef]

19. Giassi, M.; Thomas, S.; Shahroozi, Z.; Engström, J.; Isberg, J.; Tosdevin, T.; Hann, M.; Göteman, M. Preliminary results from a scaled test of arrays of point-absorbers with 6 DOF. In Proceedings of the 13th European Wave and Tidal Conference (EWTEC), Napoli, Italy, 1–6 September 2019.

20. Leijon, M.; Waters, R.; Rahm, M.; Svensson, O.; Boström, C.; Strömstedt, E.; Engström, J.; Tyrberg, S.; Savin, A.; Gravråkmo, H.; et al. Catch the wave to electricity. *IEEE Power Energy Mag.* **2009**, *7*, 50–54. [CrossRef]

21. Thomas, S.; Giassi, M.; Göteman, M.; Hann, M.; Ransley, E.; Isberg, J.; Engström, J. Performance of a Direct-Driven Wave Energy Point Absorber with High Inertia Rotatory Power Take-off. *Energies* **2018**, *11*, 2332. [CrossRef]
22. Giassi, M.; Göteman, M. Layout design of wave energy parks by a genetic algorithm. *Ocean. Eng.* **2018**, *154*, 252–261. [CrossRef]
23. Giassi, M.; Thomas, S.; Tosdevin, T.; Engström, J.; Hann, M.; Isberg, J.; Göteman, M. Capturing the experimental behaviour of a point-absorber WEC by simplified numerical models. *J. Fluids Struct.* **2020**, (accepted).
24. Göteman, M.; Engström, J.; Eriksson, M.; Isberg, J. Fast modeling of large wave energy farms using interaction distance cut-off. *Energies* **2015**, *8*, 13741–13757. [CrossRef]
25. Göteman, M. Wave energy parks with point-absorbers of different dimensions. *J. Fluids Struct.* **2017**, *74*, 142–157. [CrossRef]
26. Giassi, M.; Castellucci, V.; Göteman, M. Economical layout optimization of wave energy parks clustered in electrical subsystems. *Appl. Ocean. Res.* **2020**, *101*, 102274. [CrossRef]
27. Lamont-Kane, P.; Folley, M.; Whittaker, T. Investigating Uncertainties in Physical Testing of Wave Energy Converter Arrays. In Proceedings of the 10th European Wave and Tidal Energy Conference (EWTEC), Aalborg, Denmark, 2–5 September 2013.

*Article*

# Simple Controllers for Wave Energy Devices Compared

**Demián García-Violini** [1,*], **Nicolás Faedo** [2], **Fernando Jaramillo-Lopez** [2] and **John V. Ringwood** [2]

[1] Departamento de Ciencia y Tecnología, Universidad Nacional de Quilmes, Roque Saenz Peña 352, Bernal B1876, Argentina

[2] Centre for Ocean Energy Research, Maynooth University, W23 F2K8 Co. Kildare, Ireland; nicolas.faedo@mu.ie (N.F.); fernando.jaramillolopez@mu.ie (F.J.-L.); john.ringwood@mu.ie (J.V.R.)

* Correspondence: ddgv83@gmail.com

Received: 11 September 2020; Accepted: 1 October 2020; Published: 13 October 2020

**Abstract:** The design of controllers for wave energy devices has evolved from early monochromatic impedance-matching methods to complex numerical algorithms that can handle panchromatic seas, constraints, and nonlinearity. However, the potential high performance of such numerical controller comes at a computational cost, with some algorithms struggling to implement in real-time, and issues surround convergence of numerical optimisers. Within the broader area of control engineering, practitioners have always displayed a fondness for simple and intuitive controllers, as evidenced by the continued popularity of the ubiquitous PID controller. Recently, a number of energy-maximising wave energy controllers have been developed based on relatively simple strategies, stemming from the fundamentals behind impedance-matching. This paper documents this set of (5) controllers, which have been developed over the period 2010–2020, and compares and contrasts their characteristics, in terms of energy-maximising performance, the handling of physical constraints, and computational complexity. The comparison is carried out both analytically and numerically, including a detailed case study, when considering a state-of-the-art CorPower-like device.

**Keywords:** wave energy; wave energy converter; energy-maximising control; optimal control; impedance-matching; renewable energy systems

---

## 1. Introduction

Despite being a vast resource, wave energy conversion technology has not yet reached economic commercialisation. The main reason for the lack of proliferation of wave energy can be attributed to the fact that harnessing the irregular reciprocating motion of the sea is not as straightforward as, for example, extracting energy from the wind. This is clearly reflected in the striking absence of clear technology convergence, with over a thousand different concepts and patents proposed over the years (see, for instance, [1]).

Dynamic analysis and control system technology can impact many aspects of wave energy converter (WEC) design and operation, including device sizing and configuration, maximising energy extraction from waves, and optimising energy conversion in the power take-off (PTO) system. As a matter of fact, it is already clear (and well-established) that appropriate control technology has the capability to greatly enhance energy extraction from WECs [2,3]. In particular, the control input, supplied by means of the PTO system, and effectively realising the mechanical load on the device, plays a key role in the optimisation of the operation of wave energy devices: Ultimately, energy conversion must be performed as economically as possible, to minimise the delivered energy cost, while also maintaining the structural integrity of the device, minimising wear on WEC components, and operating across a wide range of sea conditions. This is virtually always written in terms of an energy-maximising criterion, so that the control problem for WECs can be informally posed [2] as depicted in Table 1:

**Table 1.** Control problem for WECs.

| **Design the PTO Force (Control Input) Such That:** | |
| --- | --- |
| *Maximises* | Energy absorption from incoming waves. |
| *Subject to* | WEC dynamics. |
| | Device and actuator physical limitations. |

In recent years, wave energy control researchers applied optimal control methods, where the energy-maximisation design is written in terms of an appropriate optimal control problem (OCP), and well-developed techniques (mainly originated within the theory of calculus of variations [4]) can be considered. In particular, direct optimal control techniques are often adopted [5], which discretise the variables involved in the WEC OCP, and attempt to maximise the resulting nonlinear program (NP) directly. This NP has to be solved while using numerical optimisation routines, whose complexity (both analytical and computational) depends upon a number of factors, including the specific discretisation method selected. This family of optimisation-based controllers is optimal by design, facilitated by a suitable definition of the energy-maximising control objective in the corresponding OCP. In addition, device safety can be directly addressed by adding a (feasible) set of constraints (i.e., device and actuator limitations) to the optimisation problem.

Although optimal by design, these optimisation-based controllers have their specific drawbacks, which can limit their application in 'real-world' scenarios. In particular, depending on the discretisation utilised, optimisation-based controllers may or may not be suitable for real-time implementation; the complexity of the associated NP, and the numerical routines that are required to approximate its solution, can preclude real-time operation, often especially true if nonlinearities are considered in the WEC dynamical model [6]. In addition, these controllers often lack of any 'intuitive' interpretation, given the mathematical complexity behind their derivation. In other words, very specific expertise is often required to design, synthesise, and calibrate these controllers, which is relatively unappealing for industrial practitioners.

Aiming to find simple and intuitive solutions, a number of researchers attempt to solve the WEC OCP while using fundamental theory behind maximum power transfer: the so-called impedance-matching principle [7]. In particular, this family of simple controllers attempts to provide a (physically implementable) realisation of the impedance-matching condition for maximum power transfer, by proposing simple systems, mostly characterised by well-known techniques from linear time-invariant theory. These techniques have mild computational requirements, and their implementation can be performed in real-time with almost any physical hardware platform, including commercial low-cost microcontrollers.

Naturally, this simplicity comes at a certain cost: although simple to implement, the performance of these controllers is inherently suboptimal, leading to a drop in energy absorption when compared to optimisation-based techniques. In addition, and since the impedance-matching condition does not consider any physical limitations (i.e., device and actuator limits), constraint handling is virtually always performed by means of simple mechanisms, which do not take into account optimality with respect to power absorption. In other words, the limitation mechanisms are designed independently from the energy-maximising objective, effectively providing constrained optimal solutions, rather than optimal constrained solutions. This naturally implies a loss of energy absorption under constrained conditions.

Nonetheless, despite their suboptimal performance, the main advantage of these controllers relies on their simplicity of implementation, and intuitive design, synthesis, and calibration, which makes this family of strategies highly appealing for practical applications. Additionally, it is worth noting that, when considering the simplicity of controllers based on the impedance-matching principle, this family of controllers can be implemented in almost any physical hardware platform, such as commercial low cost microcontrollers while using traditional discrete-time recursive routines, which represents one of the main features of these controllers. Motivated by the potential that is offered by this set of techniques in real-world scenarios, this paper documents a critical comparison between five (5) different controllers, developed over the period 2010–2020. It is important to note that, even though the origins of impedance-matching control originate in the 1970s [8], practical algorithms dealing with panchromatic operation and system constraints were only developed within the past decade, hence the focus on the period post 2010'. In particular, this study compares and contrasts their characteristics, in terms of energy-maximising performance, the handling of physical constraints, and computational load. The comparison is carried out both analytically and numerically, with special attention paid to the stability issue that can arise in the implementation of each selected strategy. These controllers are those published in [9–13], listed, in chronological order, in Table 2.

**Table 2.** Set of five (5) simple controllers compared in this study.

| Reference | Controller Name | Shorthand Notation |
|:---:|:---|:---:|
| [9] | *Suboptimal causal reactive controller* | C1 |
| [10] | *Simple and effective real-time controller* | C2 |
| [11] | *Multi resonant feedback controller* | C3 |
| [12] | *Feedback resonating controller* | C4 |
| [13] | *LiTe-Con* | C5 |

Note that, in Table 2, the column 'Controller name', is defined while using the title of each corresponding reference. It should be noted that, in terms of operation and performance of this set of controller under real conditions in open waters, the control algorithms reviewed in this paper are designed to operate only in the power production region of sea states and devices; under extreme wave conditions, a supervisory control system will put the system into a safe configuration, under which no power is produced.

The remainder of this paper is organised as follows. Section 1.1 introduces the mathematical notation used throughout this study. Section 2 presents the fundamentals behind control-oriented modelling of WECs, while Section 3 provides a detailed description of the fundamental principle of impedance-matching, highlighting each of its features. Section 4 outlines the theory behind the design and synthesis of each of the five (5) simple controllers that are listed in Table 2, in chronological order of publication. Section 5 presents a case study based on a full-scale CorPower-like device, where the set of selected simple controllers is assessed in terms of a number of indicators, including energy-maximising performance, constraint handling, computational complexity, and stability features. Section 6 provides a detailed discussion on the results that are presented in Section 5, specifically providing insight into each of the strenghts and weaknesses of each simple controller. Finally, Section 7 encompasses the main conclusions of this study.

*1.1. Notation*

Standard notation is considered throughout this paper, most of which is defined in this section. $\mathbb{R}^+$ ($\mathbb{R}^-$) denotes-* the set of non-negative (non-positive) real numbers. The notation $\mathbb{N}_q$ indicates the set of all positive natural numbers up to $q$, i.e., $\mathbb{N}_q = \{1, 2, \ldots, q\}$. The symbol 0 stands for any zero element, dimensioned according to the context. The notation $A^\star$, with $A \in \mathbb{C}^{n \times n}$, denote the Hermitian transpose of the matrix $A$. The notation $\mathbb{Re}\{z\}$ and $\mathbb{Im}\{z\}$, with $z \in \mathbb{C}$, stands for the real-part and the imaginary-part of $z$, respectively. The convolution between two functions $f$ and $g$, with $\{f, g\} \subset L^2(\mathbb{R})$, over the set $\mathbb{R}$, i.e., $\int_{\mathbb{R}} f(\tau)g(t - \tau)d\tau$ is denoted as $f * g$, and where $L^2(\mathbb{R}) = \{f : \mathbb{R} \to \mathbb{R} \mid \int_{\mathbb{R}} |f(\tau)|^2 d\tau < +\infty\}$ is the Hilbert space of square-integrable functions in $\mathbb{R}$. The Laplace transform of a function $f$ (provided if exists), is denoted as $F(s)$, $s \in \mathbb{C}$. With some abuse of notation[1], the same is used for the Fourier transform of $f$, written as $F(\omega)$, $\omega \in \mathbb{R}$. Finally, the expression $\deg\{p\}$ is used to denote the degree of the polynomial $p$, defined over the field $\mathbb{R}$.

## 2. Control-Oriented Modelling of WECs

This section begins by recalling well-known theory behind control-oriented linear WEC modelling (see, for instance, [8]), for a one-degree-of-freedom (DoF) wave energy device[2]. In particular, under linear potential flow theory, the equation of motion for such a WEC is generally written in terms of a dynamical system $\Sigma$, for $t \in \mathbb{R}^+$, given by the set of equations

$$\Sigma : \begin{cases} \ddot{x} = \mathcal{M}\left(f_r + f_{re} + f_{ex} - f_u\right), \\ y = \dot{x} = v, \end{cases} \tag{1}$$

where $x : \mathbb{R}^+ \to \mathbb{R}$, $t \mapsto x(t)$ is the device excursion (displacement), $v : \mathbb{R}^+ \to \mathbb{R}$ is the device velocity, $f_{ex} : \mathbb{R}^+ \to \mathbb{R}$, $t \mapsto f_{ex}(t)$ the wave excitation force (external uncontrollable input due to the incoming wave field), $f_{re}$ the linearised hydrostatic restoring force, $f_r$ the radiation force, and $\mathcal{M} \in \mathbb{R}^+/0$ is the inverse of the generalised mass matrix of the device (see [8]). Finally, the notation $f_u : \mathbb{R}^+ \to \mathbb{R}$, $t \mapsto f_u(t)$, is used for the control input, being supplied by means of a power take-off (PTO) system. As previously discussed in Section 1, the mapping $f_u$ plays a key role in the optimisation of the operation of wave energy devices: ultimately, energy conversion must be performed as economically as possible, in order to minimise the delivered energy cost, while also maintaining the structural integrity of the device, minimising wear on WEC components, and operating across a wide range of sea conditions.

Continuing with the description of Equation (1), the linearised hydrostatic force can be written as $f_{re}(t) = -s_h x(t)$, where $s_h$ denotes the hydrostatic stiffness, which depends upon the device geometry. The radiation force $f_r$ is modelled based on linear potential theory and, using the well-known Cummins' equation [14], can be written, for $t \in \mathbb{R}^+$, using the expression

$$f_r(t) = -m_\infty \ddot{x}(t) - \int_{\mathbb{R}^+} h_r(\tau) v(t - \tau) d\tau, \tag{2}$$

with $h_r : \mathbb{R}^+ \to \mathbb{R}^+$, $h_r \in L^2(\mathbb{R})$, the (causal) radiation impulse response function containing the memory effect of the fluid response, and $m_\infty = \lim_{\omega \to +\infty} A_r(\omega) \in \mathbb{R}$, where $A_r : \mathbb{R} \to \mathbb{R}$ is the radiation added-mass, defined as

$$A_r(\omega) = m_\infty - \frac{1}{\omega} \int_{\mathbb{R}^+} h_r(t) \sin(\omega t) dt. \tag{3}$$

The non-parametric term $A_r(\omega)$, together with the so-called radiation damping $B_r : \mathbb{R} \to \mathbb{R}$, given by

$$B_r(\omega) = \int_{\mathbb{R}^+} h_r(t) \cos(\omega t) dt, \tag{4}$$

fully characterise the (well-defined) Fourier transform of $h_r$, i.e., we can write $H_r : \mathbb{R} \to \mathbb{C}$ as

$$H_r(\omega) = B_r(\omega) + j\omega \left[A_r(\omega) - m_\infty\right]. \tag{5}$$

In particular, radiation damping describes the dissipative effect of the energy transmitted from the oscillating body in the form of waves (i.e., radiated waves propagate away from the body).

---

[1]   The use of the capitalised letter for Laplace or Fourier transforms is always clear from the context.
[2]   Note that all five controllers listed in Table 2 assume a one-DoF device in each corresponding analytical formulation.

The radiation added-mass represents the additional inertial effect due to the acceleration of the water, which moves together with the body. Equations (3) and (4) are commonly referred to as Ogilvie's relations [15], and they stem from the definition of the Fourier transform. Furthermore, note that the impulse response function $h_r$ completely characterises an LTI system $\Sigma_r$, describing the dynamics of radiation effects.

Finally, the equation of motion of the WEC is given by

$$\Sigma : \begin{cases} \ddot{x} = \mathcal{M}\left(-h_r * v - s_\mathrm{h}x + f_{ex} - f_u\right), \\ y = v. \end{cases} \tag{6}$$

## 3. Fundamentals of WEC Control: The Impedance-Matching Principle

One of the first and fundamental results applied within the wave energy control literature relies on a direct approach to the energy-maximising problem, inspired by impedance matching in electrical circuits, where device and actuator constraints are neglected. In particular, this principle, which is effectively utilised by the five controllers compared in this study, heavily relies on a frequency-domain analysis of the WEC dynamics, and it is detailed and discussed in the following paragraphs.

Consider the linear Cummins' formulation, as defined in Equation (6). A direct application of the Fourier transform, together with the radiation force frequency-domain equivalent introduced in Equation (5), yields

$$j\omega(M + m_\infty)V(\omega) + H_r(\omega)V(\omega) + \frac{s_h}{j\omega}V(\omega) = F_{ex}(\omega) - F_u(\omega), \tag{7}$$

where the mappings $V : \mathbb{R} \to \mathbb{C}$ and $F_u : \mathbb{R} \to \mathbb{C}$ represent the Fourier transform of the device velocity $v$ and controller input $f_u$, respectively. From (7), it directly follows that

$$V(\omega) = \frac{1}{I(\omega)}\left[F_{ex}(\omega) - F_u(\omega)\right], \tag{8}$$

where the mapping $I : \mathbb{R} \to \mathbb{C}, \omega \mapsto I(\omega)$, defined as

$$I(\omega) = B_r(\omega) + j\omega\left[A_r(\omega) + M - \frac{s_h}{\omega^2}\right], \tag{9}$$

denotes the equivalent (intrinsic) impedance of the WEC. Naturally, Equation (8) resembles well-known representations in the field of electrical/electronic engineering and circuits theory: the WEC dynamics (7) can be equivalently represented by the analogue circuit that is depicted in Figure 1a.

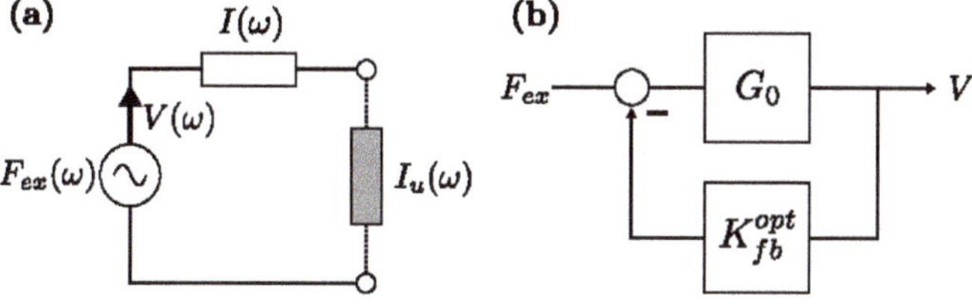

**Figure 1.** Impedance-matching principle. (**a**) Equivalent circuit for the frequency-domain analysis of Cummins' Equation (7). (**b**) Closed-loop impedance-matching formulation.

In other words, the control input $F_u(\omega)$ can be considered as a load, which has to be designed, so that maximum power transfer is achieved from the source, i.e., the wave excitation input $F_{ex}(\omega)$. From this particular point of view, this problem can be directly addressed using the so-called *impedance-matching* (or *maximum power transfer*) theorem [7], which is a well-established result within the electrical/electronic engineering community. This theorem states that the *load* impedance, $I_u$, should be designed, for alternating (a.c.) circuits, such that it exactly coincides with the complex-conjugate of the *source* impedance, $I$. In other words, the control input that maximises power transfer, for the WEC case, is given by

$$F_u(\omega) = I_u(\omega)V(\omega) = I^\star(\omega)V(\omega) = K_{fb}^{opt}(\omega)V(\omega), \tag{10}$$

where the notation $K_{fb}^{opt}$ is used to denote that the 'controller' in (10) is of a feedback-type. The result that is posed in (10) is indeed appealing, mainly due to its intrinsic simplicity, and its direct link to fundamental and well-established theory in the field of analogue circuits. Nevertheless, there are several issues that are associated with the control specifications given in (10), which prohibits the smooth application of what could potentially be an extremely appealing principle. These are listed and discussed in the following paragraphs.

To begin this discussion, note that the Laplace-transform analogue of Equation (7), when considering zero initial conditions, directly yields,

$$V(s) = G_0(s)\left[F_{ex}(s) - F_u(s)\right], \tag{11}$$

where the mapping $G_0 : \mathbb{C} \to \mathbb{C}$, defining the input-output dynamics $f_{ex} - f_u \mapsto v$, is given by

$$G_0(s) = \frac{H_r^D(s)s}{(M + m_\infty)H_r^D(s)s^2 + H_r^N(s)s + H_r^D(s)s_h}, \tag{12}$$

where the Laplace transform of the radiation impulse response, $H_r$, has been written, without any loss of generality, as $H_r(s) = H_r^N(s)/H_r^D(s)$. Given the causality property of the radiation force system $\Sigma_r$ (see, for instance, [16]), and the fact that $\Sigma_r$ is always strictly proper [8], the following relation

$$\deg\left\{H_r^N(s)\right\} < \deg\left\{H_r^D(s)\right\}, \tag{13}$$

holds. Direct observation of Equations (8) and (11) yields that, in the frequency-domain, the relation $I(\omega) = 1/G_0(\omega)$, holds. In other words, the dynamical system that is associated with the frequency-response $I(\omega)$ is inherently non-causal, as a direct consequence of the fact that the transfer function $G_0(s)$ is strictly proper (see Equation (13)). This poses a major issue with respect to the applicability of result (10): the dynamical system that is associated with the control law (10) cannot be practically implemented, due to its intrinsic non-causality. In addition to this non-causality issue, the following additional implications that are associated with the matching-principle can be identified:

- The optimal control law (10) implies a different matching-condition for each input-frequency $\omega$.
- Neither device nor actuator limitations are observed by the matching condition (10). As a matter of fact, this control strategy often requires unrealistic displacement, velocity, and control input values to successfully achieve maximum power absorption. This is a direct product of the linearising assumptions under which Equation (6) is derived.
- The stability, sensitivity, and robustness properties of the control loop associated with the impedance-matching principle of (10), depicted in Figure 1b, have been recently questioned in [17]. In particular, [17] shows that radiation damping modelling errors can be particularly detrimental in the impedance-matching condition, given that a very specific zero-pole cancellation takes place when $f_u$ is selected, as in (10).

Note that the force-to-motion (force-to-velocity in this case) frequency-response mapping, under impedance-matching condition (10), can be readily computed as

$$V(\omega) = \frac{G_0(\omega)}{1 + G_0(\omega)K_{fb}^{opt}(\omega)}F_{ex}(\omega) = \frac{1}{2B_r(\omega)}F_{ex}(\omega), \tag{14}$$

where $B_r$ is the radiation damping, as defined in Section 2. In particular, there exists an optimal real-valued scaling function $T^{opt} : \mathbb{R} \to \mathbb{R}^+$, which is given by

$$T^{opt}(\omega) = \frac{1}{2B_r(\omega)}. \tag{15}$$

Note that the image of $T^{opt}$ is effectively contained in $\mathbb{R}^+$, as a consequence of the passivity property of the radiation force, i.e., $B_r(\omega) > 0$, $\forall \omega \in \mathbb{R}/0$.

Although impedance-matching, as in (10), has some difficulties in practical application (for the reasons discussed above), it effectively describes the underlying dynamics behind maximum energy absorption, in an intuitive way. As a matter of fact, this principle underpins the family of simple controllers analysed in this study (see Table 2), which effectively attempt to provide implementable approximations of the control law derived in (10). The methodologies, which were employed by each of these five (5) controllers to approximate the impedance-matching condition (10), are discussed in detail in Section 4.

## 4. Simple WEC Controllers (in Chronological Order)

This section outlines the fundamentals behind each of the simple energy-maximising control strategies listed in Table 2, which are inherently based on the impedance-matching principle (as described in Section 3), proposed during the period between 2010–2020. In particular, the five (5) control strategies originally presented in [9–13], are recalled (in chronological order) in Sections 4.1–4.5, respectively. From now on, the shorthand notation introduced in Table 2 is used to refer to each specific strategy.

Note that this section does not simply recall results, but it also provides a critical analysis of each controller with respect to their potential success in a practical implementation, from a system dynamics perspective. In particular, the presented discussion aims to highlight the underlying simplicity of each presented controller, in terms of its applicability in the widest possible range of hardware platforms, including, for instance, low-cost microcontrollers. In the light of this, properties, such as nature of the control structure (i.e., linear, nonlinear, time-varying, etc.), as well as order, stability, and constraint handling capabilities (if any), are explicitly discussed, for each of the controllers that are listed in Table 2.

### 4.1. Suboptimal Causal Reactive Control (2011, C1)

This energy-maximising control strategy is essentially based on a velocity-profile tracking-loop (typical of approximate velocity tracking (AVT) WEC controllers [18]), schematically depicted in Figure 2, where $\hat{f}_{ex}(t)$ indicates the estimation of the wave excitation force $f_{ex}(t)$. Using the fundamental impedance-matching principle and the optimal mapping $T^{opt}(\omega)$, as described in Equation (15), the control methodology presented in [9], although being suboptimal, represents a suitable methodology for real-time application, considering that only linear time-invariant (LTI) systems are involved in the control design procedure (further discussed in the following paragraphs). It should be noted that, even though this control methodology requires the estimation of the wave excitation force (to effectively compute the velocity profile $v_{ref}$), such an estimate can be obtained by means of standard unknown-input observers. This includes, for instance, Kalman-based estimators, in combination with a harmonic description of the wave excitation input [19].

Essentially, this control technique approximates the optimal mapping $T^{opt}(\omega)$ as follows:

$$T^{opt}(\omega) = \frac{1}{2B_r(\omega)} \approx \frac{1}{2\hat{H}}, \tag{16}$$

where $\hat{H} \in \mathbb{R}^+$. In particular, the study that is presented in [9] computes the constant gain $\hat{H}$ using a second order approximation of the force-to-velocity WEC model:

$$G_0(\omega) \approx \tilde{G}_0(\omega), \tag{17}$$

where $\tilde{G}_0 : \mathbb{R} \to \mathbb{C}$ represents the frequency-response mapping of a second-order system, obtained by means of a model reduction routine that is based on Hankel singular values (see, for instance, [20]). Thus, if $G_0(\omega) \approx \tilde{G}_0(\omega)$, then the authors of the C1 controller rely on the following relation

$$B_r(\omega) \approx \tilde{B}_r = \hat{H}, \tag{18}$$

with $\tilde{B}_r \in \mathbb{R}^+/0$, being the approximation in Equation (18) valid (i.e., accurate) for a certain range of $\omega$.

Note that, in order to obtain a system $\tilde{G}_0(\omega)$ that satisfies Equation (18), while preserving the band-pass nature of the force-to-velocity mapping of a WEC system, a zero at the origin can be forced in the determination of $\tilde{G}_0(\omega)$. Nevertheless, only the order (number of eigenvalues) and stability of the approximating system can be handled using model reduction routines that are based on Hankel singular values. Note that, when the zero at the origin is guaranteed, and $\tilde{G}_0(s)$ consequently represents a band-pass structure, then:

$$\tilde{B}_r = \mathbb{Re}\left\{ \frac{1}{\tilde{G}_0(\omega)} \right\}, \tag{19}$$

where $\tilde{G}_0(\omega)$ is, as a consequence of Equations (18) and (19), a passive system. However, in practical terms, the value of $\hat{H}$ can be determined, depending on, for example, the spectral content of the sea-states characterising the operating conditions of the specific device being controlled. In other words, one can select $\hat{H}$ merely as the radiation damping of the device, i.e., $B_r(\omega)$, evaluated at the frequency associated with the most energetic waves, in the total wave power spectrum.

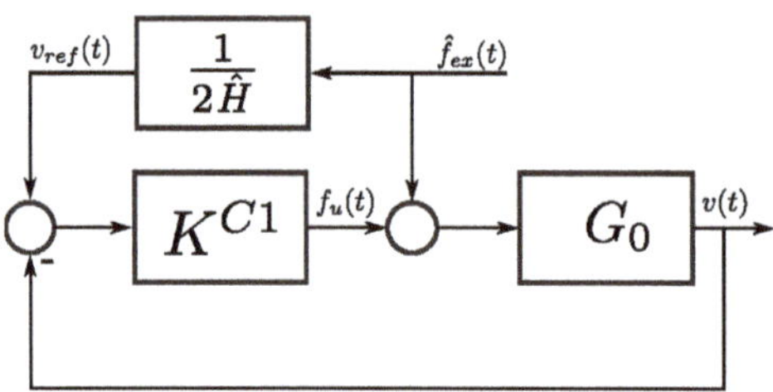


**Figure 2.** C1 control scheme presented in [9].

It should be noted that, depending on the operating conditions (i.e., sea-states under analysis), the optimal $\hat{H}$, which maximises the generated power for particular control specifications, can be potentially tuned while using a higher-order approximation of the WEC dynamics (rather than second-order, as per the original design behind the C1 controller). Furthermore, when considering that this control strategy can only interpolate the optimal condition at a finite number of frequencies[3], then only monochromatic (regular) sea-states can be efficiently addressed, in terms of maximum theoretical power production, with this energy maximising control technique.

Concerning the stability of the C1 controller, note that the system $1/2\hat{H}$ is an open-loop static mapping, and the definition of $\hat{H}$ does not affect the stability of the complete control loop;

---

[3]  Take into account that with this control strategy the optimal condition is approximated using a constant value.

therefore, $1/2\hat{H}$ can be specifically tuned for different cases, considering a number of different design criteria, driven by the operating conditions.

On the other hand, under certain assumptions, which, when considering the passivity of $G_0(s)$, are guaranteed for WEC systems, the tracking feedback controller $K^{C1}$ (see Figure 2) can be synthesised, as suggested in [9], while using the Youla-Kučera parametrisation for stable systems [21], i.e.,

$$K^{C1}(s) = \frac{Q(s)}{1 - Q(s)G_0^{-1}(s)},\tag{20}$$

where $Q(s) = F(s)G_0^{-1}(s)$ and, for example,

$$F(s) = \frac{\frac{\omega_c}{q_f}s}{s^2 + \frac{\omega_c}{q_f}s + \omega_c^2},\tag{21}$$

which represents a classical band-pass filter, such that $F(\omega_c) = 1$, and where $q_f \in \mathbb{R}^+/0$ indicates the band-width of the filter. However, note that the design of the internal tracking controller $K^{C1}$ can be performed using a wide variety of control techniques, including, for instance, $\mathcal{H}_\infty$-techniques, to cover system uncertainties, or sliding-modes methodologies, to tackle potential nonlinearities arising in the WEC dynamical system. From a stability perspective, which, as discussed in Section 3, represents a well-known issue in impedance-matching based controllers, this control strategy can be safely designed without compromising closed-loop stability.

Finally, note that this control strategy itself is purely based on LTI systems: both the computation of the reference velocity profile, $v_{ref}$, and the tracking controller, $K^{C1}$, involve LTI systems. Regarding the order of the C1 controller when the above Youla-Kučera parametrisation technique is considered, i.e., Equations (20) and (21), the order of the controller $K^{C1}(s)$ is given by $n + 1$, where $n$ denotes the order of the WEC model $G_0(s)$.

### 4.2. Simple and Effective Real-Time Controller (2013, C2)

The C2 controller, as schematically depicted in Figure 3, has one particular appealing feature: It not only provides a simple a approximation of the impedance-matching principle, but also incorporates a constraint handling mechanism. This controller, together with the C5 (see Section 4.5), are the only two simple controllers presented in the literature which are capable of handling physical constraints as part of their design.

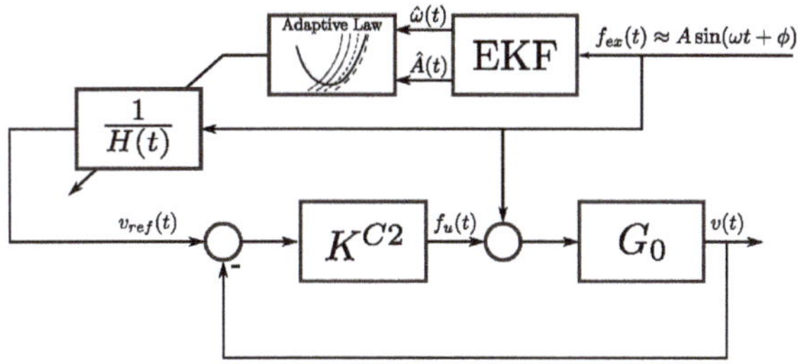


**Figure 3.** C2 control scheme presented in [10].

The C2 controller arises as an extension of the controller presented in [9], as detailed here in Section 4.1. In particular, the wave excitation force, as considered for this control strategy, is approximated using a monochromatic sinusoidal process, i.e.,

$$f_{ex}(t) \approx A(t)\sin(\omega(t)t + \phi(t)),\tag{22}$$

with parameters $\{A, \omega, \phi\} \subset \mathbb{R}$, which can vary with time. Similarly to the C1 controller, the C2 is essentially based on a velocity-profile tracking-loop (AVT controller). Unlike its predecessor, the C2 strategy features an adaptive process, where the velocity reference profile is updated in real time by means of an Extended Kalman filter (EKF), which provides an instantaneous estimate of the set of parameters $\{A, \omega, \phi\}$, fully characterising the excitation force as in (22). Such an estimation is motivated by two clear objectives, as detailed in the following. Firstly, knowledge of these estimates is used to improve the block $\frac{1}{2\hat{H}}$ that is proposed in the C1 controller (as shown in Figure 2), by means of a continuous-time adaptation procedure, aiming to approach the optimal condition $T^{opt}(\omega)$, stated by the impedance-matching principle. Secondly, the information provided by the estimation is used to normalise the wave excitation force estimate and compute a suitable velocity profile, according to pre-defined constraint specifications. To be precise, the information provided by the EKF is explicitly employed in the computation of the following time-varying scaling function:

$$\frac{1}{H(t)} = \begin{cases} \frac{1}{2B_r(\hat{\omega}(t))}, & \text{if} \quad \frac{\hat{\omega}(t)X_{lim}}{\hat{A}(t)} > \frac{1}{2B_r(\hat{\omega}(t))}, \\ \frac{\hat{\omega}(t)X_{lim}}{\hat{A}(t)}, & \text{if} \quad \text{otherwise,} \end{cases} \tag{23}$$

where $X_{lim}$ represents the maximum displacement limit. Such a mapping $H(t)$ is schematically shown in Figure 3, where $\hat{A}(t)$, $\hat{\omega}(t)$ represent the instantaneous estimates of amplitude $A$ and frequency $\omega$, both in Equation (22), respectively. Finally, the feedback tracking controller, denoted as $K^{C2}$ in Figure 3, can be computed while using the Youla-Kučera parametrisation, as per the case of the C1 controller, described in Section 4.1. To be precise, the controller $K^{C2}$ can be designed following the expressions provided in Equations (20) and (21). It is important to highlight that, although the estimation process in the C2 controller is performed in terms of an EKF, a wide variety of estimation techniques could be potentially employed to provide instantaneous estimates of $A$ and $\omega$ (see, for instance, [19]).

There are some aspects, both positive and negative, related to the C2 controller, and its applicability in realistic environments, which are worth mentioning. On the positive side, the C2 controller can effectively handle displacement limitations in terms of the design parameter $X_{lim}$. This is clearly fundamental in any practical application, where compliance with physical restrictions must be guaranteed by the energy-maximising control strategy, hence maximising energy absorption while minimising the risk of component damage. Nonetheless, on the negative side, and taking into account the challenge involved in the tuning of the corresponding estimator, and its sensitivity to design parameters [19], the inclusion of the EKF can potentially impact (negatively) on the resulting performance, as detailed in the following. In particular, when considering that the wave excitation force can be expressed as in Equation (22), methodologies that instantaneously estimate the amplitude $A$ and frequency $\omega$, virtually always assume that $A$ is approximately constant. In addition, the computation of an estimate of the frequency $\omega$, in Equation (22), inherently requires a non-linear estimation process [19]. The necessity of assuming a low variation rate for the amplitude modulation $A(t)$ (see [19]), and the fact that the design and calibration of the nonlinear estimation process that is required to compute the instantaneous input frequency is substantially challenging, can potentially degrade the quality of the approximating $f_{ex}$ defined in (22). This naturally has a direct impact on the quality of the energy-maximising solution provided by the C2 controller (which explicitly uses this estimate in order to compute the corresponding control force), both in terms of performance, and satisfaction of physical limitations (i.e., constraints).

From a stability analysis point of view, this controller presents an advantage, directly related to the simple LTI velocity tracking controller $K^{C2}$: In other words, one can always guarantee stability in the C2 controller as long as the tracking loop is stable, and the EKF strategy converges towards a bounded wave excitation estimate. However, even though the reference tracking controller is based on a LTI system, this control methodology cannot be considered LTI altogether, due to the presence of the time-varying prefilter $1/H(t)$, which is involved in the computation of the velocity profile

$v_{ref}$. Regarding the feedback controller order, when the Youla-Kučera parametrisation is considered (i.e., as in Equations (20) and (21)), the order of $K^{C2}(s)$ is given by $n + 1$, where $n$ denotes the order of the WEC model $G_0(s)$. Finally, note that convergence of the EKF estimation, towards the real wave excitation force time-trace, cannot be generally guaranteed, mainly because such a force cannot be always expressed as in Equation (22). In other words, this controller inherently assumes that the device is subject to waves that arise from a narrowbanded sea-state, which is not always the case in practice.

### 4.3. Multi Resonant Feedback Control (2016, C3)

The controller that is presented in [11], i.e., the C3 controller, is strictly based on a feedback structure. It is important to note that, in contrast to the previously presented two controllers (Sections 4.1 and 4.2), the feedback structure does not require wave excitation force estimates, which positively impacts on its associated computational complexity. Aiming to address the energy-maximisation problem in a broadband sense, the design of the C3 controller is based on a specific approximation of the impedance-matching condition using a frequency-domain discretisation, i.e., considering a finite set of frequencies as,

$$\Omega = \{\omega_1, \ldots, \omega_N\} \subset \mathbb{R}^+. \tag{24}$$

To this end, and similarly to the case that is presented in Section 4.2, the wave excitation force is approximated as a finite sum of $N$ purely sinusoidal processes, i.e.,

$$f_{ex}(t) \approx \sum_{i=1}^{N} A_i^{ex} \sin(\omega_i t + \phi_i^{ex}). \tag{25}$$

Subsequently, given the linearity associated with the WEC model $G_0(s)$, the device velocity can be consequently described as:

$$v(t) \approx \sum_{i=1}^{N} v_i(t) = \sum_{i=1}^{N} A_i \sin(\omega_i t + \phi_i). \tag{26}$$

Note that the C3 control strategy, as presented in [11], uses position as the default output of the WEC system $G_0(s)$. However, aiming to be consistent with all the (other) controllers that are considered in this study, and without any loss of generality, the velocity is used in this study as the output of the WEC system. Figure 4 schematically depicts the C3 control strategy.

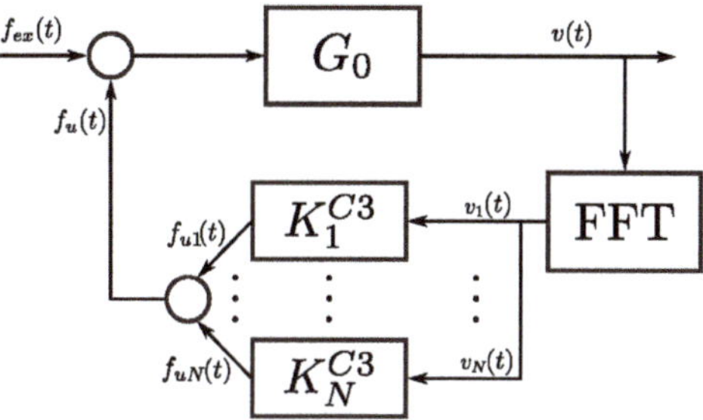

**Figure 4.** C3 control scheme presented in [11].

As a consequence of the assumption regarding the wave excitation force and velocity stated in Equations (25) and (26), the authors of [11] separate the system $G_0(s)$ into $N$ second-order subsystems, i.e., $\{G_i(s)\}_{i=1}^{N}$, such that

$$G_0(\omega_i) = G_i(\omega_i), \tag{27}$$

for all $\omega_i \in \Omega$, with $i \in \mathbb{N}_N$. The objective sought with the separation, as shown in Equation (27), is to deal with simpler individual systems $G_i(s)$ where, given the second-order nature of each subsystem $G_i(s)$, the following result

$$\mathbb{R}\mathrm{e}\left\{\frac{1}{G_i(\omega)}\right\} = B_r^i, \tag{28}$$

with $B_r^i \in \mathbb{R}^+$, for all $i \in \mathbb{N}_N$, holds. The relation that is posed in Equation (28) is a key factor in the definition of the optimal control condition exploited by the C3 controller: instead of solving the complete energy maximising control problem, i.e., considering the optimal frequency-domain impedance-matching condition for all $\omega \in \mathbb{R}$, the C3 control strategy aims to solve the impedance-matching problem for each subsystem $G_i(s)$, which is intrinsically related to each $B_r^i$, defined in Equation (28). In particular, the energy-maximising control problem is tackled while using a set $\mathbf{K}^{C3}$ of $N$ proportional-integral (PI) controllers, i.e., $\mathbf{K}^{C3} = \left\{K_i^{C3}(s)\right\}_{i=1}^{N}$, with each $K_i^{C3}$ defined as

$$K_i^{C3}(s) = \left(P_i^{C3} + \frac{I_i^{C3}}{s}\right), \tag{29}$$

where the set of parameters $\{P_i^{C3}, I_i^{C3}\}_{i=1}^{N} \subset \mathbb{R}$ is designed to approach the impedance-matching optimal condition for each subsystem $G_i(s)$.

Note that the control problem that is posed in [11] is stated as an optimisation problem, where the values $A_i$ and $\phi_i$, as expressed in Equation (26), are estimated in real-time. In other words, the determination of the instantaneous estimate of each $A_i$ and $\phi_i$, in Equation (25), plays a key role in this control strategy, and is addressed by the C3 in terms of an optimisation problem. In particular, the estimation required in [11] is tackled by means of a real-time implementation based on the fast Fourier transform (FFT), as shown in Figure 4. Note that different (time) window lengths are utilised in the case studies that are presented in [11], aiming to analyse their effect in the final energy-maximising performance.

From a general perspective, mainly concerning the applicability and the stability of this control strategy, some aspects are worth mentioning. Firstly, the use of a real-time FFT and the time window required for its implementation generates a time-delay between measurement and control actuation. It is important to highlight that time precision plays a decisive role in real-time implementation, and can intrinsically affect both the stability properties of the control loop, and its performance in terms of energy capture. By the way of example, and in order to highlight the importance of having good timing in realistic control implementations, the use of a real-time operating system (RTOS), such as, for example real-time LabView, Matlab xPC Target, RTOSs for the microcontroller architecture, or even FPGA-based systems, is a common practice for realistic control implementation. This effectively reduces the time-delay between measurement and control actuation (i.e., latency or jitter). Note that the FFT procedure can be potentially replaced by suitable recursive least square (RLS) routines, such as those that are described in [22][4]. This allows for a potentially more computationally efficient implementation than its FFT counterpart, which would be appealing for real-time applications. Secondly, the methodology that is employed in [11] to define $G_i(s)$ is not clearly stated. Note that there exists an infinite number of possible second-order systems $G_i(s)$ fulfilling Equation (27), which can lead to different control scenarios. In addition, even though a stability analysis is performed in [11] while using the classical Routh–Hurwitz criterion (see, for instance, [23]) for each controller-plant pair $(K_i^{C3}, G_i)$, it is not clear how the stability of the complete interconnection, i.e., $(\mathbf{K}^{C3}, G_0)$, can be guaranteed. Furthermore, the authors of [11] do not

---

[4]    As a matter of fact, using RLS algorithms effectively resembles optimisation-based spectral techniques based on trigonometric basis functions (see, for instance, [22]).

take into account the estimation of each $A_i$ and $\phi_i$ in the closed-loop stability analysis, which further increases the degree of uncertainty behind the practicality of this solution. Note that such an estimation process depends upon a time-dependent optimisation routine, which automatically renders the control loop as time-varying, commonly requiring converse Lyapunov theory (see, for instance, [24]) to guarantee uniform stability, which intrinsically complicates the nature of the problem.

Finally, note that this controller cannot be considered a LTI system, due to the optimisation process that is involved in the feedback path. In addition, regarding the order of the set of controllers $\mathbf{K}^{C3}$, when considering the PI structure presented in Equation (29), the total order always matches the number of elements considered in the frequency set $\Omega$, i.e., there is one PI (first-order) controller for each element of the set $\Omega$.

### 4.4. Feedback Resonating Control (2019, C4)

As in the case detailed in Section 4.3, the energy maximising control strategy presented in [12], denoted here as C4, is essentially based on a classical feedback control structure, as depicted in Figure 5. Thus, as for the C3 controller, knowledge of the wave excitation force is not required to implement the C4 control strategy, which significantly reduces the controller complexity and simplifies its implementation, as well as contributing to improving the resulting performance, as long as some guarantee of closed-loop stability can be provided.

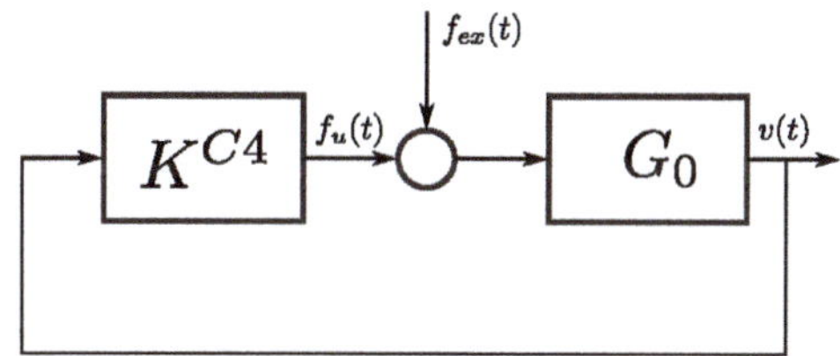


**Figure 5.** The C4 control scheme presented in [12].

In the energy-maximising control strategy proposed in [12], the controller is synthesised while using system identification algorithms, aiming to approximate the frequency-domain feedback optimal control condition, i.e., Equation (10). The control strategy is schematically depicted in Figure 5. In particular, the proposed controller has a fixed rational polynomial structure, as follows

$$K^{C4}(s) = \frac{b_2 s^2 + b_1 s + b_0}{s^2 + a_1 s + a_0}, \tag{30}$$

which structurally considers two poles and two zeros, with $\{b_0, b_1, b_2, a_0, a_1\} \subset \mathbb{R}$. Subsequently, in order to approximate the impedance-matching optimal condition over a certain frequency range, the system in Equation (30) is designed to satisfy the following relation:

$$K^{C4}(\omega) \approx K_{fb}^{opt}(\omega). \tag{31}$$

with $K_{fb}^{opt}(\omega)$ defined in Equation (10). Given that input-output stability can be generally set as a requirement in the majority of standard identification approaches available in the literature [25], the stability of the controller can be straightforwardly satisfied, which is recommended for any realistic control system implementation. However, this does not necessarily guarantee closed-loop stability for the system that is depicted in Figure 5; additional requirements need to be imposed on the structure in (30) in order to secure stable closed-loop behaviour, such as, for instance, passivity.

In addition, the simplicity of the controller structure, given its second-order nature, is worth highlighting. Furthermore, recalling that stability represents a key issue for controllers that are based on the impedance-matching principle, the C4 control methodology can be relatively easily designed in order to guarantee both controller and, more important, closed-loop, stability, although the latter is not theoretically addressed in [12]. Nevertheless, the lack of a constraint handling mechanism can render the C4 controller unsuitable for realistic implementations.

From a dynamical systems perspective, this controller is based on a single LTI system. In addition, considering the fixed structure that is presented in Equation (30), the controller order is always set to 2.

### 4.5. LiTe-Con (2020, C5)

Similarly to the case that is presented in Section 4.4, the so-called LiTe-Con [13] (referred to here as C5), is synthesised using system identification algorithms, aiming to approximate the frequency-domain energy-maximising optimal condition. In particular, based on the impedance-matching feedback law (10), in order to provide an alternative energy maximising control solution that is capable of dealing with the stability issues detailed in Section 1, the authors of [13] recast the controller solution into an equivalent feedforward structure, i.e.,

$$K_{ff}^{opt}(\omega) = \frac{\mathbb{Re}\left\{G_0(\omega)\right\} + j\mathbb{Im}\left\{G_0(\omega)\right\}}{2\mathbb{Re}\left\{G_0(\omega)\right\}},\tag{32}$$

where the mapping $f_{ex} \mapsto v$, corresponding with the feedforward structure (32), is equivalent to that presented in Equation (15). Note that, in contrast to C4, which does not require estimation of the wave excitation force, the feedforward control structure requires the estimation of the excitation force of the wave. Thus, the requirement of having an estimate of $f_{ex}(t)$, can negatively impact on the resulting performance [26]. The aspect that is related to the requirement of having an estimate of the wave excitation force can be considered as the essence of the distinction between C4 and C5. The non-parametric frequency-response mapping $K_{ff}^{opt}$ is then approximated by means of black-box system identification techniques, as further discussed in the upcoming paragraphs.

Given the inherent feedforward structure of the C5 controller, estimation of the wave excitation force is required. The control structure C5, as presented in [13], is schematically depicted in Figure 6, where $\hat{f}_{ex}$ denotes the estimate of the wave excitation force $f_{ex}$, and $K^{C5}$ is the C5 controller. Note that a constraint handling mechanism is proposed in C5, extending its applicability range to realistic environments.

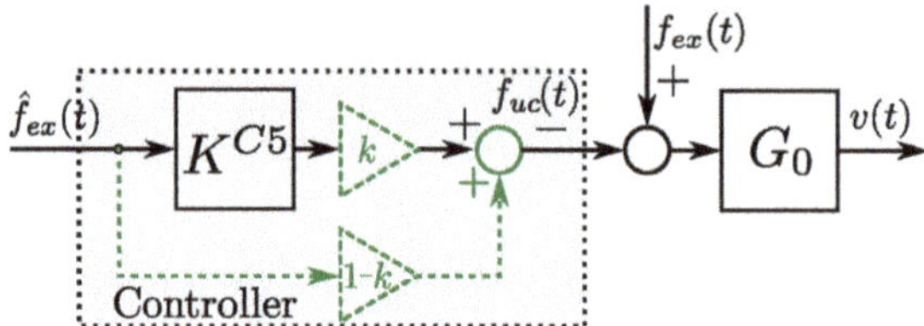


**Figure 6.** C5 control scheme presented in [13]. The constraint handling mechanism of the C5 is indicated using a shadowed-grey box, using dashed-green line to represent the internal block interconnections.

The controller methodology that is presented in [13] proposes the approximation of the optimal impedance-matching controller $K_{ff}^{opt}$, as defined in Equation (32), with a LTI-stable and implementable dynamical system $K^{C5}$, such that

$$\left. K^{C5}(s)\right|_{s=jw} \approx K_{ff}^{opt}(\omega),\tag{33}$$

where $K^{C5}(s)$ is obtained using frequency-domain system identification algorithms, such as subspace [25] or moment-matching-based [27] system identification techniques. Subsequently, the resulting control force (in the frequency domain) is expressed as:

$$F_u(\omega) = K^{C5}(\omega)\hat{F}_{ex}(\omega).\tag{34}$$

In order to implement the synthesised C5 controller in a realistic environment, a constraint handling mechanism is essential to prevent inflicting damage on the mechanical system. In particular, [13] proposes a constraint handling mechanism, using a constant value $k \in [0, 1]$, so that the control force $F_u$ in Equation (34) is redefined as:

$$F_{uc}(\omega) = \underbrace{\left[ kK^{C5}(\omega) + (1 - k) \right]}_{\text{Controller}} \hat{F}_{ex}(\omega), \tag{35}$$

where it is straightforward to check that $F_{uc}(\omega) = F_u(\omega)$ when $k = 1$. Assuming perfect matching in Equation (33), the optimal mapping in Equation (15) can be defined as:

$$T^{C4}(\omega) = k \left( \frac{\mathbb{Re}\left\{G_o(\omega)\right\}^2 + \mathbb{Im}\left\{G_o(\omega)\right\}^2}{2\mathbb{Re}\left\{G_o(\omega)\right\}} \right). \tag{36}$$

Note that, if $k = 1$ in Equation (36), then the optimal condition in Equation (15) is obtained, i.e., $T^{C4}(\omega) = T^{opt}(\omega)$. On the other hand, if $k = 0$, then $T^{C4}(\omega) = 0$. In other words, the inclusion of the term $k$ allows for a simple implementation of position and velocity constraints, effectively constraining the position and velocity between zero and their theoretical maxima. The constraint handling mechanism of the C5 is indicated in Figure 6 using a dotted-grey box, while dashed-green lines are used to represent the internal block interconnections.

Some comments that are related to this control strategy and its applicability, particularly focusing on stability features, can be mentioned. On the positive side, the feedforward nature of the C5 controller requires a simple and relatively effortless implementation in real-world applications, while also effectively taking into account device limitations in terms of a simple constraint mechanism. Nonetheless, the C5 controller requires an estimate of the wave excitation force, which can potentially have a negative impact on the energy-maximising performance of the controller [28] and computational complexity. However, in contrast to, for instance, the C2 controller (which uses requires an EKF structure), the estimation that is required by the C5 can be addressed while using relatively standard wave excitation force estimation techniques. In particular, a LTI Kalman filter[5], featuring a harmonic description for ocean waves, is considered in [28]. As demonstrated in [19], this observer provides good overall estimation quality, while inherently handling measurement noise (in an optimal sense). From a stability perspective, the complete control structure stability (WEC system, excitation force estimator, and controller) is guaranteed as long as each individual component (WEC model, estimator, and controller) is stable, i.e., under the separability principle [23].

Finally, note that this control strategy is purely based on LTI systems (even taking into account the required estimation process for the wave excitation force). Regarding the order of the controller, it directly depends upon the system identification process that is employed to compute $K^{C5}$. For instance, using a moment-based identification approach [27], where the user can select a number of interpolation frequencies (or points) to preserve steady-state response characteristics (see Section 5.1.5), the order of the controller $K^{C5}$ is twice the number of matching points.

## 5. Results

This section presents a case study, where the performance obtained with the set of five (5) controllers presented in Section 4, i.e., C1, C2, C3, C4, and C5, is assessed. Aiming to provide a set of benchmark cases in terms of maximum achievable absorbed energy, two reference measures are considered. Firstly, for the unconstrained case, the performance that is obtained with each controller is compared with the corresponding theoretical maximum, which is analytically computed while

---

[5]    Infinite horizon version [23].

using Equation (15). Such a benchmark case is denoted here as CR1. Secondly, in the constrained case, where, unlike its unconstrained counterpart, there is not an explicit analytical formulation for the maximum achievable performance, an optimisation-based controller is considered. Even though the resulting performance that is obtained with optimisation-based formulations is not a theoretical maximum per-se (due to any potential errors arising from the intrinsic necessity of discretisation and use of numerical routines), it is considered here as a surrogate reference for maximum achievable performance in a constrained scenario. In particular, a moment-based controller [29], denoted in this study as CR2, is considered in this application case for the performance comparison in the constrained case, analogously to the performance assessment that is presented in [13]. It is important to note that, for the constrained case, only the C2 and C5 control methodologies can be considered for the performance assessment presented in this section: the controllers C1, C3, and C4 do not provide any constraint handling mechanism at all, which directly compromises their application in realistic scenarios. This is further discussed in the following paragraphs.

The control strategies are applied to a state-of-the-art full-scale CorPower-like device oscillating in heave (translational motion) [30]. This type of device is often considered throughout the WEC literature, being one of the most well-established WEC system within the wave energy field (see, for example, [31] or [32]), and it is illustrated in Figure 7 with its corresponding physical dimensions specified in meters. To perform the simulations and analysis shown in this section, the considered the WEC model, as in Figure 7, is described using an seventh-order LTI state-space representation, which is denoted in the frequency-domain as $G_0(\omega)$.

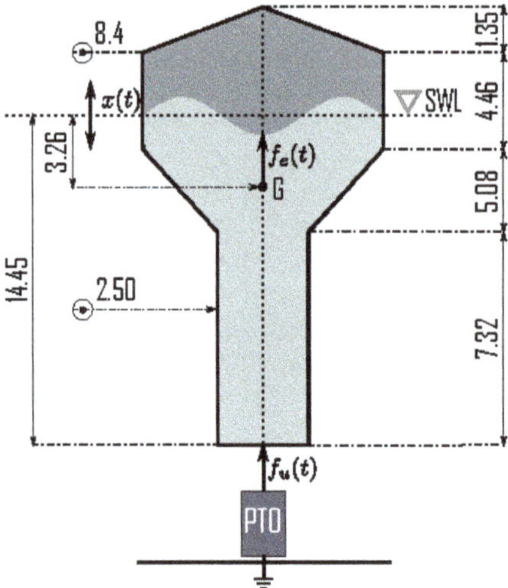


**Figure 7.** Full-scale CorPower-like device considered in this case study. Dimensions are in metres. The acronym SWL stands for still water level and the letter G is used to denote the center of gravity of the device. The lower side of the power take-off is anchored to the sea bed, which provides an absolute reference for device motion. The displacement of the device is denoted by $x(t)$.

In this application case, the time-trace of the wave excitation force, $f_{ex}(t)$, considering irregular waves, is determined from the so-called free-surface elevation, $\eta(t)$, based on a JONSWAP spectrum $S_\eta(\omega)$ [33]. The corresponding sea-state parameters, which characterise the nature of the mapping $S_\eta$, are as follows: Peak period in the interval $T_p \in [5.0, 12.0]$ s, significant wave height $H_s = 2.0$ m, and a peak-enhancement factor $\gamma = 3.3$. In Figure 8, $S_\eta(\omega)$ is depicted for $T_p = \{5, 6.4, 7.8, 9.2, 10.6, 12\}$ s, where $T_p = 5$ s and $T_p = 12$ s indicate the extremes of the considered range for $T_p$.

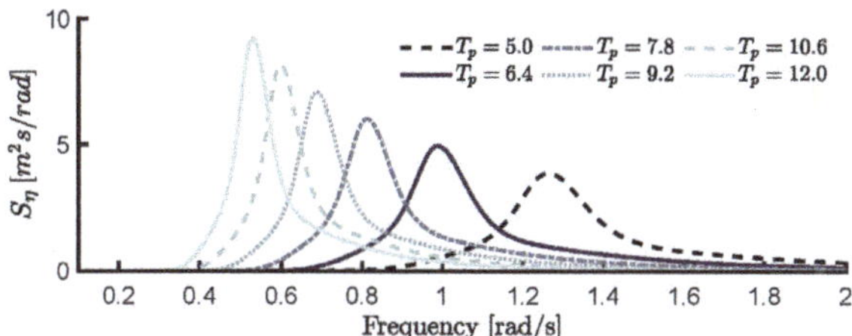

**Figure 8.** $S_\eta(\omega)$ for $T_p = \{5, 6.4, 7.8, 9.2, 10.6, 12\}$ s.

The power spectral density of the excitation force is given by the relation $|F_{ex}(\omega)| = |G_\eta(\omega)||S_\eta(\omega)|$, where $G_\eta(\omega)$, obtained using NEMOH [34], represents the frequency-response function associated with the mapping $\eta \mapsto f_{ex}$. The resulting performance is studied in both unconstrained, and constrained, scenarios, as detailed previously in this same section. In particular, in the constrained case, the maximum allowed displacement is set to $X_{max} = 1.5$ m. In this study, $\eta(t)$ is generated using a white noise signal, filtered according[6] to the spectrum $S_\eta(\omega)$ [35].

Throughout this section, performance is assessed in terms of the average absorbed power, which is evaluated as

$$P_{abs} = -\frac{1}{T} \int_0^T f_u(t)v(t)dt. \tag{37}$$

Note that, when considering the stochastic nature of the process involved, and to be statistically consistent, the results that are shown in this study are always representative average values, which are generated when considering 20 realisations of each specific sea-state.

Controllers C1, C2, and C5 require (potentially different) wave excitation force estimation procedures, as detailed throughout Section 4. For the case of C1 and C5, which rely upon an instantaneous estimate of the excitation force time-trace $f_{ex}(t)$, a linear Kalman filter is utilised, in combination with the internal model principle of control theory [36]. The design and synthesis procedure for such an observer, which is based upon a harmonic internal description of ocean waves, is well-established in the literature of WEC control, and it can be consulted elsewhere (see, for instance, [19]). Note that, in this study, the infinite-horizon Kalman gain is always utilised, i.e., the solution to the infinite-horizon algebraic Riccati equation, which directly implies that the associated estimator is LTI. The estimation requirements for controller C2 are different, and more 'sophisticated' methods than those associated with C1 and C5 are needed. In particular, this strategy relies upon having estimates of instantaneous amplitude $\hat{A}(t)$ and frequency $\omega(t)$, which, as previously discussed, inherently require a nonlinear estimation procedure, given the nature of the internal model used to describe the wave excitation force (see Equation (22)). As detailed in Section 4.2, an EKF is, as proposed in [10], considered to perform this task[7], which is designed and tuned following [10].

Although beyond the scope of this study, where 'idealised' conditions are assumed for simulation to guarantee a level playing field for the totality of the controllers studied, i.e., perfect knowledge of the WEC dynamics and noise-free measurements are considered, the reader is referred to, for instance, [17], for further detail on the impact of modelling mismatch in both feedback and feedforward structures, and [19,28], for potential performance deterioration caused by noisy measurements in wave excitation force estimators.

The remainder of this section is organised, as follows. Section 5.1 presents the design procedure of each controller recalled in Section 4. Sections 5.2.1 and 5.2.2 show the performance results for each

---

6    For a detailed discussion about the synthesis of stochastic processes, the interested reader is referred to [35].

7    An alternative to the use of an EKF can be found in [37], where estimates of $\hat{A}$ and $\hat{\omega}$ are computed based on the Hilbert–Huang transform [38].

controller, in both unconstrained, and constrained scenarios, respectively, while Section 5.3 discusses the time-domain behaviour obtained with each controller.

### 5.1. Design Procedures

This section outlines the design procedure of each controller presented in Section 4. Throughout the following paragraphs, detailed comments and insight, with respect to design process, are also included, when appropriate.

### 5.1.1. C1

The design procedure for the C1 controller, as presented in Section 4.1, can be separated into two clear stages. Firstly, the constant function $1/2\hat{H}$ has to be designed, so that a suitable velocity reference $v_{ref}$ is generated. The second stage involves the design and synthesis of a closed-loop controller, which is able to track such a velocity reference profile.

To approximate the optimal mapping $T^{opt}(\omega)$, in terms of $1/2\hat{H}$, note that the waves that are considered in this study are generated using a stochastic description with peak period $T_p \in [5, 12]$ s, i.e., waves with significant energy components in a frequency range of approximately $[0.3, 2]$ rad/s (see Figure 8). Motivated by this, the value of $\hat{H}$ is determined to be as representative as possible in such a frequency range, as depicted in Figure 9. In particular, the obtained $1/2\hat{H}$ is shown using a dashed line, while the optimal mapping $T^{opt}(\omega)$ is depicted with solid line.

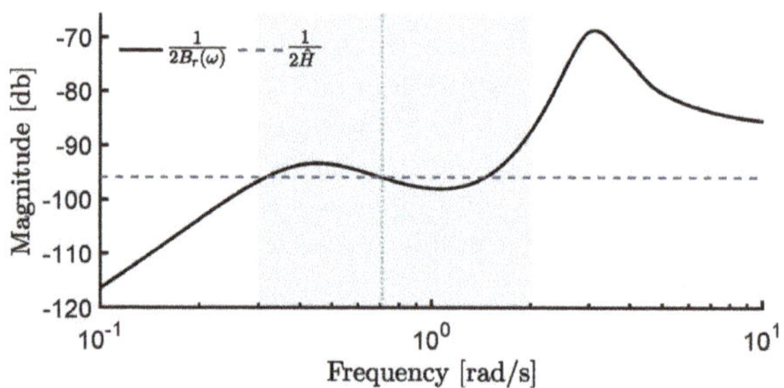


**Figure 9.** C1 design process. The optimal mapping, $T^{opt}(\omega) = 1/2B_r(\omega)$, and the obtained approximation, $1/2\hat{H}$, are depicted while using solid and dashed lines, respectively.

Note that, in Figure 9, the mean period $T_p = 8.5$ s (which is obtained as the average between the extreme of the complete range) is depicted using a dotted line. Additionally, the corresponding frequency range, i.e., $\approx [0.3, 2]$ rad/s, is depicted with a grey box. By the direct observation of Figure 9, one can appreciate that the best approximation of this optimal mapping $T^{opt}(\omega)$, provided by the constant gain $\hat{H}$, happens precisely at $2\pi/8.5$ rad/s, i.e., the C1 controller interpolates $T^{opt}(\omega)$ at $T_p = 8.5$ s.

As can be directly recalled from Section 4.1, the design procedure that is associated with the reference tracking loop is addressed while using a standard Youla-Kučera parametrisation for stable systems, which can be directly considered for the WEC case due to its passive and non-minimum phase nature [8]. To be precise, the velocity tracking loop, as depicted in Figure 2, can be synthesised using the following structure:

$$K^{C1}(s) = \frac{Q(s)}{1 - Q(s)G_0(s)^{-1}},\qquad(38)$$

where $Q(s) = F(s)G_0(s)^{-1}$ and

$$F(s) = \frac{\frac{\omega_c}{q_f}s}{s^2 + \frac{\omega_c}{q_f}s + \omega_c^2},\qquad(39)$$

with $\omega_c = 2\pi/8.5$ and $q_f = 0.1$. Thus, a LTI feedback controller $K^{C1}$, with order 8, is obtained. Figure 10 depicts the resulting mapping $v_{ref} \mapsto v$, i.e., the closed-loop frequency-response from the reference velocity, to the actual velocity of the CorPower-like device. A flat 0 dB response (both in terms of magnitude and phase) is achieved in the frequency range of interest, effectively indicating that the design criteria are being met, i.e., the output velocity is approximately following the input reference velocity, according to the spectral description of the ocean waves that are considered in this study.

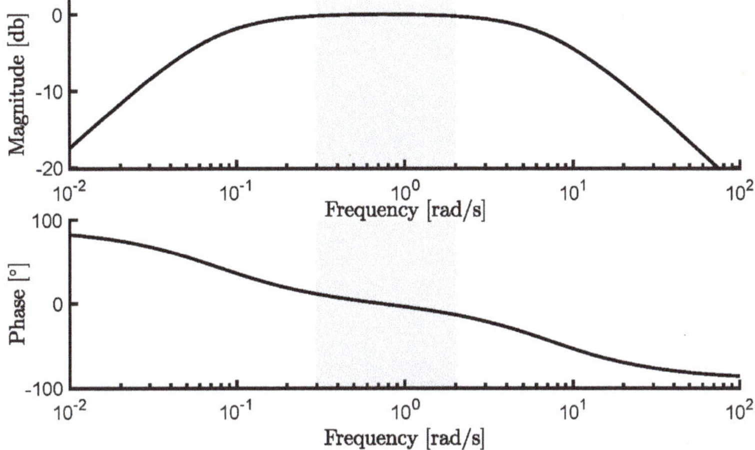

**Figure 10.** Obtained mapping between the reference velocity, and the actual velocity of the CorPower-like device, for the C1 and C2 controllers.

### 5.1.2. C2

This section describes the design procedure that is required by the C2 controller, schematically depicted in Figure 3. Similarly, as in the case presented in Section 5.1.1 for the C1 strategy, the design of the C2 controller can also be separated into two clearly distinctive stages: A velocity reference generation procedure, followed by a suitable closed-loop tracking mechanism. The reference tracking loop required by this control strategy can be analogously designed to that presented for the C1 controller in Equations (38) and (39), so the same tracking controller is utilised for the C2 controller, exhibiting the input-output behaviour previously illustrated (and described) in Figure 10, i.e., $K^{C2}(s) = K^{C1}(s)$.

The main difference between the C2 controller and its predecessor (i.e., the C1 controller) lies in the generation of the velocity profile, and its subsequent impact in final energy absorption. In the case of the C2 controller, the reference profile is generated in terms of instantaneous estimates of frequency and amplitude of the wave excitation force signal, computed by means of an EKF strategy. By way of example, Figure 11 shows estimation results for instantaneous amplitude $\hat{A}(t)$ (left axis—dashed) and frequency $\hat{\omega}(t)$ (right axis—dotted), for a particular sea-state realisation (left axis—solid), with $T_p = 8.5$ s.

Note that the sea states that are selected for this case study (see Figure 8) are based on a JONSWAP description with a peak enhancement factor of $\gamma = 3.3$, i.e., they are relatively narrowbanded. C2 generally performs better in narrowbanded seas, where a dominant frequency is present, as discussed in Section 4.2. However, and beyond the scope of this case study, the reader is reminded that this assumption might not always be fulfilled in a practical scenario. See Section 4.2 for further discussion.

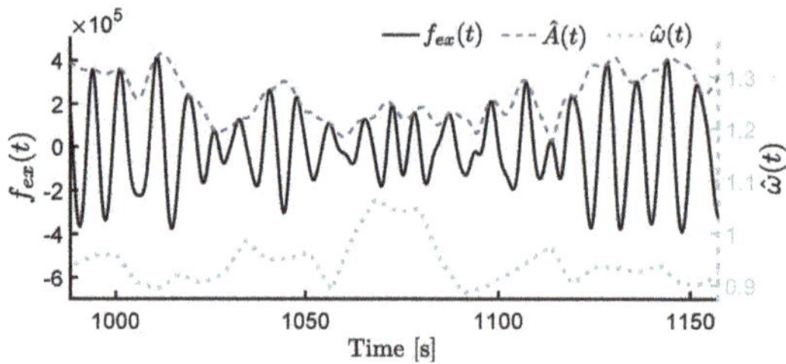

**Figure 11.** C2 design procedure. The wave excitation force and its amplitude and frequency estimation are shown with solid, dashed, and dotted lines, respectively.

### 5.1.3. C3

The design of this feedback controller begins with the definition of a finite set of frequencies $\Omega$, as detailed in Equation (24), which allows for the computation of an approximation of the device velocity in terms of a finite set of frequency components (see Equation (26)). Such a set $\Omega$ can be empirically defined, by explicitly using the frequency-domain characterisation of the stochastic process describing the different sea-states under analysis. In other words, the set $\Omega$ is generated in order to guarantee a suitable representation of excitation forces with significant frequency components in the range corresponding with $[0.3, 2]$ rad/s, i.e.,

$$\Omega = \left\{ \omega_k \in \mathbb{R}^+ \,\middle|\, \omega_k = \frac{\pi}{10} + \frac{\left(\frac{\pi}{2} - \frac{\pi}{10}\right)}{63}k, \ k \in \{0,\dots,63\} \right\}, \tag{40}$$

which represents a set with a cardinality of 64.

Subsequently, to determine estimates of each $A_i$ and $\phi_i$, in Equation (26), different FFT windows lengths, which are empirically determined using an exhaustive search methodology, are used for each different considered sea-state. In addition, each controller $K_i^{C3}(s)$ (see Equation (29)) is tuned, so that energy-maximisation is achieved, at each $\omega_i$, while bearing in mind closed-loop stability. Thus, from a dynamical systems point of view, the resulting set $\mathbf{K}^{C3}$, considering the cardinality of the set $\Omega$, represents a diagonal system of order 64.

By way of example, when considering the assumption expressed in Equation (25), which essentially inspires this control strategy, Figure 12a illustrates the approximation $\hat{f}_{ex}(t)$ (dashed line) obtained with the associated set of finite frequencies that are described in (40), for one particular realisation of the excitation force $f_{ex}(t)$ (solid line), computed according to a stochastic description with $T_p = 8.5$ s. Furthermore, Figure 12b provides the spectral representation of each of the excitation force time-traces, i.e., $\hat{f}_{ex}(t)$ and $f_{ex}(t)$, using dashed and solid lines, respectively.

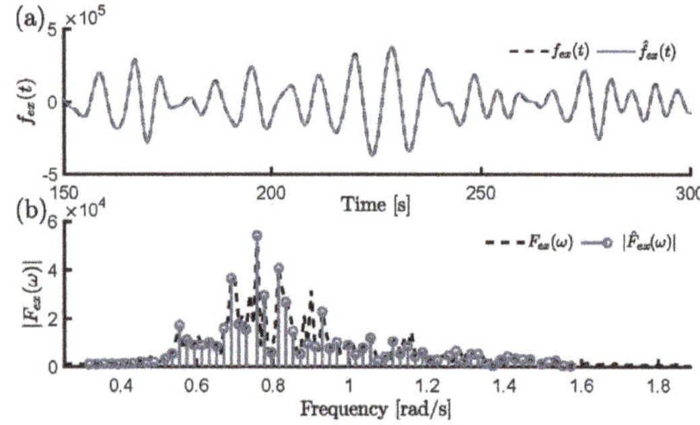

**Figure 12.** The approximated , $\hat{f}_{ex}(t)$, and the real excitation force, $f_{ex}(t)$, are shown using dashed and solid lines, respectively. In Figure 12 (**b**), the spectral representation of each time trace depicted in Figure 12 (**a**) is shown using the reference code employed for Figure 12a.

### 5.1.4. C4

The C4 control strategy, as depicted in Figure 5 and expressed in Equation (31), is essentially based on the application of system identification routines, considering a feedback control structure, as mentioned in Section 4.4. In order to satisfy Equation (31), frequency domain-based system identification routines are applied in this study. In particular, traditional least-mean-square error minimisation algorithms, for transfer function structures, are considered[8]. The identification process, required to compute $K^{C4}$, is performed using the data set defined for $K_{fb}^{opt}$ in Equation (10), which is essentially based on the WEC frequency-response mapping $G_0$. The resulting controller, for the CorPower-like device that was considered in this case study, is given by the expression

$$K^{C4}(s) = \frac{-1.195 \times 10^6 s^2 + 5.883 \times 10^4 s - 5.244 \times 10^6}{s^2 + 9.800 s + 0.216}. \tag{41}$$

Note that the focus of the frequency-domain identification algorithm is on ensuring the controller approximation in the frequency range characterising the wave inputs, i.e., $[0.3, 2]$ rad/s. To be precise, the following relation

$$K^{C4}(\omega) \approx K_{fb}^{opt}(\omega), \quad \forall \omega \in [0.3, 2], \tag{42}$$

holds. This can be clearly appreciated in Figure 13, where the frequency-response mappings, which are associated with the control-loop featuring the approximating structure $K^{C4}$, are explicitly shown. In particular, the left column of Figure 13 shows $K^{C4}(\omega)$, together with the frequency-response associated with the (theoretical) optimal feedback controller $K_{fb}^{opt}$. In addition, the right column of Figure 13 depicts the input-output frequency-response mapping when the approximating feedback controller $K^{C4}$ is considered, i.e., $T^{C4}(\omega)$, along with the corresponding (theoretical) optimal input-output frequency-response $T^{opt}(\omega)$.

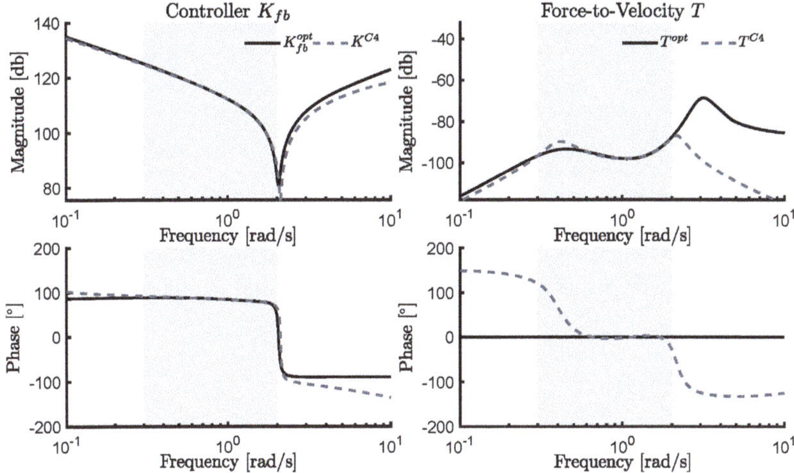


**Figure 13.** Frequency-response mappings related to the C4 controller. In particular, the left column illustrates $K^{C4}(\omega)$ (dashed line), together with the optimal feedback mapping $K_{fb}^{opt}(\omega)$ (solid line). The right column depicts the force-to-velocity mapping associated with the C4 controller, i.e., $T^{C4}(\omega)$ (dashed line), along with the optimal force-to-velocity frequency response $T^{opt}(\omega)$ (solid line).

---

[8] The interested reader is referred to [25] for a detailed discussion regarding classical system identification approaches.

### 5.1.5. C5

The C5 control strategy, as outlined in Section 4.5, is similar to that of the C4 controller, in the sense that both strategies are essentially based on performing system identification routines, aiming to approximate pre-defined optimal frequency-response mappings. However, there is one fundamental difference: While the C4 control methodology approximates a feedback structure, the C5 control strategy proposes a feedforward equivalent, which allows for constraint handling to be accommodated in a relatively straightforward manner. In particular, the C5 controller $K^{C5}$ aims to approximate the impedance-matching condition that is expressed in Equation (33), i.e.,

$$K^{C5}(\omega) \approx K^{opt}_{ff}(\omega), \quad \forall \omega \in [0.3, 2],\tag{43}$$

where, once again, the focus for the frequency-domain identification algorithm is put on the frequency range characterising the wave inputs. The identification of the feedforward structure $K^{opt}_{ff}$ is performed using a moment-matching-based identification approach, in order to ensure perfect frequency-response matching at a set of user-selected frequencies, as considered by the authors in [13]. Note that the definition of these matching points is designed to improve the fit between $K^{C5}$ and $K^{opt}_{ff}$, within the target bandwidth defined in Equation (43). According to moment-matching-based identification theory [16], the order of the resulting controller is twice the number of matching points. In this study, seven matching points are considered, selected as

$$\omega_M = \Big\{0.42, \quad 0.92, \quad 1.49, \quad 1.65, \quad 1.74, \quad 2.29, \quad 2.86\Big\},\tag{44}$$

generating an approximating structure $K^{C5}$ of order 14. Figure 14 shows the set of frequency-response mappings that are related to the C5 controller. In particular, the left column of Figure 14 illustrates $K^{C5}(\omega)$, together with the optimal feedforward mapping $K^{opt}_{ff}(\omega)$. The right column of Figure 14 depicts the force-to-velocity mapping associated with the C5 controller, i.e., $T^{C5}(\omega)$, along with the optimal force-to-velocity frequency response $T^{opt}(\omega)$. Note that both the left and right columns also show the effect of varying the constant gain $k$, used to handle physical limitations, on each respective frequency-response profile. In particular, values for $k$ in the set $\{0.5000, 0.2500, 0.1250, 0.065\}$ are considered, while using an arrow to indicate a decrease in $k$. Finally, the matching points, chosen to achieve moment-matching within the system identification procedure (i.e., the elements of the set (44)), are indicated using circular markers.

Regarding the tuning of the constraint handling mechanism, for this application case, the value of the constant $k$ (as described in Equations (35) and (36)) is determined using exhaustive (simulation-based) search, depending on each particular sea-state considered. Figure 15 shows the resulting $k$, for each $T_p \in [5, 12]$ s.

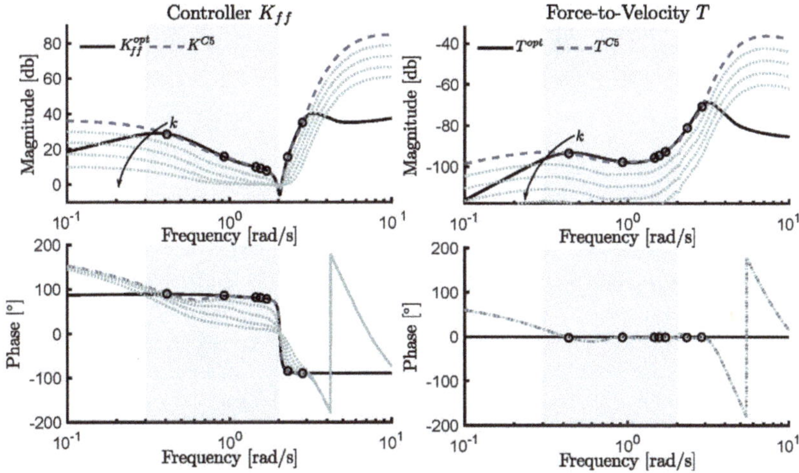

**Figure 14.** Frequency-response mappings related to the C5 controller. In particular, the left column illustrates $K^{C5}(\omega)$ (dashed line), together with the optimal feedforward mapping $K_{ff}^{opt}(\omega)$ (solid line). The right column depicts the force-to-velocity mapping associated with the C5 controller, i.e., $T^{C5}(\omega)$ (dashed line), along with the optimal force-to-velocity frequency response $T^{opt}(\omega)$ (solid line). The effect of the constraint handling mechanism for $k$ in the set $\{0.5000, 0.2500, 0.1250, 0.065\}$, is depicted with dotted lines. The matching points (used in the moment-based identification process) are depicted while using circular markers.

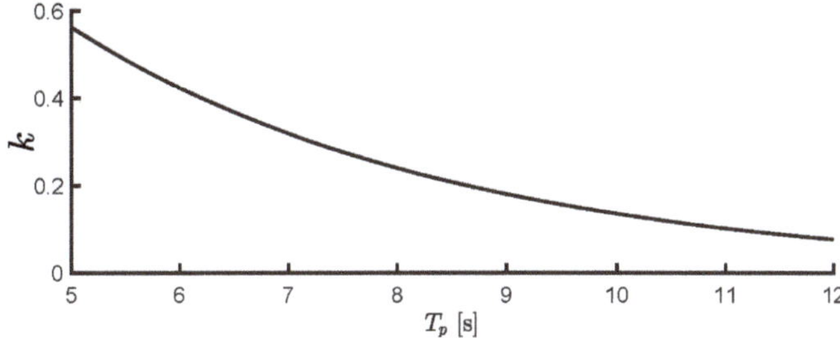

**Figure 15.** Optimised $k$ for each $T_p \in [5, 12]$ employed for the constraint handling mechanism of the C5 controller.

## 5.2. Performance Analysis: Energy Absorption

In this section, the resulting performance that is obtained which each controller presented in Section 4 is analysed and discussed, for both unconstrained and constrained scenarios. Two performance benchmarks are considered here, as mentioned in the introductory paragraph of Section 5. For the unconstrained case, the theoretical maximum absorbed energy (computed analytically, as in, for instance, [8]), denoted CR1, is considered as a reference measure. In contrast, for the constrained scenario, the performance obtained with an optimisation-based controller, denoted here as CR2, is considered as a surrogate performance reference.

Throughout this section, the performance results for each controller are shown in terms of relative values with respect to each corresponding benchmark case, i.e.,

$$\text{RGP} = \frac{P^C}{P^R}, \tag{45}$$

where RGP denotes the relative absorbed power, $P^C$ represents the power generated with the controller $C$, i.e., C1, C2, C3, C4, or C5, and $P^R$ denotes the power that is generated with the reference controller $R$, i.e., CR1 (unconstrained case) or CR2 (constrained case).

### 5.2.1. Unconstrained Control

Figure 16 shows performance results in terms of RGP, for the unconstrained case. Note that Figure 16 also includes the maximum relative RGP (dashed-black line), defined as the maximum theoretical achievable energy absorption, i.e., CR1. The RGP values that were obtained with the C1, C2, C3, C4, and C5 strategies, are denoted in Figure 16 using diamond, circle, square, triangular, and dot markers, respectively. It is important to note that, from an overall analysis, it is clear that the C2, C4, and C5 control methods obtain a high performance level, which, in practical terms, is comparable with the maximum achievable RGP. This is especially noteworthy for the case of the C4 controller, which is essentially a simple second-order system. The lowest performance level is obtained with the C3, which achieves its maximum for the central range of $T_p$. However, significant performance degradation can be observed for both lower and higher peak periods.

Note that a significant difference in terms of performance can be appreciated for the only two feedback structures analysed in this study, i.e., C3 and C4. In particular, the C4 controller significantly outperforms the C3 controller, in both terms of power absorption and controller simplicity (i.e., LTI second-order system vs. time-varying control system).

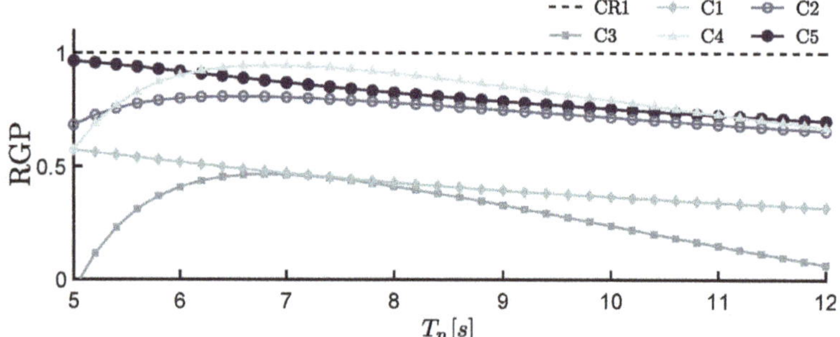


**Figure 16.** Relative generated power for the unconstrained case.

### 5.2.2. Constrained Control

In accordance with the unconstrained case, as described in Section 5.2.1, the resulting generated power for the constrained scenario is depicted in Figure 17. Analogously to Figure 16, Figure 17 also illustrates the maximum achievable performance, obtained via the surrogate reference measure CR2 (dashed-black line). Note that only the C2 and C5 control methodologies can be considered for performance assessment in the constrained case, since the other control strategies that are presented in Section 4, i.e., C1, C3, and C4, do not provide constraint handling mechanisms. The RGP obtained with the C2 and C5 strategies, is denoted using circle and dot markers, respectively. From an overall analysis, it can noted that, in general, the C5 controller performs better than the C2, for the totality of the analysed peak wave periods. In addition, both of the controllers show decreasing performance with increasing period $T_p$.

In particular, both of the controllers achieve their best performance when short $T_p$ values are considered, i.e., $T_p \approx 5$, since the oscillation amplitude of the considered waves decreases along with $T_p$ (according to the spectrum considered) and, consequently, the displacement constraint is not active. Further discussion on this topic can be found in [32].

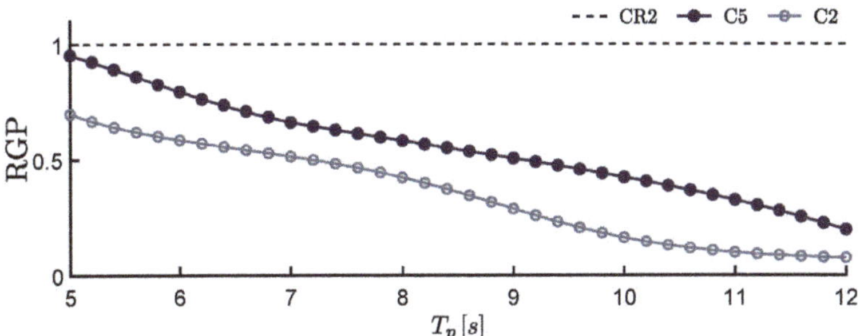



**Figure 17.** Relative generated power for the constrained case.

*5.3. Time-Domain Response*

Figure 18 illustrate a particular set of time-traces obtained with each control strategy studied, for a JONSWAP sea-state (as described in Section 5), with a peak period of $T_p = 8.5$ s, for both constrained (a) and unconstrained (b) cases. The results related to each controller, i.e., C1, C2, C3, C4, and C5, are denoted using thick-solid black, dashed, dash-dotted, thick-dotted, and solid-grey lines, respectively. In addition, unconstrained and constrained benchmark references, i.e., CR1 and CR2, are indicated with thin-dotted lines, in Figure 18a,b, respectively.

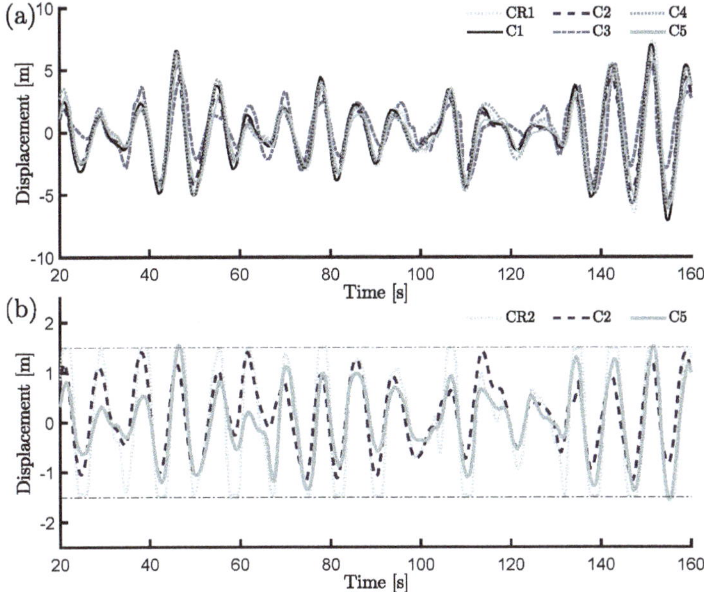



**Figure 18.** Time-traces obtained with each controller considering $T_p = 8.5$ s. The results for unconstrained and constrained cases are shown in (**a**,**b**), respectively. Note that the constraints are explicitly dealt with only the C2 and C5 controllers.

In the unconstrained case, as shown in Figure 18a, a large deviation with respect to the reference CR1 can be noted for the C3 controller, both in terms of instantaneous amplitude and phase. Note that, the effects that were observed in each corresponding time-domain behaviour, directly support the performance results (in terms of generated power) presented in Figure 16. Finally, in the constrained case, as depicted in Figure 18b, the C5 controller exploits the allowed displacement region more effectively than the C2 controller, which is directly translated into an increase in power production, supporting the results that are presented in Figure 17.

## 6. Discussion

This section presents a general discussion on the overall controller design procedure, while taking into account both capabilities and energy-maximising performance of each simple control strategy. To this end, this section provides a critical comparison between methods, using four distinguishing features (or categories): (1) computational simplicity; (2) stability; (3) constraint handling capability; and, (4) performance. In particular, a rating system is proposed, aiming to assign a score to each of the above categories, with such a score taking values between 0 (minimum) and 1 (maximum), for each controller analysed in this paper. The underlying principles behind each of this features, and its associated scoring system, are described in the following paragraphs.

- **Computational Simplicity.** With the aim of analysing computational simplicity, Table 3 presents controller *normalised run-time* (denoted as $t_R$), i.e., ratio between the time required to compute the control signal for each controller, and the length of the simulation itself (in seconds). The computations are performed using Matlab® R2019a, running on a PC comprising an Intel® Core™i3-2120 CPU @ 3.30 GHz processor with 16 GB of RAM operated by a Windows 10 Pro 64 bits version 2004 compilation OS 19041.450. Table 3 also presents the corresponding scoring that is assigned to each of the analysed controllers, based on the relation described in the following. Note that, given the wide range of normalised run-times calculated (see Table 3), a normalised logarithmic scale is used for scoring the computational simplicity. Let $t_R^{\min}$ and $t_R^{\max}$ be the minimum and maximum normalised run-times among the set of five controllers. The scoring $S : \mathbb{R} \to [0,1]$, $t_R \mapsto S(t_R)$, for each controller is defined via:

$$S(t_R) = 1 - \frac{\log_{10}\left[\frac{t_R}{t_R^{\min}}\right]}{\log_{10}\left[\frac{t_R^{\max}}{t_R^{\min}}\right]}. \tag{46}$$

It is straightforward to note that $S(t_R^{\min}) = 1$ and $S(t_R^{\max}) = 0$, i.e., in terms of run-time, the slowest controller is rated with 0 (minimum score), while the fastest is scored with 1 (maximum score).

- **Stability.** It is well-know that stability is a classical issue in controllers based on the energy-maximising impedance-matching principle, as mentioned in Section 4. When considering the discussion provided in Section 4 related to the stability of each controller, this category is rated, as follows :

  1. If the stability of the control loop can be fully guaranteed, the controller is rated with 1,
  2. if the stability of the control loop cannot be fully guaranteed, the controller is rated with 0.

- **Constraint Handling Capability.** This category analyses the capabilities of each controller in handling physical limitations, i.e., constraints. In particular, this category is rated as follows:

  1. If a controller provides a constraint handling mechanism, the controller is rated with 1,
  2. if a controller does not provide a constraint handling mechanism, the controller is rated with 0.

- **Performance.** Using the performance results that are presented for the unconstrained case (since constrained performance metrics are not available for all controllers), presented in Figure 16, this category measures the total energy-maximising performance with respect to the RGP (defined in Equation (45)), as follows:

$$\frac{1}{7} \int_5^{12} \text{RGP} \, dT_p \tag{47}$$

where the denominator of the quotient, i.e., 7, arises from computing the length (in seconds) of the range of peak periods that are considered to generate the wave inputs. Note that, if the reference measure CR1 is considered, then the expression in Equation (47) is effectively 1.

**Table 3.** Normalised run-time and computational simplicity scoring for each controller.

| Controller | Run-Time $t_R$ | Scoring $S(t_R)$ |
|---|---|---|
| C1 | $0.2 \times 10^{-3}$ | 1.00 |
| C2 | $9.8 \times 10^{-3}$ | 0.46 |
| C3 | $303 \times 10^{-3}$ | 0.00 |
| C4 | $2.5 \times 10^{-3}$ | 0.65 |
| C5 | $28 \times 10^{-3}$ | 0.32 |

The results of evaluating each control strategy, following the set of features that are defined above, can be found in both Table 4, and the four-dimensional spider plot presented in Figure 19.

**Table 4.** Score assigned to each controller considering each category.

| Controller | Computational Simplicity | Stability | Constraint Handling | Performance |
|---|---|---|---|---|
| C1 | 1.00 | 1 | 0 | 0.42 |
| C2 | 0.46 | 1 | 1 | 0.75 |
| C3 | 0.00 | 0 | 0 | 0.30 |
| C4 | 0.65 | 1 | 0 | 0.82 |
| C5 | 0.32 | 1 | 1 | 0.81 |

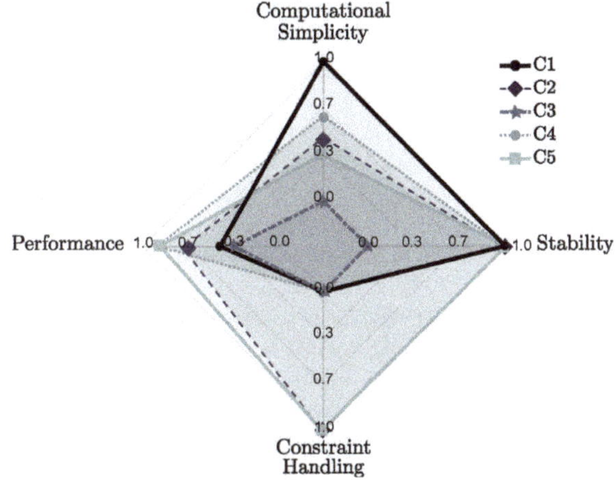


**Figure 19.** Graphical representation of the results shown in Table 4.

From the observation of Table 4, Figure 19, and the discussion in Sections 4 and 5, some comments, related to the status and perspectives of energy-maximising WEC controllers based on the impedance-matching principle, are worth highlighting. Firstly, a strong trade-off between the estimation of wave excitation forces, and capabilities of managing constraints, can be mentioned. In addition, it is noteworthy that, even with a simple feedback structure, high performance levels can be reached, as demonstrated by the C4 controller, which features a second-order LTI system. Nonetheless, the lack of a constraint handling mechanism renders the C4 controller unsuitable for realistic applications, in which large (and potentially unrealistic) displacements may be induced in unconstrained control conditions. This fact, which is also discussed here in Section 3, can be directly appreciated in Figure 18a.

From the analysis of Table 4 and Figure 19, it can be noted that the C5 controller provides a framework in which physical constraints are considered, while, at the same time, good performance levels are effectively achieved. Note that the C2 controller also provides a constraint handling mechanism, via a suitable modification of the feedforward velocity reference generation, although such a mechanism depends upon a nonlinear estimation process. In other words, constraint handling in the C2 has an inherently higher computational (and analytical) complexity than the C5 controller.

The results that are presented in Table 4 and Figure 19 give some clear directions and perspectives for future research and the development of simple controllers for wave energy systems. In particular, there is a clear lack of LTI feedback strategies that are capable of handling physical limitations. Such a scenario, if achievable, would provide controllers capable of optimising energy-absorption by using motion measurements only, dropping any requirement of wave excitation force estimation, while being able to successfully handle constraints. This would be, effectively, an ideal simple controller for WEC systems, completely covering essential requirements for WEC energy maximising controllers.

## 7. Conclusions and Future Directions

This paper documents a set of (5) controllers, all based on the impedance-matching principle, which have been developed over the period 2010–2020, and compares and contrasts their characteristics, in terms of performance, the handling of physical constraints, dynamical features, and computational load. The comparison is carried out both analytically and numerically, focusing on issues, such as stability, which can arise in the implementation of panchromatic resonant control, and the availability of suitable constraint handling mechanisms, given that the unrealistic motion often required by the impedance-matching condition. The holistic performance of the controllers is assessed under irregular waves in unconstrained, and constrained (where possible), scenarios. The assessment is carried out in terms of relative absorbed power, including a set of benchmark reference performance metrics for each case, and relative computational load.

A scoring system, which considers four specific features, is proposed in this study, aiming to perform a fair comparison between the five controllers. In particular, computational simplicity, stability, constraint handling, and resulting performance, are explicitly taken into account. From this analysis, a number generic limitations can be identified. In particular, as shown and summarised in Section 6, there is an intrinsic link between feedback-based controllers and an inability to handle constraints, while feedforward controllers, able to handle physical constraints, inevitably require wave excitation force estimates, bringing extra computational load and potential sensitivity to excitation force estimation errors.

While this study has some limitations, for example, the relative sensitivity of controllers to modelling error and measurement noise, etc. has not been considered (though the interested reader is referred to [17] in relation to the relative sensitivity of feedforward and feedback controllers), it shows that, for the unconstrained case, low-order LTI controllers can achieve almost theoretical performance levels, in realistic irregular waves. However, to date, there is not a control methodology that provides an estimator-free, low-order structure (i.e. demanding low hardware requirements), while, at the same time, being able to handle physical constraints. This points to an important potential future research direction: the development of an alternative simple, estimator-free, feedback controller, incorporating a suitable constraint handling mechanism. Whether such a controller is possible, given the generic limitations of the set of controllers studies in this paper, remains an open question.

**Author Contributions:** Conceptualization, D.G.-V., N.F., and J.V.R.; methodology, D.G.-V., N.F., and J.V.R.; formal analysis, D.G.-V., N.F., F.J.-L. and J.V.R.; writing–original draft preparation, D.G.-V. and N.F.; writing–review and editing, D.G.-V., N.F., F.J.-L. and J.V.R.; supervision, J.V.R.; project administration, J.V.R.; funding acquisition, J.V.R.. All authors have read and agreed to the published version of the manuscript.

**Funding:** This research was funded by Science Foundation Ireland under grant number SFI/13/IA/1886 and Grant No. 12/RC/2302 for the Marine Renewable Ireland (MaREI) centre.

**Conflicts of Interest:** Three of the controllers studied in this paper have been proposed by various authors of this manuscript. However, every attempt has been made to ensure a fair comparison, while no special coding attention has been given to any of the controllers/estimators in relation to computation time. The funders had no role in the design of the study; the collection, analyses, or interpretation of data; in the writing of the manuscript, or in the decision to publish the results.

# References


1. Drew, B.; Plummer, A.R.; Sahinkaya, M.N. A review of wave energy converter technology. *Proc. Inst. Mech. Eng. Part A J. Power Energy* **2009**, *223*, 887–902. [CrossRef]
2. Ringwood, J.V.; Bacelli, G.; Fusco, F. Energy-maximizing control of wave-energy converters: The development of control system technology to optimize their operation. *IEEE Control Syst.* **2014**, *34*, 30–55.
3. Korde, U.A.; Ringwood, J.V. *Hydrodynamic Control of Wave Energy Devices*; Cambridge University Press: Cambridge, UK, 2016.
4. Liberzon, D. *Calculus of Variations and Optimal Control Theory: A Concise Introduction*; Princeton University Press: Princeton, NJ, USA, 2011.
5. Faedo, N.; Olaya, S.; Ringwood, J.V. Optimal control, MPC and MPC-like Algorithms for Wave Energy Systems: An Overview. *IFAC J. Syst. Control* **2017**, *1*, 37–56. [CrossRef]
6. Faedo, N. Optimal Control and Model Reduction for Wave Energy Systems: A Moment-Based Approach. Ph.D. Thesis, Department of Electronic Engineering, Maynooth University, Kildare, Ireland, 2020.
7. Floyd, T.L.; Pownell, E. *Principles of Electric Circuits*; Prentice Hall: Upper Saddle River, NJ, USA, 2000.
8. Falnes, J. *Ocean Waves and Oscillating Systems: Linear Interactions Including Wave-Energy Extraction*; Cambridge University Press: Cambridge, UK, 2002.
9. Fusco, F.; Ringwood, J. Suboptimal causal reactive control of wave energy converters using a second order system model. In Proceedings of the 21st 2011 International Offshore and Polar Engineering Conference. International Society of Offshore and Polar Engineers (ISOPE), Maui, HI, USA, 19–24 June 2011; pp. 687–694.
10. Fusco, F.; Ringwood, J.V. A simple and effective real-time controller for wave energy converters. *IEEE Trans. Sustain. Energy* **2013**, *4*, 21–30. [CrossRef]
11. Song, J.; Abdelkhalik, O.; Robinett, R.; Bacelli, G.; Wilson, D.; Korde, U. Multi-resonant feedback control of heave wave energy converters. *Ocean Eng.* **2016**, *127*, 269–278. [CrossRef]
12. Bacelli, G.; Nevarez, V.; Coe, R.G.; Wilson, D.G. Feedback Resonating Control for a Wave Energy Converter. *IEEE Trans. Ind. Appl.* **2019**, *56*, 1862–1868. [CrossRef]
13. García-Violini, D.; Peña-Sanchez, Y.; Faedo, N.; Ringwood, J.V. An Energy-Maximising Linear Time Invariant Controller (LiTe-Con) for Wave Energy Devices. *Trans. Sustain. Energy* **2020**, *11*, 2713–2721. [CrossRef]
14. Cummins, W.E. The impulse Response Function and Ship Motions. *Schiffstechnik* **1962**, *47*, 101–109.
15. Ogilvie, T.F. Recent progress toward the understanding and prediction of ship motions. In Proceedings of the 5th Symposium on Naval Hydrodynamics, Bergen, Norway, 10–12 September 1964; Volume 1, pp. 2–5.
16. Peña-Sanchez, Y.; Faedo, N.; Ringwood, J.V. Moment-based parametric identification of arrays of wave energy converters. In Proceedings of the 2019 American Control Conference, Philadelphia, PA, USA, 10–12 July 2019.
17. Ringwood, J.V.; Mérigaud, A.; Faedo, N.; Fusco, F. An Analytical and Numerical Sensitivity and Robustness Analysis of Wave Energy Control Systems. *IEEE Trans. Control Syst. Technol.* **2019**, *28*, 1337–1348. [CrossRef]
18. Hals, J.; Falnes, J.; Moan, T. A comparison of selected strategies for adaptive control of wave energy converters. *J. Offshore Mech. Arct. Eng.* **2011**, *133*, 031101–031113. [CrossRef]
19. Peña-Sanchez, Y.; Windt, C.; Josh, D.; Ringwood, J.V. A Critical Comparison of Excitation Force Estimators for Wave Energy Devices. *IEEE Trans. Control Syst. Technol.* **2019**, *28*, 2263–2275. [CrossRef]
20. Antoulas, A.C. *Approximation of Large-Scale Dynamical Systems*; SIAM: Philadelphia, PA, USA, 2005.
21. Sánchez-Peña, R.S.; Sznaier, M. *Robust Systems Theory and Applications*; Wiley: New York, NY, USA, 1998.
22. Astrom, K.J.; Wittenmark, B. *Adaptive Control*, 2nd ed.; Addison-Wesley Publishing Company: Boston, MA, USA, 1994.
23. Goodwin, G.C.; Graebe, S.F.; Salgado, M.E. *Control System Design*; Prentice Hall: Upper Saddle River, NJ, USA, 2001; Volume 240.
24. Khalil, H.K.; Grizzle, J.W. *Nonlinear Systems*; Prentice Hall: Upper Saddle River, NJ, USA, 2002; Volume 3.
25. Ljung, L. *System Identification—Theory for the User*; Prentice Hall: Upper Saddle River, NJ, USA, 1999.

26. Fusco, F.; Ringwood, J. A model for the sensitivity of non-causal control of wave energy converters to wave excitation force prediction errors. In Proceedings of the 9th European Wave and Tidal Energy Conference (EWTEC), Southampton, UK, 5–9 September 2011; School of Civil Engineering and the Environment, University of Southampton: Southampton, UK, 2011.

27. Faedo, N.; Peña-Sanchez, Y.; Ringwood, J.V. Finite-Order Hydrodynamic Model Determination for Wave Energy Applications Using Moment-Matching. *Ocean Eng.* **2018**, *163*, 251–263. [CrossRef]

28. García-Violini, D.; Peña-Sanchez, Y.; Faedo, N.; Windt, C.; Ringwood, J.V. Experimental implementation and validation of a broadband LTI energy-maximising control strategy for the Wavestar device. *IEEE Trans. Control Syst. Technol.* **2020**, response to reviewers submitted.

29. Faedo, N.; Scarciotti, G.; Astolfi, A.; Ringwood, J.V. Energy-maximising control of wave energy converters using a moment-domain representation. *Control Eng. Pract.* **2018**, *81*, 85–96. [CrossRef]

30. Todalshaug, J.H.; Ásgeirsson, G.S.; Hjálmarsson, E.; Maillet, J.; Möller, P.; Pires, P.; Guérinel, M.; Lopes, M. Tank testing of an inherently phase-controlled wave energy converter. *Int. J. Mar. Energy* **2016**, *15*, 68–84. [CrossRef]

31. Giorgi, G.; Ringwood, J.V. Analytical representation of nonlinear Froude-Krylov forces for 3-DoF point absorbing wave energy devices. *Ocean Eng.* **2018**, *164*, 749–759. [CrossRef]

32. Faedo, N.; García-Violini, D.; Peña-Sanchez, Y.; Ringwood, J.V. Optimisation-vs. non-optimisation-based energy-maximising control for wave energy converters: A case study. In Proceedings of the 2020 European Control Conference (ECC), Saint Petersburg, Russia, 12–15 May 2020; pp. 843–848.

33. Hasselmann, K. Measurements of wind wave growth and swell decay during the Joint North Sea Wave Project (JONSWAP). *Deutches Hydrogr. Inst.* **1973**, *8*, 95.

34. LHEEA, NEMOH-Presentation. Laboratoire de Recherche en Hydrodynamique Énergetique et Environnement Atmospherique. 2017. Available online: https://goo.gl/yX8nFu (accessed on 1 August 2019).

35. Papoulis, A. *Probability, Random Variables and Stochastic Processes*; McGraw-Hill: New York, NY, USA, 1991.

36. Francis, B.A.; Wonham, W.M. The internal model principle of control theory. *Automatica* **1976**, *12*, 457–465. [CrossRef]

37. Garcia-Rosa, P.B.; Kulia, G.; Ringwood, J.V.; Molinas, M. Real-time passive control of wave energy converters using the Hilbert-Huang transform. *IFAC-PapersOnLine* **2017**, *50*, 14705–14710. [CrossRef]

38. Huang, N.E.; Shen, Z.; Long, S.R.; Wu, M.C.; Shih, H.H.; Zheng, Q.; Yen, N.C.; Tung, C.C.; Liu, H.H. The empirical mode decomposition and the Hilbert spectrum for nonlinear and non-stationary time series analysis. *Proc. R. Soc. Lond. Ser. A Math. Phys. Eng. Sci.* **1998**, *454*, 903–995. [CrossRef]

Journal of
*Marine Science and Engineering*

*Article*

# A Real-Time Detection System for the Onset of Parametric Resonance in Wave Energy Converters

**Josh Davidson * and Tamás Kalmár-Nagy**

Department of Fluid Mechanics, Faculty of Mechanical Engineering, Budapest Univeristy of Technology and Economics, 1111 Budapest, Hungary; kalmarnagy@ara.bme.hu
* Correspondence: davidson@ara.bme.hu

Received: 10 September 2020; Accepted: 13 October 2020; Published: 20 October 2020

**Abstract:** Parametric resonance is a dynamic instability due to the internal transfer of energy between degrees of freedom. Parametric resonance is known to cause large unstable pitch and/or roll motions in floating bodies, and has been observed in wave energy converters (WECs). The occurrence of parametric resonance can be highly detrimental to the performance of a WEC, since the energy in the primary mode of motion is parasitically transferred into other modes, reducing the available energy for conversion. In addition, the large unstable oscillations produce increased loading on the WEC structure and mooring system, accelerating fatigue and damage to the system. To remedy the negative effects of parametric resonance on WECs, control systems can be designed to mitigate the onset of parametric resonance. A key element of such a control system is a real-time detection system, which can provide an early warning of the likely occurrence of parametric resonance, enabling the control system sufficient time to respond and take action to avert the impending exponential increase in oscillation amplitude. This paper presents the first application of a real-time detection system for the onset of parametric resonance in WECs. The method is based on periodically assessing the stability of a mathematical model for the WEC dynamics, whose parameters are adapted online, via a recursive least squares algorithm, based on online measurements of the WEC motion. The performance of the detection system is demonstrated through a case study, considering a generic cylinder type spar-buoy, a representative of a heaving point absorber WEC, in both monochromatic and polychromatic sea states. The detection system achieved 95% accuracy across nearly 7000 sea states, producing 0.4% false negatives and 4.6% false positives. For the monochromatic waves more than 99% of the detections occurred while the pitch amplitude was less than 1/6 of its maximum amplitude, whereas for the polychromatic waves 63% of the detections occurred while the pitch amplitude was less than 1/6 of its maximum amplitude and 91% while it was less than 1/3 of its maximum amplitude.

**Keywords:** paramertic resonance; wave energy conversion; real-time detection; recursive least squares

---

## 1. Introduction

Parametric resonance is an instability phenomenon caused by the time-varying parameters of a system [1]. Whereas normal resonance causes oscillations in a system to grow linearly with time, parametric resonance causes an exponential increase in the oscillation amplitude. This phenomena was first identified by Faraday [2], who observed the generation of surface waves when a liquid-filled container was oscillated vertically, thereby varying the restoring force usually provided by the constant gravitational acceleration only. The time-varying wetted surface of a floating body subject to incident waves can produce temporal variations in the system's parameters and thus trigger parametric resonance. The first noted observation of parametric resonance in floating bodies dates back to Froude in the 19th century [3], who described that large roll motions occur when the natural period of a ship's roll is twice the natural period of the heave or pitch modes of motion. This parametric coupling between heave, pitch and roll degrees of freedom (DoFs) has been investigated in offshore engineering fields, such as shipping (see, for example, [4,5] and the references therein) and in offshore spar-type structures [6–10].

The large unstable motions which accompany the occurrence of parametric resonance can be extremely hazardous for the safe operation of marine structures and vessels. For example, container ships have lost cargo overboard due to large parametric roll motions. Research in the marine engineering field has therefore investigated control methods for the suppression and stabilisation of parametric pitch/roll [11,12]. Similarly to ships and other offshore floating structures, the large amplitude unstable motions—due to parametric resonance—have been observed in wave energy converters (WECs). As such, there is motivation to design control systems to mitigate the onset of parametric resonance in WECs.

### 1.1. Parametric Resonance in WECs

The concept of resonance is very well known in the study of wave energy conversion, since a WEC is usually designed to resonate with the incident waves for maximum power extraction. By comparison, parametric resonance has received far less attention. A major barrier in studying parametric resonance is the complexity of the mathematical models required to capture this nonlinear phenomenon, compared to the computationally efficient linear/frequency domain hydrodynamic models traditionally utilised in WEC research and analysis. Therefore, the first reported observations of parametric resonance in WECs stem from physical scale model wave tank testing, dating back to work on the frog in the late 1980s by Bracewell [13]. Numerous further observations of parametric resonance were subsequently reported from physical scale model wave tank testing of WECs [14–26], which in many cases was unanticipated when first observed, since the occurrence of parametric resonance was not predicted by the simple numerical modelling results that preceded the wave tank experiments.

Therefore, over the past decade, in line with the availability of computationally efficient, nonlinear hydrodynamic models [27], increased attention towards numerical modelling of parametric resonance in WECs can be noted. The first publication in this direction was at the end of the 2000s by Babarit et al. [28], which investigated the use of a nonlinear hydrodynamic model to capture the parametric roll resonance in the SeaREV, observed during the earlier experimental tests in Durand et al. [14]. The nonlinear hydrodynamic model calculates the Froude–Krylov force on the exact wetted surface of the WEC at each time step, rather than using the time-invariant mean wetted surface (as is the case for linear hydrodynamic models). This type of nonlinear Froude–Kylov force model has subsequently been used in [29–34] to investigate parametric resonance in WECs. Other nonlinear hydrodynamic modelling approaches include only modelling the restoring force nonlinearly, rather than the entire Froude–Krylov force [35–40]. Modelling parametric resonance has also been explored through high-fidelity CFD simulations [41].

In some circumstances, the parametric excitation can arise due to the effect of the mooring system on the overall WEC dynamics. Therefore, several modelling approaches have been employed to capture this mooring induced effect. For the case of a taut moored WEC, Nicoll et al. [42] utilised the pendulum equation to model the parametric excitation, whereas Orszagova et al. [21,25,43] utilised a 2nd order Taylor series for the mooring system dynamics to successfully capture the parametric resonance phenomenon.

### 1.2. Suppression Control Methods for Parametric Resonance in WECs

The various control methods designed to suppress the occurrence of parametric resonance in WECs can be broadly classified into two main categories: passive and active control. The passive control approaches typically utilise fins or strakes to increase the hydrodynamic damping in the pitch/roll DoFs, which has been shown to reduce the occurrence of parametric resonance in offshore spar platforms [44]. This type of passive control method was first investigated for a WaveBob-like, two-body heaving point absorber in Beatty et al. [20] using numerical simulations, which confirmed the ability of the strakes to reduce pitch and roll amplitudes, allowing increased WEC power output. Follow-up work by Ortiz [45] investigated an alternative passive control approach, which utilises the mooring dynamics to reduce the occurrence of parametric resonance. The use of fins has been investigated experimentally for the stabilisation of the OWC Spar Buoy in Gomes et al. [46]. The wave flume experiments on a scale model device confirmed the ability of the fins to reduce, though not eliminate, the amplitude of the pitch and roll oscillations due to the occurrence of parametric resonance. A different passive control approach is shown in Cordonnier et al. [47] and Gomes et al. [48], based on the optimisation of the initial WEC design. A procedure to identify the frequency and amplitude ranges in which parametric resonance will occur is included into the numerical optimisation routines for the design of the geometry and mass distributions. A penalty is then added in the optimisation routine to the configurations where parametric resonance occurs due to the decreased performance. Cordonnier et al. [47] showed that the optimal geometry when taking parametric resonance into consideration has a larger width than the original geometry, which does not consider parametric resonance during the design optimisation.

The active control approaches involve a mechanism to influence the WEC dynamics and mitigate the onset of parametric resonance. Villegas and van der Schaaf [49] proposed an active control system for the WaveBob, after parametric pitch and roll were detected in physical experiments and were observed to hinder the performance of the WEC. The control system acts on the WEC dynamics, through the effect of the power take-off (PTO) force between the outer torus and inner spar of the WaveBob. The PTO force couples the motion of the two bodies, influencing the resonant peaks in the frequency responses of the heave, pitch and roll DoFs. The active control system proposed by Villegas and van der Schaaf applies a notch filter, designed to eliminate any PTO forces at the frequencies for which parametric resonance occurs. Experimental results for a model scale device in a wave tank validate the effectiveness of this approach in [49]. Maloney [50] also considered a WaveBob-like WEC, but proposed including an internal, elastically supported reaction mass within the inner spar, which can be utilised to control the natural frequency of the spar through the use of a variable inertia system. They term this three-body system the VISWECand utilise the variable inertia system to eliminate parametric excitation in roll, in addition to its primary function of tuning the response of the spar to increase the output power.

Looking forward, more advanced active control mechanisms may switch on only when required and remain inactive during normal operation. A key component for such a system is a monitoring and detection system, to give an early warning. Such early warning systems have already been developed for ships [51–57], enabling the crew to take corrective measures such as speed and/or course changes to avoid the occurrence of parametric roll. However, to date, no such early warning system has been developed for a WEC.

### 1.3. Objectives and Outline of the Paper

The objective of the present paper is to provide the first implementation of a real-time detection system for the onset of parametric resonance in WECs, capable of providing an early warning for a control system to take preventive measures. The details of the proposed detection system are introduced in Section 2. Next, the performance of the detection system is assessed in an illustrative test case. The details of the test case and the implementation of the detection system are presented in Section 3. The test case results are presented in Section 4 and the performance of the detection system is evaluated and discussed. Finally, a number of conclusions are drawn in Section 5.

## 2. A Real-Time Detection System for Early Warning of Parametric Resonance in WECs

The proposed monitoring and detection system is based on a method developed for the detection of parametric roll in ships, presented in Holden et al. [51]. The method works by performing real-time parameter identification of a linear time-varying model, based on the onboard measurements of the angular displacement and velocity of the ship's roll. The detection system regularly assesses the stability of the identified model by monitoring the eigenvalues of the model system, giving a warning that roll resonance is probable when the model becomes unstable.

Owing to the simplicity of this detection method, it was selected for our investigation as the first real-time detection system for the onset of parametric resonance in WECs. The details of the linear time-varying model are given in Section 2.1, the real-time parameter identification in Section 2.2 and then the detection criteria in Section 2.3.

### 2.1. The Linear Time-Varying Model

Since the detection system is based on real-time measurements from the WEC, the system requires a discrete time model, taking as input the discrete measured signals at each sampling instant. The discrete time model employed here for the evolution of the displacement $z_1$ and velocity $z_2$, subject to an external forcing $\mathbf{u}$, can be expressed as:

$$\mathbf{z}_{k+1} = \mathbf{A}_k \mathbf{z}_k + \mathbf{u}_k,$$

$$\begin{bmatrix} z_{1,k+1} \\ z_{2,k+1} \end{bmatrix} = \begin{bmatrix} A_{1,k} & A_{2,k} \\ A_{3,k} & A_{4,k} \end{bmatrix} \begin{bmatrix} z_{1,k} \\ z_{2,k} \end{bmatrix} + \begin{bmatrix} u_{1,k} \\ u_{2,k} \end{bmatrix}, \tag{1}$$

where the parameters of the matrix $\mathbf{A}$ are unknown and time-varying.

### 2.2. Real-Time Parameter Identification

A recursive least-squares algorithm is employed to regularly update the parameter values of $\mathbf{A}$, based on the real-time measurements of $z_1$ and $z_2$. Recursive least squares is employed in [51] to identify the roll dynamics of a ship and has also been used for online identification of the model parameters for the WEC dynamics in an adaptive controller [58,59]. The recursive least squares algorithm in [58,59] assumes that the excitation force is known a priori, which requires the use of a robust estimation algorithm to be used in conjunction with the adaptive controller (see Pena et al. [60] for a critical comparison of excitation force estimators for WECs). However, the approach used in [51], which is followed in the present study, makes the assumption that the external forcing parameters in Equation (1), $u_1$ and $u_2$, are independent, zero-mean, Gaussian white noise processes. Under this assumption the recursive least squares identification takes the form

$$\theta_{i,k} = \theta_{i,k-1} + P_{i,k}\Phi_{i,k-1}e_{i,k}, \tag{2}$$

$$P_{i,k}^{-1} = P_{i,k-1}^{-1} + \Phi_{i,k-1}^2, \tag{3}$$

where $i \in \{1,2\}$ and

$$\begin{bmatrix} \theta_{1,k} \\ \theta_{2,k} \end{bmatrix} = \begin{bmatrix} -(A_{1,k} + A_{4,k}) \\ A_{1,k}A_{4,k} - A_{2,k}A_{3,k} \end{bmatrix}, \tag{4}$$

$$\Phi_{i,k} = -z_{i,k} - z_{i,k-1}, \tag{5}$$

$$e_{i,k} = -z_{i,k} - \Phi_{i,k-1}\theta_{i,k-1}. \tag{6}$$

### 2.3. Detecting Instability

The detection algorithm monitors the eigenvalues of $\mathbf{A}$ to identify whether the system is possibly becoming unstable. For a discrete-time, linear time-invariant system, if the eigenvalues of the matrix $\mathbf{A}$ lie outside of the unit circle, then the system is unstable. However, for a discrete-time, linear time-varying system, such as Equation (1), instability cannot be concluded when the eigenvalues lie outside of the unit circle. Therefore, by monitoring when the eigenvalues of $\mathbf{A}$ moving outside of the of the unit circle, the present detection system provides an early warning for the probable occurrence of parametric resonance (the analytical conditions for the stability/instability of a discrete system are given in, for example, [61]). The eigenvalues $\lambda_k$ are determined by solving

$$\lambda_k^2 + \theta_{1,k}\lambda_k + \theta_{2,k} = 0. \tag{7}$$

## 3. Test Case

The test case considers evaluating the proposed detection system in a numerical simulation of a generic heaving point-absorber-type WEC. The performance of the proposed detection system is assessed with the following two main criteria:

- Correctly warning when parametric resonance occurs (correct positive) and not giving a false warning when parametric resonance does not occur (correct negative).
- How early the system detects the onset of parametric resonance and sends a warning.

The specific device considered in this test case is introduced in Section 3.1. The range of input waves in which the device was tested is detailed in Section 3.2. The details of the numerical model are presented in Section 3.3. The simulation details and the implementation of the early warning system are provided in Section 3.4. The results are then presented in Section 4.

### 3.1. The Device

The device in this study is a cylindrical spar-buoy, providing a generic representation of a heaving point absorber WEC. The selection of this type of WEC stems from the numerous reports of parametric pitch/roll for axisymmetric heaving point absorbers [17–20,29–34,39,41,45,46,48–50]. The particular geometry used in this study is a representation of a classic spar platform from the literature [62–65], comprising a simple vertical cylinder whose natural period of pitch motion is about twice the heave resonant period. The 2:1 ratio of heave to pitch natural frequency makes the floating cylinder prone to parametric excitation in pitch. The simple geometry was selected in order for the test case to remain generic. Likewise, no power take-off mechanisms or wave enengy extraction capability are considered on the device, with the focus of the tests being on the ability of the detection system to give an early warning of the onset of parametric resonance from measurements of the pitch motion. Hong et al. [62] tested the floating cylinder at model scale, under various regular wave conditions in the Samsung Ship Model Basin. A numerical model of the cylinder was used by Jingrui et al. [63] to study the combination resonance response. Gavasoni et al. [64] employed a numerical model for the investigation of nonlinear vibration modes and instability. Giorgi et al. [65] demonstrated the potential yaw instability which may arise when including nonlinear kinematics in a nonlinear hydrodynamic model.

This particular spar-buoy was chosen since it is well known to exhibit parametric resonance and there exists a validated numerical model able to capture the existence of parametric resonance (see Section 3.3). However the spar-buoy's geometry is much larger than would be used in wave energy applications, with a heave natural period $T_{n,3} = 29$ s and pitch natural period $T_{n,5} = 57$ s. Thus the geometrical properties will be normalised in the analysis to provide insights for performance on smaller scale geometries, such as WECs, following the normalisation utilised in [65], where the frequencies are normalised against the pitch natural frequency $w_{n,5}$ and the wave heights are normalised against the metacentric height $\overline{GM}$.

### 3.2. Input Waves

The early warning detection system was evaluated in both monochromatic and polychromatic waves. For the monochromatic waves, a total of 4050 simulations were performed, spanning an array of 81 frequencies equally spaced between 1.6 and 2.4 times the natural pitch frequency $\omega_{n,5}$, and 50 wave heights from 2% to 100% of the metacentric height, $\overline{GM}$. For the polychromatic waves, the same number and range of wave heights were used; however, a larger frequency range was employed. The wide bandwidth of the polychromatic wave spectra meant there was a broad range of peak frequency values for which significant energy was present in the expected frequency range where parametric resonance was likely to occur. Thus, the frequency range for the polychromatic waves spanned from $0.6\omega_{n,5}$ to $3.4\omega_{n,5}$, with a resolution of $0.05\,\omega_{n,5}$, resulting in 2800 simulations. The reason why more monochromatic simulations were performed than polychromatic (4050 compared to 2800) is that that the time taken for each polychromatic simulation was twice as long as for the monochromatic simulations.

For the monochromatic waves, each simulation was run for a time equal to 100 times the pitch natural period, $T_{n,5}$. The wave amplitude was linearly ramped up over the first $5T_{n,5}$ of the signal to reduce transient effects. An example of the monochromatic wave signal is shown in Figure 1a. The length of each simulation for the polychromatic waves was twice as long as for the monochromatic waves, i.e., $200T_{n,5}$, since parametric resonance may take longer to develop due to the irregular nature of the polychromatic waves. The polychromatic waves are based on a JONSWAP spectrum, defined by a significant wave height and peak frequency. Each spectrum comprises 100 frequencies $\omega_j$, each with a randomly assigned phase $\phi_j$ uniformly distributed between 0 and $2\pi$. The spectra have a frequency range from $0\omega_{n,5}$ to $9\omega_{n,5}$ with an average frequency resolution $\Delta\omega$ of $0.09\omega_{n,5}$. The frequencies are not equally spaced on harmonics (to avoid the time series being periodic and repeating itself, as demonstrated in [66]), with a random shift between $\pm\frac{\Delta\omega}{2}$ applied to each frequency. Note that the same set of random phases and frequency shifts is applied to each spectrum. Examples of the polychromatic wave signal and its power spectral density (PSD) are shown in Figure 1b,c, normalised against the metacentric height and pitch natural frequency respectively.

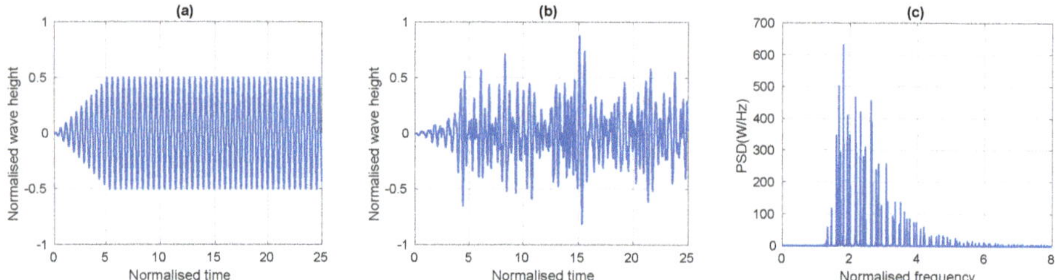

**Figure 1.** Example of the input waves: (**a**) monochromatic, (**b**) polychromatic, (**c**) power spectral density (PSD) of the polychomatic waves.

*3.3. Numerical Model*

The hydrodynamic model for the wave structure interaction follows the validated model for the same device from [62–64]. The model is able to articulate the occurrence of parametric resonance in the floating cylinder, through the coupled equations for heave $x_3$ and pitch $x_5$

$$F_{I,3}(\ddot{x}_3) + F_{D,3}\left(\dot{x}_5\right) + F_{R,3}\left(x_3, x_5, t\right) = F_{E,3}\left(t\right), \tag{8}$$

$$F_{I,5}(\ddot{x}_5) + F_{D,5}\left(\dot{x}_5\right) + F_{R,5}\left(x_3, x_5, t\right) = F_{E,5}\left(t\right), \tag{9}$$

where:

- $F_{I,i}$ is the inertia, detailed in Section 3.3.1.
- $F_{R,i}$ is the hydrostatic restoring force, detailed in Section 3.3.2.
- $F_{D,i}$ is the hydrodynamic damping force, detailed in Section 3.3.3.
- $F_{E,i}$ is the wave excitation force, detailed in Section 3.3.4.

The values for all of the parameters in the model are listed in Table 1 in Section 3.3.5.

### 3.3.1. Inertia

The inertial forces are given by the product of the inertia and acceleration:

$$F_{I,3} = (M + m_3)\ddot{x}_3, \tag{10}$$

$$F_{I,5} = (I_5 + m_5)\ddot{x}_5, \tag{11}$$

where $M$ is the cylinder mass, $m_3$ is the hydrodynamic added mass in heave, $I_5$ is the pitch moment of inertia and $m_5$ is the hydrodynamic added moment of inertia in pitch.

### 3.3.2. Hydrostatic Restoring Force/Moment

The hydrodystatic restoring force and moment terms have time-varying parameters, which enables the existence of parametric resonance in the numerical model. In [63,64], the hydrodystatic restoring force and moment are given by

$$F_{R,3}\left(x_3, x_5, t\right) = \rho g A_C \left[ x_3 - x_5^2 \frac{L_{MS}}{2} \right], \tag{12}$$

$$F_{R,5}\left(x_3, x_5, t\right) = \rho g A_C L_D \left[ \overline{GM} x_5 - \frac{1}{2} x_3 x_5 + \frac{1}{2} \eta(t) x_5 \right], \tag{13}$$

where $\rho$ is the density of water, $g$ is the gravitational constant, $A_C$ is the WEC horizontal cross-sectional area, $L_{MS}$ is distance from the WEC centre of mass to the still water level, $L_D$ is the WEC draft, $\overline{GM}$ is the WEC metacentric height and $\eta(t)$ is the wave elevation.

### 3.3.3. Hydrodynamic Damping

Following [63,64], the hydrodynamic damping is measured from free decay experiments in [62] and is given by

$$F_{D,3}\left(\dot{x}_3\right) = C_3 \dot{x}_3, \tag{14}$$

$$F_{D,5}\left(\dot{x}_5\right) = C_5 \dot{x}_5, \tag{15}$$

where $C_i$ are the damping coefficients identified from the experimental data.

### 3.3.4. Wave Excitation

For an input wave with amplitude $A$, frequency $\omega$ and phase $\phi$, the wave elevation is

$$\eta(t) = A \sin\left(\omega t + \phi\right). \tag{16}$$

The excitation force is given by

$$F_{E,i}\left(t\right) = H_{E,i}(\omega) A \sin\left(\omega t + \phi + \phi_{E,i}(\omega)\right), \tag{17}$$

where $H_{E,i}(\omega)$ and $\phi_{E,i}(\omega)$ are the hydrodynamic excitation force coefficients for the amplitude gain and phase shift, respectively. The hydrodynamic excitation force coefficients are frequency dependent and their values for the given spar-buoy geometry are calculated using the open-source linear potential flow boundary element method software Nemoh [28]. The excitation force coefficient values, as a function of frequency, are displayed in Figure 2.

For the case of polychromatic waves, the excitation force comprises a linear superposition of the contributions from each of the 100 frequency components.

$$F_{E,i}\left(t\right) = \sum_{j=1}^{100} H_{E,i}(\omega_j) A_j \sin\left(\omega_j t + \phi_j + \phi_{E,i}(\omega_j)\right). \tag{18}$$

### 3.3.5. Model Parameters

The values for the parameters used in the model are listed in Table 1 following the values provided in [62–64] for the the validated model of the same device. Note again, the geometry is much larger than would likely be used for WECs, as detailed in Section 3.1. The values for the excitation force coefficients are plotted in Figure 2.

**Table 1.** Model parameter values used in the test case.

| Parameter | Value | Parameter | Value | Parameter | Value |
|---|---|---|---|---|---|
| $A_C$ | 1087 m$^2$ | $\rho$ | 1000 kg/m$^3$ | $M$ | 2.15$\times$10$^8$ kg |
| $L_D$ | 198.1 m | $g$ | 9.81 m/s$^2$ | $m_3$ | 1.37 $\times$ 10$^7$ kg |
| $\overline{GM}$ | 10.1 m | $C_3$ | 1.19 $\times$ 10$^6$ kg/s | $I_5$ | 1.12 $\times$ 10$^{12}$ kgm |
| $L_{MS}$ | 109.1 m | $C_5$ | 7.54 $\times$ 10$^9$ kgm/s | $m_5$ | 7.26 $\times$ 10$^{11}$ kg |

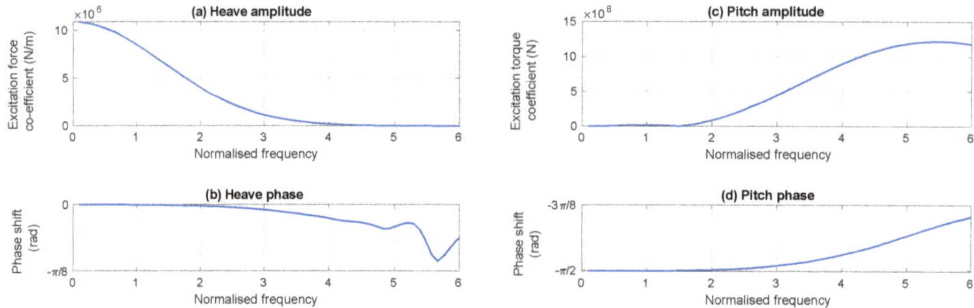

**Figure 2.** The excitation force coefficients (**a**) $H_{E,3}$, (**b**) $\phi_{E,3}$, (**c**) $H_{E,5}$ and (**d**) $\phi_{E,5}$.

### 3.4. Simulation Details and the Implementation of an Early Warning Detection System

#### 3.4.1. Simulation Details

The model was implemented in MATLAB and Simulink. A Runge–Kutta 4th order solution scheme was employed, with a fixed time step $dt$ equal to $0.01T_{n5}$. If the pitch angle exceeded 90°, the simulation was aborted and a maximum pitch angle of 90° was recorded.

#### 3.4.2. Recursive Least Squares Implementation

The detection system monitors the pitch motion of the device; thus $z_1 = x_5$ and $z_2 = \dot{x}_5$. The recursive least squares algorithm is executed at each time step, after the ramping up of the input wave is completed, i.e., $t > T_{n,5}$ (see Section 3.2). The decision to postpone starting the detection until $t = 5T_{n,5}$ is to eliminate the possibility that the increase of pitch motion during this time, due to the ramping up of the wave amplitude, results in an increase of the eigenvalues which are calculated under the assumption of a Gaussian wave amplitude distribution. However, investigating this after the test cases were simulated reveals that the magnitudes of the eigenvalues oscillate less around their initial values during the ramping of the input wave amplitude, $t < 5T_{n,5}$, compared to when the wave amplitude is maximal.

The values of $\theta_{i,0}$ were initialised from the results of precursory trial runs, in numerous wave conditions where no parametric resonance was present. In all of these trial runs the values of both $\theta_{1,k}$ and $\theta_{2,k}$ were observed to converge to $-0.5$, regardless of the wave conditions (height, frequency, monochromatic/polychromatic) and of the initial values chosen. An illustrative subset of the precursory trial runs is shown in Figure 3, which plots the evolution of the $\theta_{i,k}$ values, demonstrating their convergence to $-0.5$ for different wave heights (Figure 3a), wave frequencies (Figure 3b) and initial conditions (Figure 3c). Therefore the values of $\theta_{1,0}$ and $\theta_{2,0}$ were both initialised to -0.5 for all of the simulations in the case study. Note that from Equation (7), the initial values correspond to eigenvalues of $\lambda_1 = 1$ and $\lambda_2 = -0.5$.

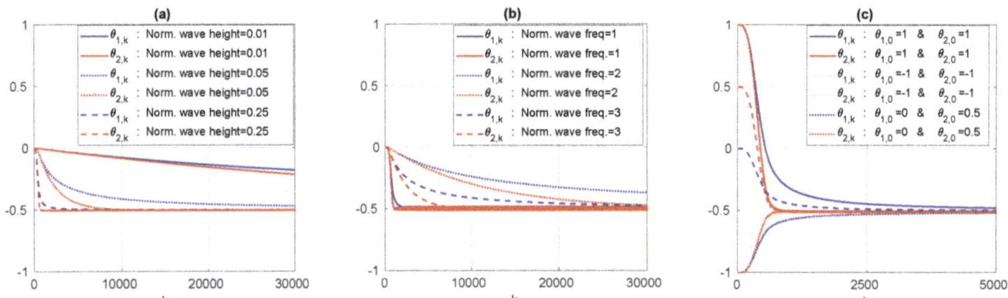

**Figure 3.** The online adaptation of $\theta_{i,k}$ using the recursive least squares algorithm. For monochromatic waves with: (**a**) a normalised wave frequency of 1.6 and varying wave height; (**b**) a normalised wave amplitude of 0.01 and varying wave frequency; and (**c**) a normalised wave frequency of 3, a normalised wave height of 0.1 and varying initial values.

### 3.4.3. Detection System Implementation

The principle of the detection system is to provide a warning of the probable onset of parametric resonance if one of the eigenvalues moves outside of the unit circle. Therefore, the absolute values of the eigenvalues are monitored at each time step and a warning is generated by the detection system when the following condition is met:

$$|\lambda_i| > 1 + \epsilon, \tag{19}$$

where $\epsilon$ is a user-specified threshold value. For the present test case, $\epsilon = 1$; thus, the detection system generates a warning if the magnitude of an eigenvalue exceeds 2. The pragmatic choice of $\epsilon = 1$ stemmed from the observations of the preliminary test runs, where small fluctations were present in the eigenvalues for normal operational conditions, $0.95 < \lambda_1 < 1.05$, but for the case where parametric resonance occurred, the eigenvalues rapidly diverged to values greater than 1000.

## 4. Results

The monochromatic waves are presented in Section 4.1 and then the polychromatic waves in Section 4.2. A brief discussion of the results and their implications is then provided in Section 4.3.

### 4.1. Monochromatic Waves

To evaluate the performance of the real-time early warning detection system, first each simulation had to be post-processed to determine the instances in which parametric resonance occurred (Section 4.1.1). Once the regions of parametric resonance were identified, then the performance of the early warning detection system could be assessed on its ability to correctly identify the onset of parametric resonance in a timely manner (Section 4.1.2).

### 4.1.1. Post Process Identification of Parametric Resonance

For normal operating conditions, in monochromatic waves, the frequencies of the pitch motion and the input waves are the same. However, when parametric resonance occurs, the frequency of the pitch motion becomes half of the input wave frequency. Therefore, the occurrence of parametric resonance in monochromatic waves can be easily identified during post processing by comparing the frequency of the pitch motion to the input wave frequency.

For each of the 4050 monochromatic wave simulations, a fast fourier Transform (FFT) was performed on the time series of the pitch displacement. If the peak amplitude of the PSD occurred at a frequency less than 75% of the input wave frequency, then it is clear that parametric resonance occurred for that input wave condition. Figure 4 shows the wave frequencies and amplitudes for which parametric resonance occurs (the white region) and where it does not (the grey region). Three points are marked in red on Figure 4, which are subjected to the same wave height but with different frequencies, such that point A lays in the stable (no parametric resonance) region, point B on the border between the two regions and point C in the parametric resonance region. The pitch displacement time series and the PSD for points A, B and C are plotted in Figure 5A–C respectively. The pitch displacement for point A is seen to remain constant, with a small amplitude below 0.1 degrees and a frequency equal to the input wave frequency. For point B, the amplitude slowly grows throughout the simulation and the frequency has two peaks, one equal to the wave frequency and the other at half the wave frequency. For point C, the pitch displacement is seen to grow exponentially (to a value above $10°$) and at a frequency equal to half the wave frequency.

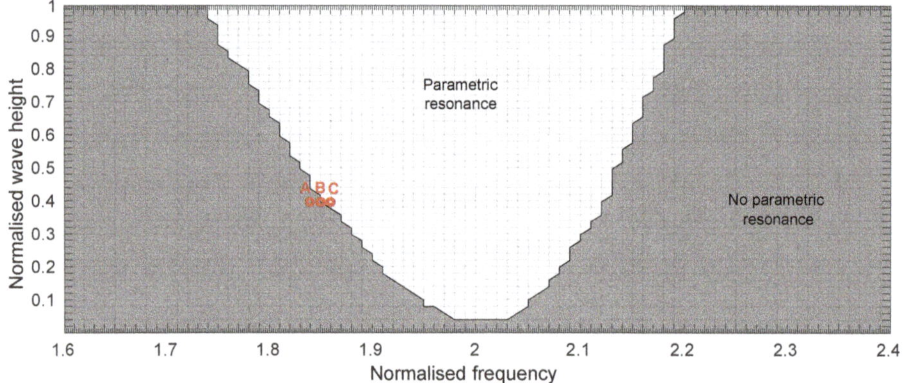


**Figure 4.** The amplitude and frequency range spanned by the monochromatic waves test cases. The white region denotes where parametric resonance was observed. Note the inclusion of grid lines in this figure to highlight the frequency and amplitude resolution, with a result simulated at every grid intersection. For reference, as discussed in Section 3.1, the wave height is normalised against $\overline{GM}$ and the wave frequency by $\omega_{n,5}$.

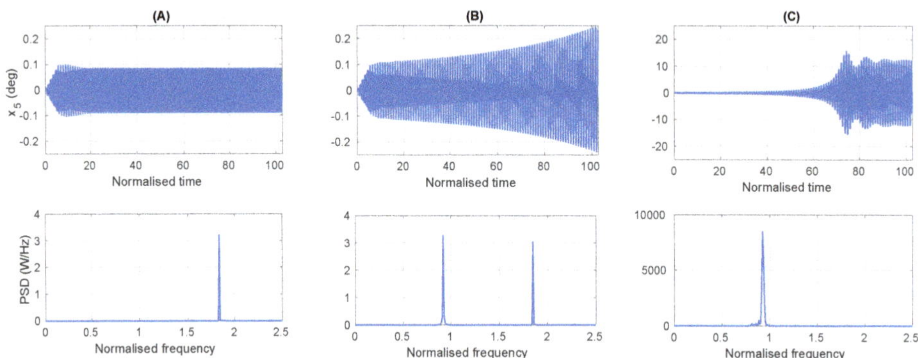


**Figure 5.** The pitch displacement time series and resulting PSD for the points (A) A, (B) B and (C) C in Figure 4.

4.1.2. Performance of the Early Warning Detection System

Figure 6 provides the results for the first assessment criteria: *correctly warning when parametric resonance occurs (correct positive) and not giving a false warning when parametric resonance does not occur (correct negative).* Of the 4050 simulations, the detection system made 4 false negatives and 81 false positives. The false negatives all occurred on the border between the stable and parametric resonance regions, for low wave amplitudes, around twice the pitch natural frequency. The false positives occurred for large wave amplitudes at high frequencies, with three occurring on the border and the remainder occurring in a connected region in the corner of the graph with the highest wave amplitudes and frequencies. The behaviour of the false negatives is explored in Figure 7, which relates to the points A, B and C, marked in red on Figure 6, and the behaviour of the false positives is explored in Figure 8, which relates to the points D, E and F.

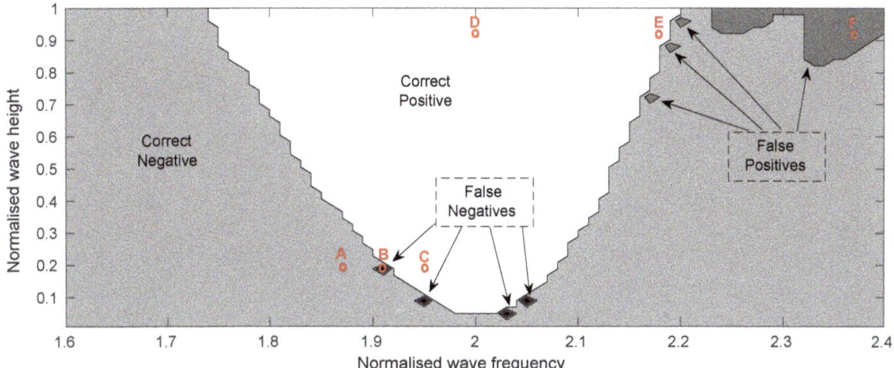


**Figure 6.** The performance of the early warning detection system in monochromatic waves; correct positives (white), correct negatives (grey), false positives (dark grey) and false negatives (black).

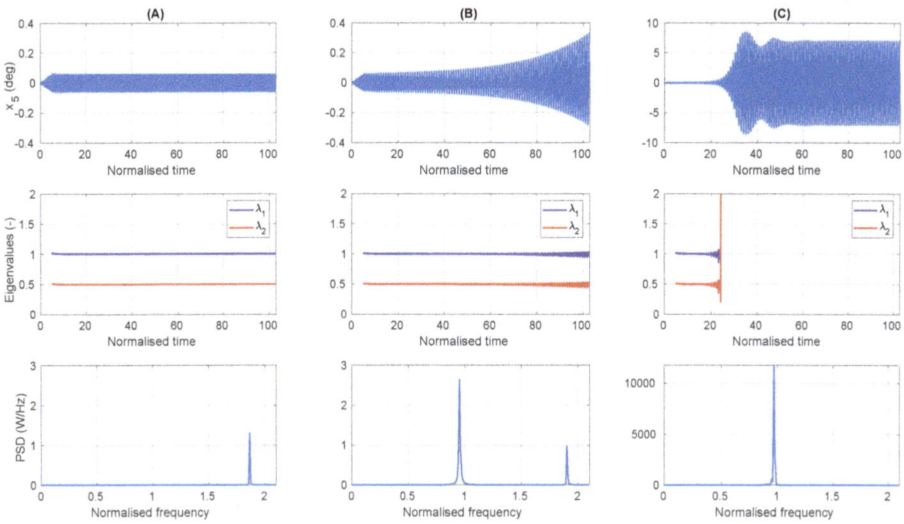


**Figure 7.** The pitch displacement time series (**top**), the magnitudes of the eigenvalues of the matrix **A** (**middle**) and the PSD of the pitch displacement (**bottom**), for the points (A) A, (B) B and (C) C in Figure 6.

Figure 7A relates to point A, which lies outside of the parametric resonance region. The top graph shows the pitch displacement is extremely small, about $0.05°$, and the bottom graph shows that the pitch frequency is equal to the input wave frequency, as is expected for normal operating conditions with no parametric resonance. The magnitudes of the eigenvalues, in the middle graph, are seen to remain constant and within the unit circle, signifying that the system is stable. For Figure 7C, since point C lies within the parametric resonance region, the bottom graph shows that the pitch frequency is now equal to half the input wave frequency. The top graph shows that the amplitude of the pitch displacement is approximately 100 times greater than for point A, even though the input wave amplitude is the same for at both points. The magnitudes of the eigenvalues in the middle graph are seen to diverge well outside of the unit complex circle at the same time as the amplitude of the pitch displacement starts to exponentially increase, highlighting the correct performance of the detection system. Since point B is near the border just inside the parametric resonance region, the exponential growth of the pitch displacement, seen in Figure 7B, occurs at a slower rate, only reaching a value of 0.2° by the end of the simulation. The frequency of the pitch displacement has a small peak at the wave frequency, but a larger one at half the wave frequency; thus this simulation is categorised in the parametric resonance region. The magnitudes of the eigenvalues are seen to increase slightly towards the end of the simulation, thus if the simulation ran for a longer time, then the eigenvalues would have likely diverged and the detection system would have correctly identified this case, rather than producing a false negative.

Points D, E and F correspond to the same large input wave amplitude and all trigger the parametric resonance warning system, although point F is a false positive. Figure 8D relates to point D, which is in the top middle of the parametric resonance region, resulting in the pitch motion becoming unstable and exceeding 90° (at which time the simulation is terminated) within approximately 20 wave periods. Figure 8E relates to point E, which is just within the boundary of the parametric resonance region, resulting in large pitch displacements of approximately 30 degrees at half the input wave frequency, which cause the eigenvalues to diverge out of the unit circle. For Figure 8F, which relates to the false positive at point F, the pitch displacement is stable throughout the simulation, with an amplitude less than 0.5° and a frequency equal to the wave frequency. However, although the magnitudes of the eigenvalues do not diverge, the largest eigenvalue has a magnitude of about 2.5, thereby triggering the parametric resonance detection system.

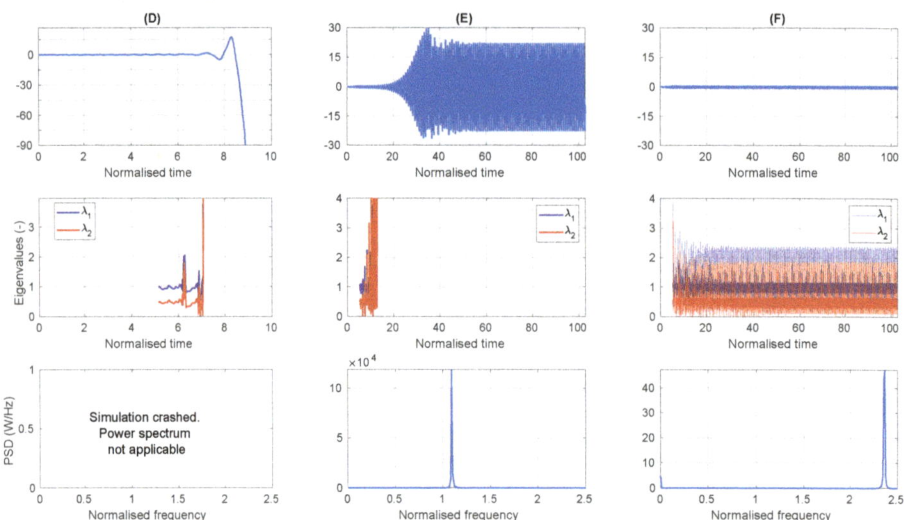

**Figure 8.** The pitch displacement time series (**top**), the magnitudes of the eigenvalues of the matrix **A** (**middle**) and the PSD of the pitch displacement (**bottom**), for the points (D) D, (E) E and (F) F in Figure 6.

Figure 9a–c highlights the performance of the detection system for the second criteria: *how early the system detects the onset of parametric resonance and sends a warning*. Figure 9a shows the amplitude of the pitch displacement when the warning system detected parametric resonance, which for the majority of the cases occurred when the pitch displacement was less than 3°. While for some cases, the pitch amplitude was as high as 7° when the warning system detected parametric resonance, these cases correspond to extremely unstable conditions where the pitch displacement diverged within a few periods after the initial ramping of the input wave and exceeded 90° (such as in Figure 8D). The maximum amplitude of the pitch displacement when parametric resonance occurred is shown in Figure 9b[1]. A quantitative comparison of Figure 9a,b reveals that more that 99% of the detections occurred while the pitch amplitude was less than 1/6 of its maximum amplitude. Finally, Figure 9c shows the normalised time between when the warning system detects the onset of parametric resonance and when the pitch displacement reaches it maximum value. For the majority of cases, the parametric resonance was detected between 5 to $10T_{n5}$. For the extremely unstable regions in the top middle of the parametric resonance region, less warning time was available, with the detection occurring between 2 to $5T_{n5}$. On the border between the stable and parametric resonance regions, very early warning was given, more than $10T_{n5}$, due to the slower exponential growth rate of the pitch displacement amplitude (for example see Figure 8E).

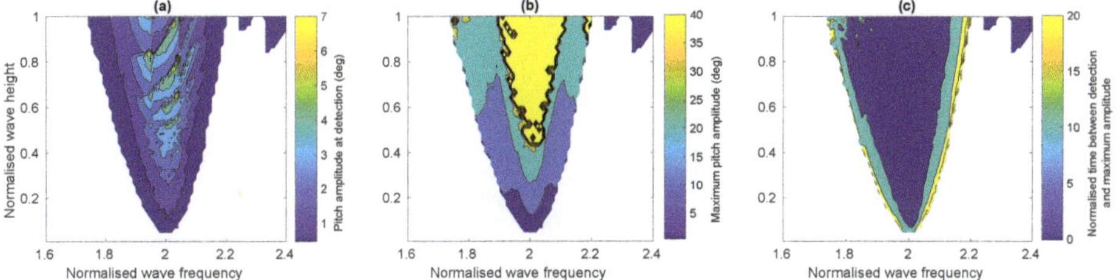

**Figure 9.** The performance of the early warning detection system in monochromatic waves. (**a**) The amplitude of the pitch displacement when the warning system detected parametric resonance. (**b**) The maximum amplitude of the pitch displacement when parametric resonance occurred (clipped at 40°, above which simulations always crashed). (**c**) The time between the when the warning system detected parametric resonance and when the maximum pitch displacement occurred.

*4.2. Polychromatic*

In this section, the performance of the real-time early warning detection system is evaluated for polychromatic input waves. Following the same layout as for the monochromatic waves, first each simulation was post-processed to determine the instances in which parametric resonance occurred (Section 4.2.1) and then the performance assessment was conducted (Section 4.2.2).

### 4.2.1. Post Process Identification of Parametric Resonance

Identifying the occurrence of parametric resonance in polychromatic waves is not as straightforward as for monochromatic waves. For a polychromatic wave spectrum, there is energy in the wave frequencies below the peak of the spectrum; thus, if there is pitch motion at half the peak frequency of the input waves, it can not be instantly identified whether this pitch motion response is due to parametric excitation or direct excitation from the spectral wave components at that lower frequency. Similarly, there is energy at wave frequencies above the peak of the spectrum, so once again it can not be instantly identified whether large pitch motion at low frequencies is due to direct excitation from the wave components at that frequency, or from parametric excitation caused by the

---

[1]  Note: to limit the scale, the maximum amplitude is clipped at 40°, since for any simulation in which the amplitude exceeded this value, the pitch displacement grew to 90° and the simulation was terminated.

wave components at twice that frequency. Therefore, a different metric is applied for the polychromatic waves, based on the energy increase in the pitch motion observed during the simulation.

The energy in the pitch motion (calculated from the root mean square value of $x_5$), measured in the first $5T_{n5}$ after the input wave height is ramped to maximum amplitude (i.e., from $5T_{n5}$ to $10T_{n5}$), is compared to the energy in the last $5T_{n5}$ of the simulation (i.e., from $195T_{n5}$ to $200T_{n5}$). The energy increase in the pitch motion as a function of the normalised wave frequency and height is shown in the contour plot in Figure 10. The contour lines have a resolution of 0.25 and it can be seen that for low amplitude waves and for frequencies well away from $2\omega_{n,5}$, the energy is relatively constant over the length of the simulation, with the increase in energy laying between the 0.5 and 1.5 contour lines. However, for the region of larger amplitude waves centred around $2\omega_{n,5}$, where parametric resonance was expected, the increase in energy has a large positive gradient with respect to the wave height and frequency once the increase in energy values exceeds 1.5. Based on the results shown in Figure 10, if the energy increase from the start to the end of the simulation is more than 2, then the simulation is added to the parametric resonance region, as depicted in Figure 11.

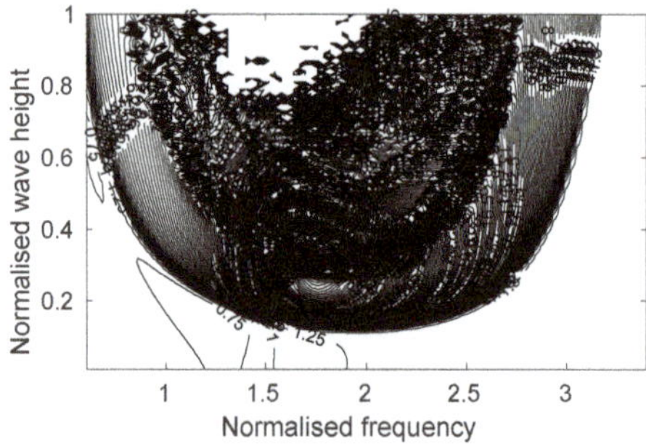

**Figure 10.** The energy increase in the pitch displacement from the start to the end of the simulation as a function of the normalised wave frequency and height. The contour lines have a resolution of 0.25.

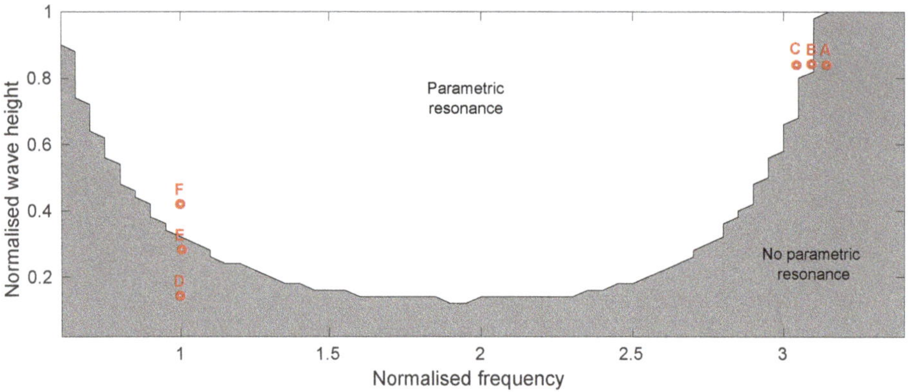


**Figure 11.** The amplitude and frequency range spanned by the polychromatic waves' test cases. The region in which parametric resonance was observed is shown in white.

As an example, three points marked in red, A, B and C, were selected, spanning the borderline between the stable and parametric resonance regions; their time series and PSD are plotted in Figure 12. Point A, which is just outside of the parametric resonance region, is seen to have a fairly constant pitch displacement amplitude, whose frequency components are predominantly in a spectrum which has a peak frequency equal to the peak frequency of the input wave spectrum. However, some frequency components exist, centred at half the peak frequency, whose amplitude is about half of the amplitude at the peak frequency. For point B, just inside parametric resonance region, the pitch motion still contains a spectrum of frequency components with a peak frequency equal to the peak frequency of the input wave spectrum and whose amplitudes are approximately the same as for point A. However, the frequency component at half the peak frequency of the input wave spectrum has an amplitude more than 15 times greater than the amplitude at the peak wave frequency. For point C, only the frequency component at half the peak frequency of the input wave spectrum can be seen, whose amplitude is 90 times greater than the amplitude at the peak wave frequency.

To highlight the difference between the classification of normal resonance and parametric resonance, points D, E and F are selected at increasing wave heights on the pitch natural frequency, and their time series and PSD are plotted in Figure 13. The wave height is doubled between points D and E, which results in a doubling of the pitch motion, as is expected for normal resonance. However, between points D and F, the wave height is increased by a factor of three, but the pitch motion increases approximately tenfold, highlighting the nonlinear parametric resonant response.

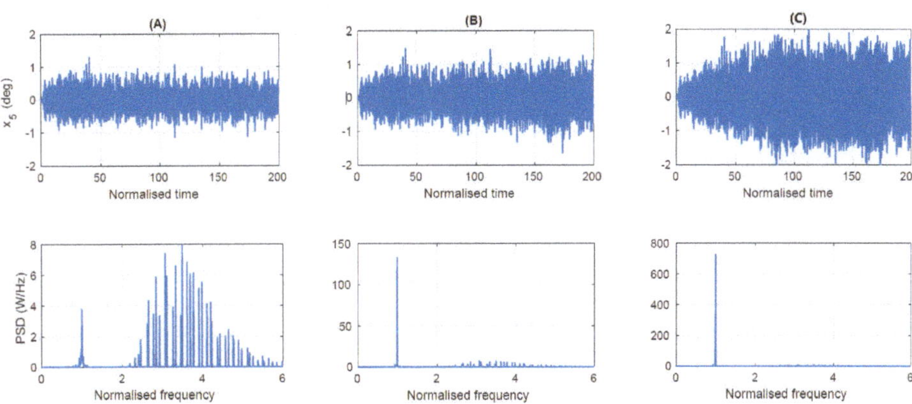

**Figure 12.** The pitch displacement time series and resulting PSD for the points (A) A, (B) B and (C) C in Figure 11.

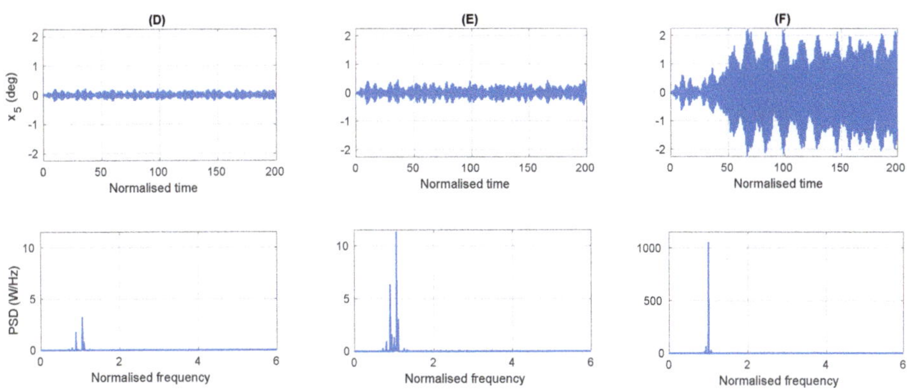

**Figure 13.** The pitch displacement time series and resulting PSD for the points (D) D, (E) E and (F) F in Figure 11.

### 4.2.2. Performance of the Early Warning Detection System

Figure 14 provides the results for the first assessment criteria: correctly warning when parametric resonance occurs (correct positive) and not giving a false warning when parametric resonance does not occur (correct negative). Of the 2800 simulations, the detection system had 25 false negatives and 235 false positives. The false negatives occurred along the border between the stable and the parametric resonance regions, for low wave amplitudes. The false positives, on the other hand, spanned an area from the border into the stable region, at large wave amplitudes, for both low and high frequencies.

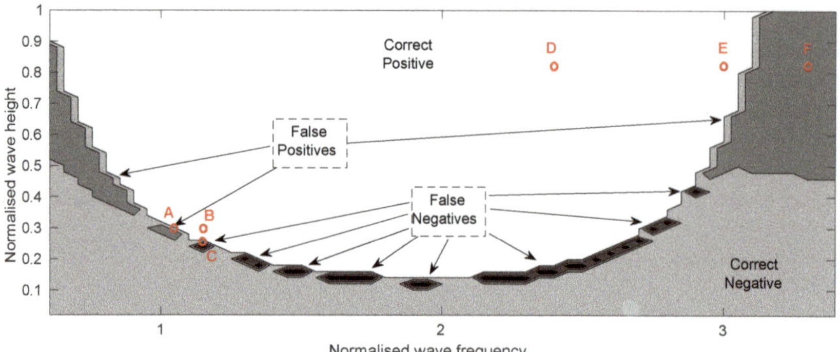


**Figure 14.** The performance of the early warning detection system in polychromatic waves; correct positives (white), correct negatives (grey), false positives (dark grey) and false negatives (black).

Figure 15A–C shows the time series of the pitch displacement and the magnitudes of the eigenvalues of the matrix **A**, along with the PSD of the pitch displacement, for points A–C, respectively. While point A corresponds to a false positive and point C corresponds to a false negative, their time series look quite similar. The increase of energy between the start and end of the simulation was 190% for point A and 232% for point B; thus, they both sit very close on either side of the selected threshold for which the parametric resonance region was defined in this study. At the ends of both simulations, large pitch oscillations occurred, with a maximum amplitude of 0.57° for point A, which caused the largest eigenvalue to grow to an absolute value of 264 and trigger the detection system. For point B, the pitch displacement amplitude only grew to 0.48° and the eigenvalues only grew to 1.44. For a comparison to the behaviour of these two points that sit on the borderline, point B sits just inside the parametric resonance region, with the same amplitude as point A and the same frequency as point C, but the energy increased by more than 800% from the start to the end of the simulation, and the maximum pitch displacement was about three times greater than for points A and C.

Yhe false positives at large wave amplitudes, Figure 16D–F, correspond to the points D–F, respectively, which are all subjected to an input spectrum with the same wave height. Point D is well within the parametric resonance region, point E is just inside the parametric resonance region and point F is in the stable region. Point D is seen to have an oscillatory, large amplitude pitch displacement, with rapid exponential growth and decay rates, and a single dominant frequency at half the peak frequency of the input wave spectrum. Point E also has a single dominant frequency at half the peak frequency of the input wave spectrum. However, the maximum amplitude of the pitch displacement is about 20% of the amplitude at point D and grows much slower and remains fairly constant once it reaches its maximum. For point F, the pitch displacement has a much smaller amplitude, which does not grow, and its frequency content is distributed in a spectrum with the same peak as the input waves. However, the magnitudes of the eigenvalues for point F are seen to diverge in much the same way as for points D and E.

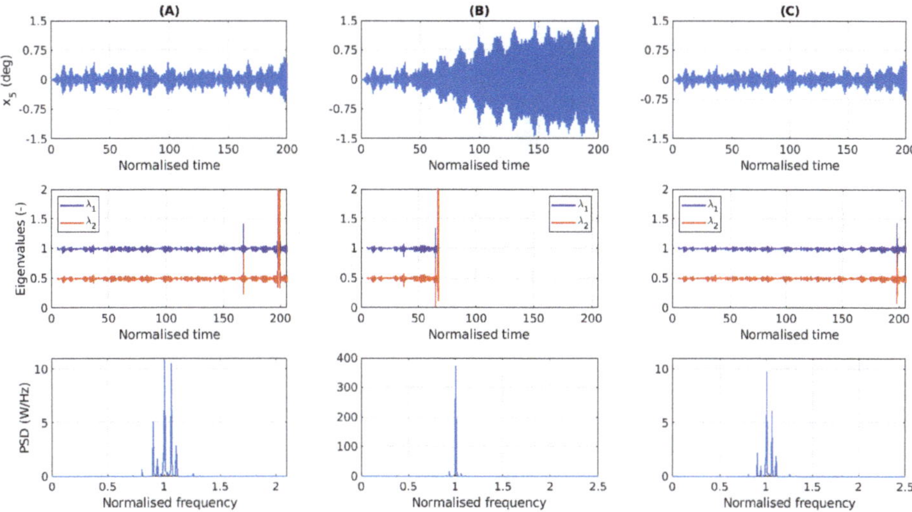

**Figure 15.** The time series for the pitch displacement (**top**), the magnitudes of the eigenvalues of the matrix **A** (**middle**) and the PSD of the pitch displacement (**bottom**), for the points (A) A, (B) B and (C) C in Figure 14.

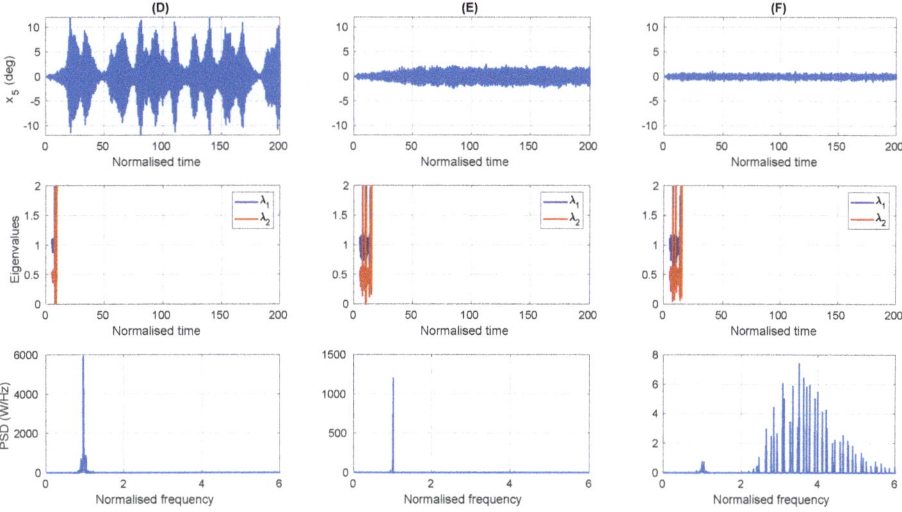

**Figure 16.** The time series for the pitch displacement (**top**), the magnitudes of the eigenvalues of the matrix **A** (**middle**) and the PSD of the pitch displacement (**bottom**), for the points (D) D, (E) E and (F) F in Figure 14.

Figure 17a–c highlights the performance of the detection system for the second criteria: how early the system detects the onset of parametric resonance and sends a warning. Figure 17a shows that for the majority of the cases, the detection system sends a warning when the pitch displacement amplitude is less than 1 degree, and in all cases before the amplitude reaches $2°$. For the polychromatic waves, the maximum pitch amplitude is generally less than 20 degrees, except in a highly unstable region (normalised wave heights above 0.8 and normalised frequencies between 1.25 and 2) where the simulations are terminated at a pitch amplitude of $90°$. Near the parametric resonance border and in the region of false positives, the maximum displacement is less than $5°$. A quantitative comparison of Figure 17a,b reveals that more that 91% of the detections occurred while the pitch amplitude was less than 1/3 of its maximum amplitude and 67% while the pitch amplitude was less than 1/6 of its maximum amplitude. Finally, Figure 17c shows the normalised time between detection and the maximum amplitude, which is in excess of $10T_{n5}$ for all cases within the parametric resonance region, and in many cases is more than $100T_{n5}$. The reason for the large times is that after the pitch amplitude grows larger, due to parametric resonance, it will then have a degree of variation around the new mean amplitude because of the irregular nature of the input waves. Therefore the maximum amplitude can occur at any time after the initial growth to the larger mean amplitude. For example, in Figure 16E, the pitch amplitude grows to a mean maximum level of about $2°$ within $60T_{n5}$; however, at $110T_{n5}$ the amplitude has a spike up to $2.4°$ which is the maximum value for the simulation.

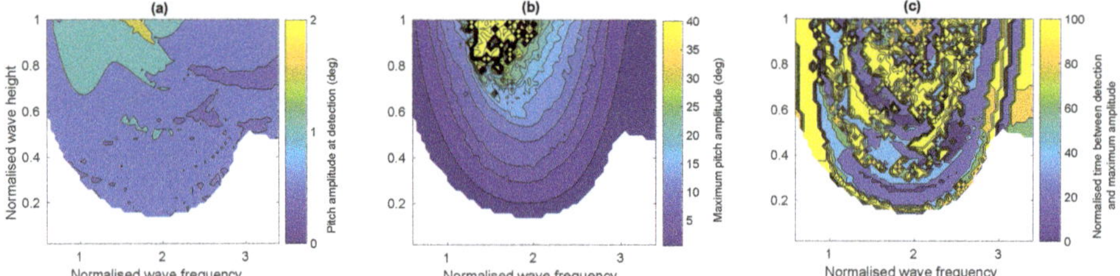

**Figure 17.** The performance of the early warning detection system in polychromatic waves. (**a**) The amplitude of the pitch displacement when the warning system detected parametric resonance. (**b**) The maximum amplitude of the pitch displacement when parametric resonance occurred (clipped at $40°$, above which simulations always crashed). (**c**) The time between the when the warning system detected parametric resonance and when the maximum pitch displacement occurred.

*4.3. Discussion*

Overall, the results demonstrate the ability of the proposed system to effectively detect the onset of parametric resonance. The detection occurred in most cases while the pitch oscillations were small fractions of their maximum amplitudes, with many periods of early warning given for a control system to mitigate the impending growth of the pitch oscillations. For the extremely unstable conditions, much less warning time was available; however, these conditions occurred in the middle of the parametric resonance region. Thus, for such waves conditions to develop the sea state must first transition through the wave conditions in the border of the parametric resonance region, where ample warning for the potential instability of the pitch oscillations is provided.

The proposed system rarely failed to detect the occurrence of parametric resonance, with very few false negatives observed, especially for the monochromatic wave case. For the polychromatic wave case, the slightly higher percentage of false negatives might be attributed to the arbitrary classification of the parametric resonance region. As described in Section 4.2.1, if the energy of the pitch oscillations doubled from the start compared to the end of the simulation, then the simulation was classified as belonging to the parametric resonance region. From Figure 10, choosing a 2-fold increase in the energy of the pitch oscillations as the cut-off for the parametric resonance region seems like a reasonable choice, representing the knee of the curve, above which the pitch oscillation amplitude readily increases. However, choosing a slightly larger value around the knee of the curve, such as a 3-fold increase in the energy for example, could also be valid and might have resulted in a lower number of false negatives, but possibly at the cost of an increased number of false positives. Nevertheless, the false negatives produced by the detection system under the current classification system are not particularly troublesome, relating to pitch displacements of less than $1°$.

The main advantages of the proposed system relate to its simplicity in terms of computational modelling and required inputs. The detection system does not require a complex, high-fidelity, computationally expensive model. Instead, the simple black-box, linear model in Equation (1) has only four parameters that are adaptively tuned to any type of WEC system/geometry. This allows the detection system to be arbitrarily applied to different type of WECs, without requiring bespoke modelling to be tailored for each individual application, system and/or DoF. Indeed, another advantage of the detection system is its ability to treat each DoF independently, since it updates the parameters of a single DoF model using measurements of that single DoF. Thus, if multiple DoFs are to be considered, each DoF can be treated individually with its own detection system, independent of the other DoFs. However, the system requires that the dynamics of the DoF to be modelled can be approximated by a linear second-order differential equation, which is a reasonable assumption in most cases since the DoFs for which parametric resonance is to be monitored are not the primary mode of motion and amplitudes are relatively small in normal operating conditions. Future work will investigate different types of WECs and DoFs.

Perhaps the biggest advantage is the relatively simple and accessible set of inputs required by the detection system: the position and velocity of a single DoF. Measurement of these variables is relatively straightforward, using accelerometers, inclinometers, etc. However, the robustness of the detection system to noise and error in these measuring instruments should be investigated in future work. Nevertheless, the ability of the detection system to work without requiring more challenging inputs, such as prediction of the input waves or estimation of the excitation force, is a major advantage. The current inputs simplify the system requirements and remove the uncertainty/error involved with prediction and estimation methods.

Being a first implementation, for a real-time detection system for the onset of parametric resonance in WECs, the proposed system is not without its flaws and has room for improvement. The main disadvantage observed from the results was the number of false positives. For the monochromatic waves, the false positives occurred for the high frequency and large amplitude input waves, whereas for the polychromatic waves the false positives occurred for both high and low frequency input waves with large amplitudes.

One possible remedy to reduce the amount of false positives is to increase the detection threshold, i.e., the value of $\epsilon$ in Equation (19). For example, as seen in Figure 8F, increasing the value of $\epsilon$ from 1 to 3 would eliminate the false positive. Although increasing the detection threshold can in theory increase the number of false negatives, in practice when parametric resonance was detected the magnitudes of the eigenvalues typically diverged very rapidly to values exceeding 1000. For many false positives, the magnitudes of the eigenvalues were less than 10. Thus, it is unlikely that increasing the detection threshold to any single digit number would cause a currently correct positive detection to become a false negative. However, a false negative is more detrimental to the WEC system than a false positive, so prudently selecting the value of $\epsilon$ should be investigated on a case-by-case basis for different WEC systems and DoFs.

Alternatively, to reduce the nuumber of false positives, the underlying cause should be identified if possible and the detection algorithm improved accordingly. In particular, a possible explanation for the growth of the eigenvalues, which trigger the false detection, is due to the input excitation not being Gaussian white noise, as is assumed in Equation (1). The Gaussian white noise assumption stems from the original application of this detection method in [51], the roll motion of ships, for which in head seas there is no direct wave excitation to the roll DoF. In contrast, the present case study examined the pitch motion, which is directly excited by the input waves. Both the monochromatic and polychromatic waves were coloured, rather than a white spectrum; thus the assumption of a Gaussian white noise excitation was violated since the input spectrum was not flat, with equal amplitude across all frequencies. Therefore, future work will investigate a means to improve upon the assumption of Gaussian white noise input whilst hopefully maintaining the benefit of not explicitly requiring an estimation of the input waves/excitation force.

## 5. Conclusions

This paper demonstrated the first implementation of a real-time detection system for the early warning of the onset of parametric resonance in a WEC, by applying a method developed for the monitoring and detection of parametric roll motion in ships. The results from a test case, considering the parametric pitch motion in a heaving spar-buoy, show that the detection system performed very well, achieving 95% accuracy across a range of monochromatic and polychromatic sea states. In addition, the detection system provides an early warning while the pitch motion is a small fraction of its impending maximum amplitude. However, one potential shortcoming is the assumption of the external excitation to the system being described by Gaussian white noise, which is less valid for the pitch motion in a WEC compared to the roll motion in a ship. Assuming the input to be Gaussian white noise resulted in the system incorrectly detecting instability when the input waves were of large amplitudes and high frequencies, resulting in the warning system providing 10 times more false positives than false negatives. The practical implementation of the system imposes a low overhead, only requiring measurement of the position and velocity of the DoF to be monitored and enough computational resources to perform a recursive least squares algorithm for two time-varying parameters and then finding the roots of a quadratic equation.

**Author Contributions:** Conceptualization, J.D.; Formal analysis, J.D.; Investigation, J.D.; Supervision, T.K.-N.; Writing—original draft, J.D.; Writing—review & editing, T.K.-N. Both authors have read and agreed to the published version of the manuscript.

**Funding:** This project has received funding from the European Union's Horizon 2020 research and innovation programme under the Marie Sklodowska-Curie grant agreement number 867453. The research reported in this paper was supported by the BME Water Sciences and Disaster Prevention TKP2020 IE grant of NKFIH Hungary (BME IE-VIZ TKP2020) and by the BME NC TKP2020 grant of NKFIH Hungary.

**Conflicts of Interest:** The authors declare no conflict of interest.

## References


1. Fossen, T.; Nijmeijer, H. *Parametric Resonance in Dynamical Systems*; Springer: New York, NY, USA, 2011.
2. Faraday, M. On a peculiar class of acoustical figures; and on certain forms assumed by groups of particles upon vibrating elastic surfaces. In *Philosophical Transactions of the Royal Society of London*; The Royal Society: London, UK, 1831; pp. 299–340.
3. Froude, W. *On the Rolling of Ships*; Institution of Naval Architects: London, UK, 1861.
4. Galeazzi, R. Autonomous Supervision and Control of Parametric Roll Resonance. Ph.D. Thesis, Department of Naval Architecture and Offshore Engineering, Technical University of Denmark, Lyngby, Denmark, 2009.
5. Shigunov, V.; El Moctar, O.; Rathje, H.; Germanischer Lloyd, A. Conditions of parametric rolling. In Proceedings of the 10th International Conference on Stability of Ships and Ocean Vehicles, St Petersberg, Russia, 22–26 June 2009.
6. Koo, B.; Kim, M.; Randall, R. Mathieu instability of a spar platform with mooring and risers. *Ocean Eng.* **2004**, *31*, 2175–2208. [CrossRef]
7. Neves, M.A.; Sphaier, S.H.; Mattoso, B.M.; Rodri´guez, C.A.; Santos, A.L.; Vileti, V.L.; Torres, F.G. On the occurrence of Mathieu instabilities of vertical cylinders. In Proceedings of the 27th International Conference on Offshore Mechanics and Arctic Engineering, Estorill, Portugal, 15–20 June 2008.
8. Li, B.B.; Ou, J.P.; Teng, B. Numerical investigation of damping effects on coupled heave and pitch motion of an innovative deep draft multi-spar. *J. Mar. Sci. Technol.* **2011**, *19*, 231–244.
9. Yang, H.; Xu, P. Effect of hull geometry on parametric resonances of spar in irregular waves. *Ocean Eng.* **2015**, *99*, 14–22. [CrossRef]
10. Jang, H.; Kim, M. Mathieu instability of Arctic Spar by nonlinear time-domain simulations. *Ocean Eng.* **2019**, *176*, 31–45. [CrossRef]
11. Umeda, N.; Hashimoto, H.; Minegaki, S.; Matsuda, A. An investigation of different methods for the prevention of parametric rolling. *J. Mar. Sci. Technol.* **2008**, *13*, 16–23. [CrossRef]

12. Galeazzi, R.; Pettersen, K.Y. Parametric resonance in dynamical systems. In *Parametric Resonance in Dynamical Systems*; Chapter Controlling Parametric Resonance: Induction and Stabilization of Unstable Motions; Fossen, T.I., Nijmeijer, H., Eds.; Springer: New York, NY, USA, 2012; pp. 305–327.

13. Bracewell, R.H. Frog and PS Frog: A Study of Two Reactionless Ocean Wave Energy Converters. Ph.D. Thesis, University of Lancaster, Lancashire, UK, 1990.

14. Durand, M.; Babarit, A.; Pettinotti, B.; Quillard, O.; Toularastel, J.; Clément, A. Experimental validation of the performances of the SEAREV wave energy converter with real time latching control. In Proceedings of the 7th European Wave and Tidal Energy Conference, Porto, Portugal, 11–13 September 2007.

15. Payne, G.S.; Taylor, J.R.; Bruce, T.; Parkin, P. Assessment of boundary-element method for modelling a free-floating sloped wave energy device. Part 2: Experimental validation. *Ocean Eng.* **2008**, *35*, 342–357. [CrossRef]

16. Raftery, M.W. Harnessing Ocean Surface Wave Energy to Generate Electricity. Master's Thesis, Stevens Institute of Technology, Hoboken, NJ, USA, 2008.

17. Sheng, W.; Flannery, B.; Lewis, A.; Alcorn, R. Experimental studies of a floating cylindrical OWC WEC. In Proceedings of the 31st International Conference on Ocean, Offshore and Arctic Engineering, Rio de Janeiro, Brazil, 1–6 July 2012.

18. Gomes, R.; Henriques, J.; Gato, L.; Falcao, A. Testing of a small-scale floating OWC model in a wave flume. In Proceedings of the 4th International Conference on Ocean Energy, Dublin, Ireland, 17–19 October 2012.

19. Gomes, R.; Henriques, J.; Gato, L.; Falcão, A.d.O. Wave channel tests of a slack-moored floating oscillating water column in regular waves. In Proceedings of the 11th European Wave and Tidal Energy Conference, Nantes, France, 6–11 September 2015.

20. Beatty, S.J.; Roy, A.; Bubbar, K.; Ortiz, J.; Buckham, B.J.; Wild, P.; Stienke, D.; Nicoll, R. Experimental and numerical simulations of moored self-reacting point absorber wave energy converters. In Proceedings of the 25th International Ocean and Polar Engineering Conference, Kona, HI, USA, 25–30 June 2015.

21. Orszaghova, J.; Wolgamot, H.; Taylor, R.E.; Draper, S.; Rafiee, A.; Taylor, P. Simplified dynamics of a moored submerged buoy. In Proceedings of the 32nd International Workshop on Water Waves and Floating Bodies, Dalian, China, 23–26 April 2017.

22. Sergiienko, N.Y. Three-Tether Wave Energy Converter: Hydrodynamic Modelling, Performance Assessment and Control. Ph.D. Thesis, University Adelaide, Adelaide, Australia, 2018.

23. Gomes, R.; Henriques, J.; Gato, L.; Falcão, A. Experimental Tests of a 1: 16th-Scale Model of the Spar-Buoy OWC in a Large Scale Wave Flume in Regular Waves. In Proceedings of the 37th International Conference on Ocean, Offshore and Arctic Engineering, Madrid, Spain, 17–22 June 2018.

24. Kurniawan, A.; Grassow, M.; Ferri, F. Numerical modelling and wave tank testing of a self-reacting two-body wave energy device. *Ships Offshore Struct.* **2019**, *14*, 344–356. [CrossRef]

25. Orszaghova, J.; Wolgamot, H.; Draper, S.; Eatock Taylor, R.; Taylor, P.; Rafiee, A. Transverse motion instability of a submerged moored buoy. *Proc. R. Soc. A* **2019**, *475*, 20180459. [CrossRef]

26. Gomes, R.; Henriques, J.; Gato, L.; Falcão, A. Time-domain simulation of a slack-moored floating oscillating water column and validation with physical model tests. *Renew. Energy* **2020**, *149*, 165–180. [CrossRef]

27. Davidson, J.; Costello, R. Efficient Nonlinear Hydrodynamic Models for Wave Energy Converter Design—A Scoping Study. *J. Mar. Sci. Eng.* **2020**, *8*, 35. [CrossRef]

28. Babarit, A.; Mouslim, H.; Clément, A.; Laporte-Weywada, P. On the numerical modelling of the nonlinear behaviour of a wave energy converter. In Proceedings of the 28th International Conference on Offshore Mechanics & Arctic Engineering, Honolulu, HI, USA, 31 May–5 June 2009.

29. Tarrant, K.R. Numerical Modelling of Parametric Resonance of a Heaving Point Absorber Wave Energy Converter. Ph.D. Thesis, Trinity College Dublin, Dublin, Ireland, 2015.

30. Tarrant, K.; Meskell, C. Investigation on parametrically excited motions of point absorbers in regular waves. *Ocean Eng.* **2016**, *111*, 67–81. [CrossRef]

31. Giorgi, G.; Ringwood, J.V. A compact 6-DoF nonlinear wave energy device model for power assessment and control investigations. *IEEE Trans. Sustain. Energy* **2018**, *10*, 119–126. [CrossRef]

32. Giorgi, G.; Ringwood, J.V. Articulating Parametric Nonlinearities in Computationally Efficient Hydrodynamic Models. *IFAC-PapersOnLine* **2018**, *51*, 56–61. [CrossRef]

33. Giorgi, G.; Ringwood, J.V. Parametric motion detection for an oscillating water column spar buoy. In Proceedings of the 3rd International Conference on Renewable Energies Offshore, Lisbon, Portugal, 8–10 October 2018.

34. Giorgi, G.; Gomes, R.P.; Bracco, G.; Mattiazzo, G. The Effect of Mooring Line Parameters in Inducing Parametric Resonance on the Spar-Buoy Oscillating Water Column Wave Energy Converter. *J. Mar. Sci. Eng.* **2020**, *8*, 29. [CrossRef]

35. Zou, S.; Abdelkhalik, O.; Robinett, R.; Korde, U.; Bacelli, G.; Wilson, D.; Coe, R. Model Predictive Control of parametric excited pitch-surge modes in wave energy converters. *Int. J. Mar. Energy* **2017**, *19*, 32–46. [CrossRef]

36. Yerrapragada, K.; Ansari, M.; Karami, M.A. Enhancing power generation of floating wave power generators by utilization of nonlinear roll-pitch coupling. *Smart Mater. Struct.* **2017**, *26*, 094003. [CrossRef]

37. Abdelkhalik, O.; Zou, S. Time-varying linear quadratic gaussian optimal control for three-degree-of-freedom wave energy converters. In Proceedings of the 12th European Wave and Tidal Energy Conference, Cork, Ireland, 27 August–2 September 2017.

38. Abdelkhalik, O.; Zou, S.; Robinett, R.; Bacelli, G.; Wilson, D.; Coe, R. Control of three degrees-of-freedom wave energy converters using pseudo-spectral methods. *J. Dyn. Syst. Meas. Control* **2018**, *140*, 074501. [CrossRef]

39. Davidson, J.; Karimov, M.; Szelechman, A.; Windt, C.; Ringwood, J. Dynamic mesh motion in OpenFOAM for wave energy converter simulation. In Proceedings of the 14th OpenFOAM Workshop, Duisburg, Germany, 23–26 July 2019.

40. Zou, S.; Abdelkhalik, O. Time-varying linear quadratic Gaussian optimal control for three-degree-of-freedom wave energy converters. *Renew. Energy* **2020**, *149*, 217–225. [CrossRef]

41. Palm, J.; Bergdahl, L.; Eskilsson, C. Parametric excitation of moored wave energy converters using viscous and non-viscous CFD simulations. In *Advances in Renewable Energies Offshore*; Taylor & Francis Group: Abingdon, UK, 2019; pp. 455–462.

42. Nicoll, R.S.; Wood, C.F.; Roy, A.R. Comparison of physical model tests with a time domain simulation model of a wave energy converter. In Proceedings of the 31st International Conference on Ocean, Offshore and Arctic Engineering, Rio de Janeiro, Brazil, 1–6 July 2012.

43. Orszaghova, J.; Wolgamot, H.; Draper, S.; Taylor, P.H.; Rafiee, A. Onset and limiting amplitude of yaw instability of a submerged three-tethered buoy. *Proc. R. Soc. A* **2020**, *476*, 20190762. [CrossRef]

44. Rho, J.B.; Choi, H.S.; Shin, H.S.; Park, I.K. A study on Mathieu-type instability of conventional spar platform in regular waves. *Int. J. Offshore Polar Eng.* **2005**, *15*, 104–108.

45. Ortiz, J.P. The Influence of Mooring Dynamics on the Performance of Self Reacting Point Absorbers. Master's Thesis, University of Victoria, Victoria, BC, Canada, 2016.

46. Gomes, R.; Malvar Ferreira, J.; Ribeiro e Silva, S.; Henriques, J.; Gato, L. An experimental study on the reduction of the dynamic instability in the oscillating water column spar buoy. In Proceedings of the 12th European Wave and Tidal Energy Conference, Cork, Ireland, 27 August–2 September 2017.

47. Cordonnier, J.; Gorintin, F.; De Cagny, A.; Clément, A.; Babarit, A. SEAREV: Case study of the development of a wave energy converter. *Renew. Energy* **2015**, *80*, 40–52. [CrossRef]

48. Gomes, R.; Henriques, J.; Gato, L.; Falcao, A. An upgraded model for the design of spar-type floating oscillating water column devices. In Proceedings of the 13th European Wave and Tidal Energy Conference, Naples, Italy, 1–6 September 2019.

49. Villegas, C.; van der Schaaf, H. Implementation of a pitch stability control for a wave energy converter. In Proceedings of the 10th European Wave and Tidal Energy Conference, Southampton, UK, 5–9 September 2011.

50. Maloney, P. Performance Assessment of a 3-Body Self-Reacting Point Absorber Type Wave Energy Converter. Master's Thesis, University of Victoria, Victoria, BC, Canada, 2019.

51. Holden, C.; Perez, T.; Fossen, T.I. Frequency-motivated observer design for the prediction of parametric roll resonance. *IFAC Proc. Vol.* **2007**, *40*, 57–62. [CrossRef]

52. Shaw McCue, L.; Bulian, G. A numerical feasibility study of a parametric roll advance warning system. *J. Offshore Mech. Arct. Eng.* **2007**, *129*. [CrossRef]

53. Galeazzi, R.; Blanke, M.; Poulsen, N.K. Parametric roll resonance detection on ships from nonlinear energy flow indicator. *IFAC Proc. Vol.* **2009**, *42*, 348–353. [CrossRef]

54. Galeazzi, R.; Blanke, M.; Poulsen, N.K. Early detection of parametric roll resonance on container ships. *IEEE Trans. Control Syst. Technol.* **2012**, *21*, 489–503. [CrossRef]

55. Galeazzi, R.; Blanke, M.; Falkenberg, T.; Poulsen, N.K.; Violaris, N.; Storhaug, G.; Huss, M. Parametric roll resonance monitoring using signal-based detection. *Ocean Eng.* **2015**, *109*, 355–371. [CrossRef]

56. Caamaño, L.S.; González, M.M.; Casás, V.D. On the feasibility of a real time stability assessment for fishing vessels. *Ocean Eng.* **2018**, *159*, 76–87. [CrossRef]

57. Caamaño, L.S.; Galeazzi, R.; Nielsen, U.D.; González, M.M.; Casás, V.D. Real-time detection of transverse stability changes in fishing vessels. *Ocean Eng.* **2019**, *189*, 106369. [CrossRef]

58. Davidson, J.; Genest, R.; Ringwood, J. Adaptive control of a wave energy converter simulated in a numerical wave tank. In Proceedings of the 12th European Wave and Tidal Energy Conference, Cork, Ireland, 27 August–2 September 2017.

59. Genest, R.; Davidson, J.; Ringwood, J.V. Adaptive control of a wave energy converter. *IEEE Trans. Sustain. Energy* **2018**, *9*, 1588–1595.

60. Peña-Sanchez, Y.; Windt, C.; Davidson, J.; Ringwood, J.V. A Critical Comparison of Excitation Force Estimators for Wave-Energy Devices. *IEEE Trans. Control. Syst. Technol.* **2019**. [CrossRef]

61. Thompson, J.M.T.; Stewart, H.B. *Nonlinear Dynamics and Chaos*; John Wiley & Sons: Hoboken, NJ, USA, 2002.

62. Yong-Pyo, H.; Dong-Yeon, L.; Yong-Ho, C.; Sam-Kwon, H.; Se-Eun, K. An experimental study on the extreme motion responses of a spar platform in the heave resonant waves. In Proceedings of the 15th International Offshore and Polar Engineering Conference, Seol, Korea, 19–24 June 2005.

63. Zhao, J.; Tang, Y.; Shen, W. A study on the combination resonance response of a classic spar platform. *J. Vib. Control* **2010**, *16*, 2083–2107. [CrossRef]

64. Gavassoni, E.; Gonçalves, P.B.; Roehl, D.M. Nonlinear vibration modes and instability of a conceptual model of a spar platform. *Nonlinear Dyn.* **2014**, *76*, 809–826. [CrossRef]

65. Giorgi, G.; Davidson, J.; Habib, G.; Bracco, G.; Mattiazzo, G.; Kalmar-Nagy, T. Nonlinear Dynamic and Kinematic Model of a Spar-Buoy: Parametric Resonance and Yaw Numerical Instability. *Mar. Sci. Eng.* **2020**, *8*, 504. [CrossRef]

66. Wacher, A.; Nielsen, K. Mathematical and numerical modeling of the AquaBuOY wave energy converter. *Math. Case Stud.* **2010**, *2*, 16–33.

Journal of
*Marine Science
and Engineering*

*Article*

# Real-Time Wave Excitation Forces Estimation: An Application on the ISWEC Device

**Mauro Bonfanti [1],*, Andrew Hillis [2], Sergej Antonello Sirigu [1], Panagiotis Dafnakis [1], Giovanni Bracco [1], Giuliana Mattiazzo [1] and Andrew Plummer [2]**

[1]  Marine Offshore Renewable Energy Lab (MOREnergy Lab), DIMEAS, Politecnico di Torino, Corso Duca degli Abruzzi 24, 10129 Turin, Italy; sergej.sirigu@polito.it (S.A.S.); panagiotis.dafnakis@polito.it (P.D.); giovanni.bracco@polito.it (G.B.); giuliana.mattiazzo@polito.it (G.M.)

[2]  Department of Mechanical Engineering, University of Bath, Bath 44210, UK; a.j.hillis@bath.ac.uk (A.H.); A.R.Plummer@bath.ac.uk (A.P.)

*  Correspondence: mauro.bonfanti@polito.it; Tel.: +39-01-1090-5921

Received: 3 October 2020; Accepted: 19 October 2020; Published: 21 October 2020

**Abstract:** Optimal control strategies represent a widespread solution to increase the extracted energy of a Wave Energy Converter (WEC). The aim is to bring the WEC into resonance enhancing the produced power without compromising its reliability and durability. Most of the control algorithms proposed in literature require for the knowledge of the Wave Excitation Force (WEF) generated from the incoming wave field. In practice, WEFs are unknown, and an estimate must be used. This paper investigates the WEF estimation of a non-linear WEC. A model-based and a model-free approach are proposed. First, a Kalman Filter (KF) is implemented considering the WEC linear model and the WEF modelled as an unknown state to be estimated. Second, a feedforward Neural Network (NN) is applied to map the WEC dynamics to the WEF by training the network through a supervised learning algorithm. Both methods are tested for a wide range of irregular sea-states showing promising results in terms of estimation accuracy. Sensitivity and robustness analyses are performed to investigate the estimation error in presence of un-modelled phenomena, model errors and measurement noise.

**Keywords:** Kalman Filter; Neural Network; wave excitation forces; estimation; Optimal Control; Wave Energy Converter

---

## 1. Introduction

Among the different renewable sources, ocean waves represent one of the most powerful and in the last few decades have been widely investigated. Despite this, wave energy remains a relatively untapped resource. The application of Wave Energy Converter (WEC) systems to such irregular sources requires a robust control logic capable of maximizing the extracted energy with acceptable efficiencies. At present, the Optimal Control problem represents an active area of research. The work of [1] reports a comprehensive review of the advances in Optimal Control, Model Predictive Control (MPC) and MPC-like techniques applied to the wave energy field. Within the context of Optimal Control, accurate knowledge about the future Wave Excitation Force (WEF) is essential to compute the optimal control signal. In literature, several approaches have been proposed to address the problem of the WEF estimation with promising results. An exhaustive classification and comparison of several estimation techniques is presented in the work of [2]. Some examples are cited hereafter. In [3,4] a Kalman Filter (KF) observer is employed on a linear Point-Absorber (PA) WEC. It is assumed that the WEF can be modelled as a linear superposition of fixed and finite harmonic components. Similarly, In [5] the wave force estimation and prediction problem for arrays of WECs is approached , comparing both global and independent estimators and forecasters. In the work of [6] two approaches are presented: the first approach is based on a KF coupled with a random-walk model of the WEF; the second performs a receding horizon—unknown input estimation. In [7] a modified form of the well-known Fast Adaptive actuator Fault Estimation called Fast Adaptive Unknown Input Estimation (FAUIE) is applied to a non-linear PA. In [8] is studied a direct approach by measuring the pressures at discrete points on the buoy surface, in addition to the buoy heave position, to obtain the estimation of the WEF by an Extended Kalman Filter (EKF). Finally, In [9] studied the WEF estimation of WaveSub (by Marine Power Systems Ltd. (Swansea, England)), a multiple Degree of Freedom (DoF) non-linear WEC. The estimation problem is tackled with a stochastic and a periodic KF, using only quantities which are measurable in practice. A model-free approach is proposed by [10] emoploying a Neural Network (NN) framework to estimate the WEF on a PA. Similarly, In [11] is studied the estimation of the wave elevation using the measurements from a nearby buoy employing a Non-linear AutoRegressive with eXogenous input network (NARX). All the mentioned studies are promising in terms of estimation performances and robustness of the observer. However, most of the work cited refer to single Degree of Freedom (DoF) WECs and few studies on non-linear multiple DoF systems have been conduced so far.

The aim of this work is to estimate the WEFs on a non-linear three Degrees of Freedom (3-DoF) WEC using only readily measureable quantities to perform the estimation. The study is applied to the Inertial Sea Wave Energy Converter (ISWEC) device designed for the Mediterranean Sea. In this context, three different approaches have been applied to the ISWEC device in previous works. In [12] is built an unknown state observer with a second order augmented state space representation of the ISWEC for the estimation of the pitch excitation torque. The gain of the observer has been found with an LQR optimization. In [13] is presented a method to estimate the sea state Power Spectral Density (PSD) of the wave climate by using the device motion. The heave motion measurements are used to estimate the PSD of the incoming wave and the results compared with the wave PSD measured by a wave measurement system. Then, in [14] a feedforward NN is proposed to relate the ISWEC motion to the WEFs acting on surge, heave and pitch DoFs. In the current study two approaches are applied to the non-linear 3-DoF model of the ISWEC to estimate the WEFs acting on surge, heave and pitch DoFs:

1.  Model-Based approach: a KF observer modelling the WEF as an unknown input with a harmonic nature. This estimation framework is studied in [3,4] and named Kalman Filter with Harmonic Oscillator (KFHO) in [2];
2.  Model-Free approach: a feedforward NN to relate the ISWEC motion and gyroscopic reaction to the WEFs. The same approach reported in [14] is considered.

The main challenges lie in considering a non-linear 3-DoF model of the WEC and estimate the WEFs along three degrees of motion. In the KF context, tuning the system for the best performance in presence of measurement noise is crucial as well as identifying un-modelled phenomena and decoupling them from the signal to be estimated. Moreover, since the ISWEC device is slack-moored to the seabed, an accurate acquisition of the absolute displacements of the WEC is not trivial; in this context, measurement errors of the absolute diplacement could affect the estimation performances. For what concerns the NN, it is key to assess both the ability to estimate data that are not considered in the training process and the accuracy in presence of model uncertanities. The aim is to use the minimum amount of measurements to model the unknown exctitation forces in respect to the ISWEC kinematic. Outcomes of this investigation are the estimators performances in different irregular sea-states and different measurement frameworks in presence of sensors noises and plant uncertainties. The results of the 3-DoF KF estimator and the NN model are compared for different sensors framework in order to analyse the influence of the measurements available. Then, measurements noises and variations of the main plant parameters are introduced. Numerical simulations under different irregular sea conditions aim to compare the estimation results of the two approaches as well as the sensitivity to changes in the plant parameters relative to the case study presented. The novelty of this work is to perform an excitation force estimation for a 3-DoF non-linear WEC decoupling the mooring forces or, more in general, undesired phenomena that will appear in real operating condition. Moreover, the NN represents a novel framework to address the wave extitation force estimation problem, able to handle strong non-linearities. In the NN context, there are very few examples of application to a multi-DOF non-linear WEC system in literature. Finally, investigating the influence of available sensors constitutes a key aspect of this work, which aims to highlight the importance of having reliable measurements in the context of WEF estimation.

This paper is organized as follows. First, in Section 2 the ISWEC architecture is described together with the non-linear 3-DoF time domain model. Second, Section 3 presents the linear 3-DoF ISWEC model in conjunction with the KF formulation used to estimate the WEFs. Section 4 describes the NN approach to estimate the WEFs as well as the network architecture. In Section 5 the parameters of both KF and NN are tuned considering the operating conditions of the ISWEC. Then, in Section 6, the numerical results are presented and discussed. Strengths and weaknesses of each method are compared and numerical experiments are used to assess estimation accuracy, robustness to different wave conditions, sensor noises and sensitivity to model errors. Finally, Section 7 presents conclusions and future works.

## 2. ISWEC Device

The ISWEC system has been considered as the case study in this paper. The first concept is dated 2005, conceived by the renewable group of the Department of Mechanical and Aerospace Engineering of the Politecnico di Torino (Italy). In 2012, a 1:8 prototype was realized and tested at INSEAN wave tank in Rome, crucial to draw the main guidelines for the ISWEC design. In 2012, the design of a 60 kW rated power ISWEC full-scale prototype (shown in Figure 1) started and the device has been installed in Pantelleria island (Sicily, Italy) in Summer 2015. The project aimed to evaluate the device energy production capabilities and efficiency. In 2018, a new collaboration with the Italian company Eni S.p.a. led to the construction of a second full-scale prototype in the early 2019, installed in the Adriatic Sea (Italy). This second full-scale prototype maintains the architecture and operating principle of the Pantelleria device. The development of more accurate numerical models together with ad hoc optimization algorithms validated and supported by the experimental data of the Pantelleria device enabled efficiency and reliability improvements to the overall device. In this section the device working principle and the non-linear 3-DoF numerical model refer to the prototype installed in Adriatic Sea.

(a)   (b)

**Figure 1.** ISWEC device installed in Pantelleria island (Sicily, Italy): (**a**) ISWEC Pantelleria hull; (**b**) ISWEC Pantelleria gyroscope.

### 2.1. ISWEC Working Principle

The major disadvantage of PA-like devices is that they harness wave energy damping directly from the motion of an oscillating buoy, exploiting the relative motion of mechanical parts in contact with the harsh marine environment. The idea of the ISWEC is to harvest wave energy damping from the floater motion through a gyroscope system which is protected within the hull, reducing the risk of corrosion and biofouling, with consequent increase of reliability. Morevoer, also the power conversion and conditioning systems are completely enclosed into the floater. The only part that have continuity from the inside out is the electric cable, following the idea of a "deploy and plug" device [15]. A schematic representation of the device concept is shown in Figure 2. It consists of a sealed hull carrying inside two gyroscopic units. The power conversion principle relies on the gyroscopic effect that converts the pitching angular motion of the hull into an inner precession oscillation of the gyroscopes. ISWEC is a directional WEC since, in normal working conditions, it is able to align itself with the wave direction. As shown in Figure 2b, the gyroscope is composed of a spinning flywheel that rotates around the $z_1$ axis with a speed $\dot{\varphi}$. The flywheel is supported by a frame that allows the rotation of the gyroscope around its precession axis $x_1$. The precession motion is excited through the dynamic coupling between the hull motion around the $y_1$ axis and the flywheel angular velocity around the $z_1$ axis. A mechanical gearbox and an electrical generator, connected to the gyroscopic frame, compose the electrical Power Take-Off (PTO). Two electrical generators extract electricity acting as a linear damper braking the precession motion of the gyroscope. Notwithstanding the energy required to keep the flywheel in rotation, the possibility to regulate its speed along with the tuning of the PTO torque allows the adaptation of the natural resonant frequency of the system to the incoming wave. An accurate description of the internal components and their working principle can be found in the works of [15–17].

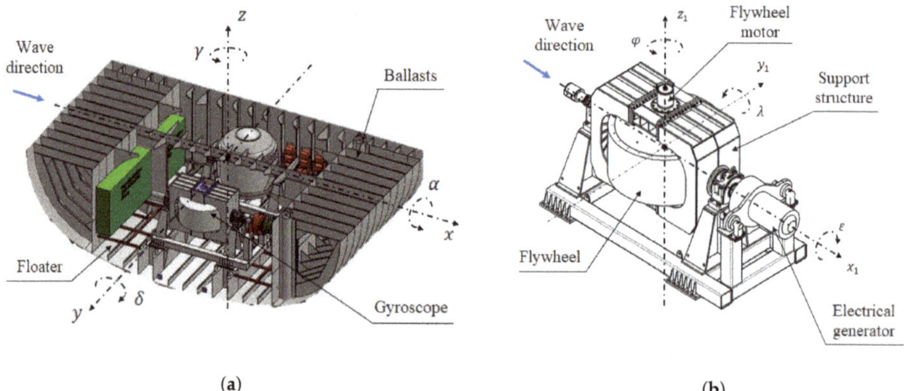


(a)                                                                    (b)

**Figure 2.** ISWEC device: architectures. (**a**) ISWEC hull; (**b**) ISWEC gyroscope.

### 2.2. ISWEC Non-Linear Time Domain Model

The ISWEC mathematical model is obtained by coupling the hull hydrodynamics and the gyroscope dynamics. The non-linear system equations have been implemented in the MATLAB®/Simulink® environment employing the Simscape™ Multibody™ toolbox, which is particularly suitable for multi-physics system modelling. The derivation and the experimental validation of the model equations of the complete system is not within the scope of this work and details can be found in the studies of [18–21].

WEC modelling is the major area of interest within the field of Wave Energy. Multitudes of WEC models are proposed in literature with different levels of confidence and uncertainties [22,23]. In this work, the fluid-structure interaction model is based on linear potential flow theory under the assumptions of irrotational flow, inviscid and incompressible fluid, harmonic oscillations of the hull for each DoF and zero-forward-speed conditions [24]. Then, according to the well-known Cummins' equation [25], the dynamic behaviour of a floating body can be derived in the time domain. Some non-linear effects are considered: the non-linear viscous forces, the drift forces in the surge direction, the mooring action and the gyroscopic reaction on the hull. As stated in the previous paragraph, the ISWEC device extracts energy from the sea exploiting only the motion around the pitch axis. Moreover, the hull is symmetrical with respect to its longitudinal and transversal plane. Under these assumptions, a planar 3-DoF model of the hull has been considered in this work. The reference plane is identified by the vertical gravity axis $z$ and the horizontal direction of the incoming wave $x$ as shown in Figure 2a. Let $X_f(t) \in \mathbb{R}^{n_D}$ be the vector containing the $n_D$ DoFs of the hull:

$$X_f(t) = \begin{bmatrix} x(t) & z(t) & \delta(t) \end{bmatrix}^T \tag{1}$$

Then, $\dot{X}_f(t) \in \mathbb{R}^{n_D}$ and $\ddot{X}_f(t) \in \mathbb{R}^{n_D}$ are the first- and second-time derivative of $X_f$ respectively. In a planar reference frame, $x(t)$ represents the surge motion, $z(t)$ the heave motion and $\delta(t)$ the pitch motion. Following the notation of Equation (1), the time-domain equation of the hull dynamics can be written as follows:

$$M\ddot{X}_f(t) + F_r(t) + F_\beta(t) + KX_f(t) = F_w(t) + F_d(t) + F_m(t) + F_g(t) \tag{2}$$

where $M \in \mathbb{R}^{n_D \times n_D}$ is the mass matrix including the added mass contribution evaluated for infinite oscillation frequency, $F_r(t) \in \mathbb{R}^{n_D}$ are the radiation forces, $F_\beta(t) \in \mathbb{R}^{n_D}$ the non-linear viscous forces, $K \in \mathbb{R}^{n_D \times n_D}$ the linear hydrostatic stiffness, $F_w(t) \in \mathbb{R}^{n_D}$ the wave excitation forces, $F_d(t) \in \mathbb{R}^{n_D}$ the non-linear wave drift forces, $F_m(t) \in \mathbb{R}^{n_D}$ the mooring line actions and $F_g(t) \in \mathbb{R}^{n_D}$ the gyroscopic reactions on the hull. For the sake of clarity, the subscripts $x$, $z$ and $\delta$ will be used in the next sections to specify the DoF to which the force or parameters refer and the subscript $j$ to indicate the $j$-th DoF.

### 2.2.1. Radiation Convolution Term

The radiation forces arise from the motion of the hull through the water that results in inertia and friction components. These contributions can be obtained by solving the convolution integral of the impulse response function $h_r(t)$ [26]:

$$F_r(t) = \int_0^t h_r(t - \tau)\dot{X}_f(\tau)d\tau \tag{3}$$

As the computation of the convolution integral can be very time consuming, it is convenient to express this term with a state-space representation:

$$\begin{aligned} \dot{\zeta}(t) &= A_r\zeta(t) + B_r\dot{X}_f(t) \\ F_r(t) &= C_r\zeta(t) + D_r\dot{X}_f(t) \end{aligned} \tag{4}$$

The vector $\zeta(t) \in \mathbb{R}^{n_R}$ represents the state vector that approximates the radiation force contributions and $n_R$ is the approximation order. The state space matrices $A_r \in \mathbb{R}^{n_R \times n_R}$, $B_r \in \mathbb{R}^{n_R \times n_D}$, $C_r \in \mathbb{R}^{n_D \times n_R}$ and $D_r \in \mathbb{R}^{n_D \times n_D}$ have been identified following the well-known Perez and Fossen and approach [26,27].

### 2.2.2. Non-Linear Viscous Forces

For evaluating the non-linear viscous forces $F_\beta$, [28] proposed a method for the identification of the viscous force for the pitch DoF:

$$F_{\beta\delta}(t) = \beta\dot{\delta}(t)|\dot{\delta}(t)| \tag{5}$$

The method relies on a full CFD approach through which a pitch free decay test is simulated. Once the time series is obtained, it is possible to identify the viscous damping coefficient $\beta$ with two different methodologies: computing the viscous damping coefficients through the logarithmic decrement method and/or integrating the Cummins' equation of motion fitting the damping coefficients that minimize the difference with CFD results. In the surge direction, The work of [29] evaluated the viscous force according to the drag force contribution of the Morison equation, assuming the hypothesis of low forward speed:

$$F_{\beta x} = \frac{1}{2}\rho C_d A\dot{x}(t)|\dot{x}(t)| \tag{6}$$

In Equation (6), $\rho$ represents the water density, $C_d$ the drag coefficient and $A$ the wetted area of the hull. The drag coefficient formulation can be found in literature for simple geometries. The viscous damping in the heave direction is not considered since the hull motion along the vertical axis does not contribute to the power extraction in this WEC. Moreover, it is decoupled for the surge and pitch DoFs due to the symmetry of the hull on the transversal plane.

### 2.2.3. Drift Forces

The drift force is required in order to describe properly the hydrodynamic behaviour of the hull along the surge DoF. The derivation of this action for the ISWEC device can be found in the work of [29]. Here the time series of wave drift force for irregular sea state were derived through the Newman approximation [30]:

$$F_{dx}(t) = 2\left(\sum_{h=1}^{W} \eta_h \sqrt{f_d(\omega_h)}cos(\omega_h t + \theta_h)\right)^2 \tag{7}$$

where the subscript $h$ indicates the $h$-th harmonic component of the wave spectrum, $W$ is the number of frequencies of the spectrum discretization, $\eta_h$ the harmonic amplitude of the wave spectrum, $f_d(\omega_h)$ the drift force coefficient for each frequency component $\omega_h$ of the wave spectrum and $\theta_h$ the phase angle.

### 2.2.4. Mooring System

The mooring system is modelled through a quasi-static approach following the formulation proposed by [29]. The static equilibrium of the system is studied by varying the $x$ coordinate and the $z$ coordinate of the connection point of the mooring to the hull. Then, computing the equilibrium condition of the mooring line for all the different possible positions of the device, the mooring tensions are identified. In the numerical model, $F_m$ is obtained with MATLAB® look-up tables that map the mooring forces in respect to the hull planar motion. As shown in Figure 3, the ISWEC mooring system consists of a slack catenary type with multiple mooring lines, jumpers and clump-weights. Two bridles connect the hull to a central joint to prevent the roll motion of the device. To guarantee the weather-vaning of the device in respect to the wave direction, the mooring connection points are placed towards the bow, with respect to the centre of gravity of the device. On each catenary, a sub-surface buoy and clump-weight are installed to enhance the elastic recall of the system and avoid snatches [31].

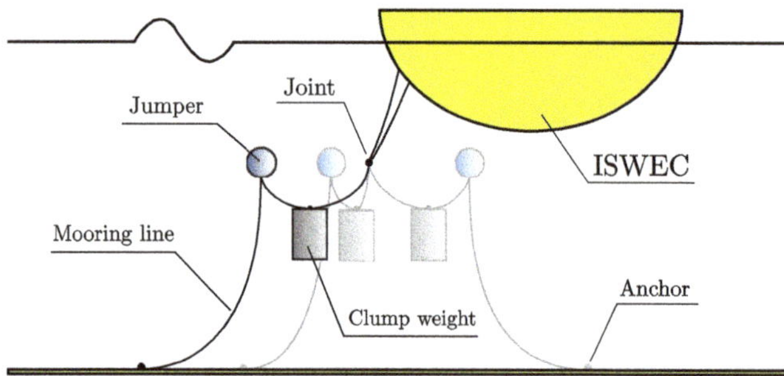


**Figure 3.** ISWEC mooring system.

### 2.2.5. Gyroscope Reactions

The gyroscope dynamics can be described through the Newton's law and derived from the conservation of the flywheel angular momentum. The non-linear numerical model considers all the contributions of the gyroscope reaction along the $x$, $z$ and $\delta$ DoFs. Through a linearization of the angular momentum of the gyroscope around the $y_1$ axis, the expression of the gyroscopic reaction discharged on the pitch DoF of the hull can be determined [32]:

$$F_{g\delta}(t) = J\dot{\varphi}\dot{\varepsilon}(t)cos(\varepsilon(t)) \tag{8}$$

where $J$ is the flywheel moment of inertia, and $\dot{\varepsilon}(t)$ and $\varepsilon(t)$ are the gyroscope speed and angular position around the precession axis $y_1$, respectively. This torque acts on the pitch axis representing the third component of the $F_g$ term in Equation (2). In this work, the gyroscope reaction is considered as a known input of both KF and NN as it can be computed through the measurements of the gyroscope kinematics. The complete derivation of the gyroscope dynamic equation can be found in [32].

### 2.2.6. Wave Excitation Force Modelling

The linear wave theory describes the irregular water surface as superposition of harmonic waves with different frequencies, phases and directions [33,34]. In this work, only unidirectional waves are considered since the ISWEC device is described only in its longitudial plane. Then, the wave signal can be approximated through a Fourier series of any arbitrary number $N$ of harmonic wave components [35]:

$$\eta(t) = \sum_{n=1}^{N} A_n cos(\omega_n t + \alpha_n) \tag{9}$$

where $A_n$ is the amplitude, $\omega_n$ the angular frequency and $\alpha_n$ the $n$-th phase associated with the $n$-th harmonic. Traditionally, Power Spectral Densities (PSD) of real waves are modelled analytically and parametrized according to their spectral properties. In literature, different analytical PSD functions are proposed. Given the wave spetrum, the amplitude of the sinusoidal $n$-th wave component of $\eta(t)$ is obtained by the following relation [35]:

$$\eta_n = \sqrt{2S_{\eta\eta}(\omega_n)\Delta\omega} \tag{10}$$

where $\Delta\omega$ is the PSD frequency resolution and $S_{\eta\eta}$ is the value of spectral energy density. Therefore, the WEF associated to the $j$-th DoF can be calculated given the geometry of the floater and sea-state characteristics [36]:

$$F_{wj}(t) = \sum_{n=1}^{N} |f_j(\omega_n)|\eta_n cos(\omega_n t + \theta_n + \angle f_j(\omega_n)) \tag{11}$$

where $f_j(\omega_n)$ is the Froude-Krylov and diffraction coefficient associated to the $j$-th DoF and the $n$-th wave frequency. In this work, the Joint North Sea Wave Project (JONSWAP) spectrum [37] is considered to define 35 different irregular sea-states to evaluate the estimation performances. However, only four sea-states are employed to tune the KF and NN models according to the operating conditions of the ISWEC device. The JONSWAP spectrum can be identified using three parameters: Significant wave Height ($H_s$), Energy Period ($T_e$) and Peak Shape ($\gamma$). These spectral parameters are reported in Table 1 for each sea-states used for the tuning process, sorted in ascending order of wave Power density ($P_d$).

Figure 4 gives a qualitative representation of annual wave energy of the sea-states considered. Data refer to a typical annual distribution of the wave energy in the Adriatic Sea (Italy) acquired during an experimental campaign started in 2018. The data are normalized on their maximum value. The squared blue marker represents the four waves used for the tuning process.

**Table 1.** Tuning wave data.

| Id | $T_e$ (s) | $H_s$ (m) | $\gamma$ | $P_d$ (kW/m) |
|----|-----------|-----------|----------|--------------|
| 1  | 4.00      | 0.75      | 3.30     | 1.10         |
| 2  | 5.00      | 1.25      | 3.30     | 3.83         |
| 3  | 5.85      | 1.75      | 3.30     | 8.87         |
| 4  | 6.75      | 2.25      | 3.30     | 16.74        |

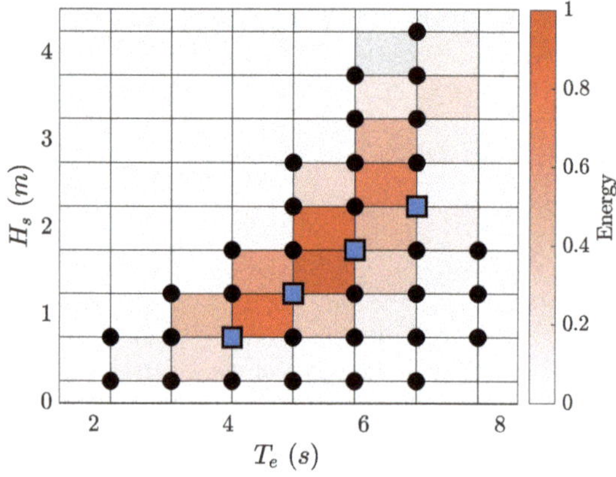

**Figure 4.** Normalized annual wave energy of the 35 irregular sea-states, indicated with black dots and squares.

### 2.2.7. Sensor and Acquisition System Model

The ISWEC device installed in the Adriatic Sea is equipped with a Data AcQuisition (DAQ) system to record experimental signals of the physical quantities of interest. The motion of the three DoFs of interest of the hull has to be acquired. Moreover, to complete the estimation the precession motion of the gyroscope as well as the flywheel speed are required. The linear accelerations, angular orientations and rates of the hull are provided by an Inertial Unit of Measurement (IMU) Xsens MTi-30 AHRS [38] fixed inside the floater. The sensor is rigidly fixed inside the hull, appropriately oriented in respect to the hull reference system. An internal data processor uses the velocity and orientation increments and, through a strapdown integration algorithm, gives in output the orientations on the three rotational DoFs. The measurements considered from the IMU are: the linear acceleration along the surge and heave directions ($\ddot{x}(t)$ and $\ddot{z}(t)$), the angular position and velocity of the hull ($\delta(t)$ and $\dot{\delta}(t)$). Two digital encoders Heidenhain Ecn 413 [39] are mounted on the gyroscope and flywheel shaft, respectively. The measurements available from the encoders used for computing the gyroscope reaction $F_g(t)$ are: the flywheel speed ($\dot{\varphi}(t)$), the angular position and speed of the gyroscope ($\varepsilon(t)$ and $\dot{\varepsilon}(t)$). The data acquisition is managed by a National Instrument compactRIO NI cRIO-9030 [40] which is a dual core 1.33 GHz real time control unit. These sensors and hardware, together with temperature, umidity, voltage and current measurements compose the actual ISWEC DAQ.

Unlike the majority of PA type WECs, the ISWEC device is slack-moored to the seabed and a precise measure of its absolute position is not trivial. In this context, a Differencial GPS (DGPS) is considered to acquire the position and elevation of the WEC. The differential positioning technique enhance the GPS position accuracy of geo-location by comparing their data with those recorded in the same time interval by other multiple GPS receivers. The measurements given by the DGPS are: the position $x(t)$ and the elevation $z(t)$. The measures of IMU and DGPS can be combined to perform the well-known sensor fusion technique [41,42]. Sensor fusion represents a foundamental part of localization and position tracking and can be applied to estimate the absolute positions and velocities with high accuracy. In this context, Motion Reference Units (MRU) are meant for measuring positions, velocities, accelerations, angular rates and orientations representing an all-in-one solution for acquiring the WEC motion. The MRU are more precise than IMU units and returns absolute positions and velocities relative to a specified equilibrium point. These sensors are employed in modern navigation systems and dynamic positioning applications. The measurements available from the MRU are: the positions ($x(t)$ and $z(t)$), velocities ($\dot{x}(t)$ and $\dot{z}(t)$) and acceleration ($\ddot{x}(t)$ and $\ddot{z}(t)$) along the surge and heave directions, the angular position and velocity of the hull ($\delta(t)$ and $\dot{\delta}(t)$). Table 2 reasumes all the frameworks considered in this work and their noise standard deviations $\sigma$.

**Table 2.** Measurement frameworks available. Four cases are considered: Full Measurement (FM), Motion Reference Unit (MRU), Inertial Measurement Unit with Differencial GPS (IMU+DGPS) and Inertial Measurement Unit (IMU).

| Measures | FM | MRU | | IMU+DGPS | | IMU | |
|---|---|---|---|---|---|---|---|
| | Data | Data | $\sigma$ | Data | $\sigma$ | Data | $\sigma$ |
| $x$ (m) | ● | ○ | 0.05 | ○ | 0.5 | – | – |
| $\dot{x}$ (m/s) | ● | ○ | 0.03 | – | – | – | – |
| $\ddot{x}$ (m/s$^2$) | ● | ○ | 0.001 | ○ | 0.01 | ○ | 0.01 |
| $z$ (m) | ● | ○ | 0.05 | ○ | 0.5 | – | – |
| $\dot{z}$ (m/s) | ● | ○ | 0.03 | – | – | – | – |
| $\ddot{z}$ (m/s$^2$) | ● | ○ | 0.001 | ○ | 0.01 | ○ | 0.01 |
| $\delta$ (rad) | ● | ○ | $5 \times 10^{-4}$ | ○ | 0.01 | ○ | 0.01 |
| $\dot{\delta}$ (rad/s) | ● | ○ | $10^{-4}$ | ○ | 0.002 | ○ | 0.002 |
| $\varepsilon$ (rad) | ● | ○ | 0.02 | ○ | 0.02 | ○ | 0.02 |
| $\dot{\varepsilon}$ (rad/s) | ● | ○ | 0.001 | ○ | 0.001 | ○ | 0.001 |
| $\dot{\varphi}$ (rad/s) | ● | ○ | 0.001 | ○ | 0.001 | ○ | 0.001 |

The Full Measurements (FM) configuration represent an ideal measurement framework in which all measures are available. Filled black dots in Table 2 indicates that the FM framework considers measurements without noise. This framework is the referance case where maximum estimation performances are expected. The angular acceleration along the pitch DoF $\ddot{\delta}(t)$ is not considered since none of the available sensors provide it. Encoders data are considered available for all the configurations. It is worth pointing out that missing measurements in IMU+DGPS and IMU frameworks can be obtained by numerical derivation/integration of position, velocity and acceleration signals. However, in this work, only raw measurements obtained directly from the sensors are considered.

### 2.3. ISWEC Model Block Diagram

The block diagram of the ISWEC plant is shown in Figure 5. The input of the system are the *Wave Excitation Force* $F_w(t)$ and the *Wave Drift Fore* $F_d(t)$. First, the Equation (2) is integrated inside the block *Hull*. The output of this block are the hull positions $X_f(t)$, velocities $\dot{X}_f(t)$ and accelerations $\ddot{X}_f(t)$ that represent the input of the *Gyroscope* block. Here the dynamic equation of the gyroscope is solved computing the gyroscope motion $\varepsilon(t)$ and $\dot{\varepsilon}(t)$ around the precession axis and the gyroscope reaction $F_g(t)$ that is feedback to the *Hull*. The *Mooring* block computes the recall $F_m(t)$ in respect to the hull position $X_f(t)$. Then, the block *Power Take-Off* computes the generator torque $T_\varepsilon(t)$ and sends feedbacks back to the *Gyroscope*. The control algorithm implemented by the *Power Take-Off* is described in detail in [15]. The motion of the hull is acquired through the *IMU* [38] and the gyorscope and flywheel angular speeds with two *Encoders* [39]. The measurements are tainted with noise $w_{IMU}(t)$ and $w_E(t)$ and sampled at 10 Hz with the cRIO unit [40] to obtain the discrete data with pedix $d$.

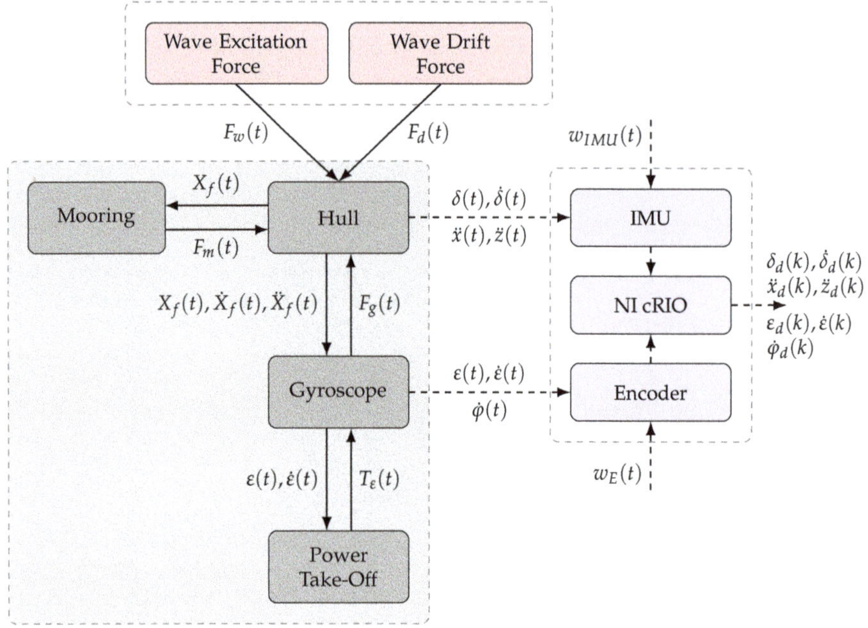


**Figure 5.** ISWEC device block diagram. Forces and motions in continuos line, signals and disturbances in dashed line.

## 3. Wave Excitation Force Estimation with Kalman Filter

One solution to estimate the WEF could be measuring directly the pressure acting on the wetted surface of the hull [8]. However, this method could be expensive as it requires a large number of sensors. Furthermore, since the pressure measured on the wetted surface is the combination of all the hydrodynamics forces it could be challenging to distinguish the contribution of the WEF. Another method relies on the measurement of the hull dynamics and an estimation is performed. The KF is ubiquitous in many applications to estimate the current state of a linear dynamic system from a set of measurements affected by uncorrelated Gaussian noise with known covariance. Under these assumptions, this estimator is defined as optimal because it estimates the system states minimizing the covariance of the estimation error. In this work, the estimation procedure is carried out by modelling the WEF as an unknown state to be estimated [4]. In order to obtain the KF formulation the Equation (2) is taken into account. First, the non-linear model is simplified considering the drift forces included into the $F_{wx}(t)$ contribution. The viscous damping along the pitch direction is linearized and the surge damping action is neglected in order to obtain a 3-DoF linear state space model. Then, the state space model of the ISWEC is discretized in the time domain to derive the KF formulation. The mooring forces are identified and the filter is designed to decouple them from the WEF. These steps are described in detail in the following sub-sections.

### 3.1. Linear 3-DoF ISWEC Model

The non-linear model described in the previous chapter is simplified to achieve a compromise between the computational effort required to solve the KF algorithm and the accuracy of the estimation. The non-linear viscous force along the pitch DoF defined by Equation (5) has to be linearized for inclusion in the estimation process. Four simulations are performed in four different sea-states using the non-linear 3-DoF numerical model and wave data of Table 1. The pitching rate amplitude time series is extracted and its mean is computed for each simulation. The mean is substituted into Equation (5) and $\beta_{lin}$ is defined as the product between the viscous damping $\beta$ and the mean pitching amplitude:

$$F_{\beta\delta}(t) \cong \beta\ddot{\delta}\dot{\delta}(t) = \beta_{lin}\dot{\delta}(t) \tag{12}$$

For the surge direction, Figure 6 demonstrates that the viscous force along the surge DoF can be neglected in respect to the WEF.

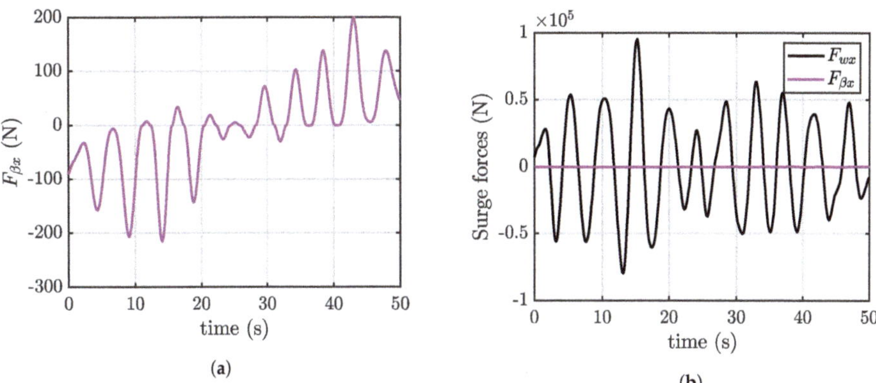

**Figure 6.** Viscous damping in surge direction magnitude (**a**) and comparison with WEF (**b**) (Wave Id 2). (**a**) Viscous damping in surge direction; (**b**) WEF and viscous damping in surge direction comparison.

The term $F_\beta(t)$ can be rewritten as:

$$F_\beta(t) \cong \begin{bmatrix} 0 & 0 & 0 \\ 0 & 0 & 0 \\ 0 & 0 & \beta_{lin} \end{bmatrix} \dot{X}_f(t) = B_v \dot{X}_f(t) \tag{13}$$

Substituting the Equation (13) in (2), Equation (2) can be expressed by the following linear continuous-time state-space model:

$$\begin{aligned} \dot{X}(t) &= AX(t) + BF_g(t) + B[F_w(t) + F_m(t)] \\ Y(t) &= CX(t) + DF_g(t) + D[F_w(t) + F_m(t)] \end{aligned} \tag{14}$$

where $X(t) \in \mathbb{R}^{n_S}$ and $Y(t) \in \mathbb{R}^{m_S}$ are the states and measurements vectors defined as:

$$X(t) = \begin{bmatrix} X_f(t) & \dot{X}_f(t) & \zeta \end{bmatrix}^T$$

$$Y(t) = \begin{bmatrix} X_f(t) & \dot{X}_f(t) & \ddot{X}_f(t) \end{bmatrix}^T \tag{15}$$

$A \in \mathbb{R}^{n_S \times n_S}, B \in \mathbb{R}^{n_S \times n_D}, C \in \mathbb{R}^{m_S \times n_S}$ and $D \in \mathbb{R}^{m_S \times n_D}$ are given by:

$$A = \begin{bmatrix} 0 & I & 0 \\ -KM^{-1} & -B_vM^{-1} & -C_rM^{-1} \\ 0 & B_r & A_r \end{bmatrix}, \quad B = \begin{bmatrix} 0 \\ M^{-1} \\ 0 \end{bmatrix}$$

$$C = \begin{bmatrix} I & 0 & 0 \\ 0 & I & 0 \\ -KM^{-1} & -B_vM^{-1} & -C_rM^{-1} \end{bmatrix}, \quad D = \begin{bmatrix} 0 \\ 0 \\ M^{-1} \end{bmatrix} \tag{16}$$

Here, $n_S = 2n_D + n_R$ is the number of states, $m_S$ the number of outputs, $I$ and $0$ stands for identity and zero matrices according to the problem dimensions. The system (14) represents the most general framework for a linear multi-DoF WEC model. It can be noted that, without loss of generality, gyroscope reaction $F_g(t)$ is the controlled input (e.g., the PTO action), the WEFs $F_w(t)$ the exogenous input to be estimated and the mooring forces $F_m(t)$ could be interpreted as an unknown unmodelled phenomena to be decoupled from the $F_w(t)$ estimation. The outputs of the linear model (14) are position, velocity and acceleration of the WEC body resulting in a FM configuration. However, matrices $C$ and $D$ can be chosen according to the measurements available on the system as well as the requirements of the observer as defined in Table 2.

### 3.2. Kalman Filter Problem Statement

In this work it is assumed that the excitation force $F_w(t)$ has a harmonic nature and it can be described as a linear combination of different wave components with a finite number of frequencies $n_W$ [2–4]. In this form, the WEF can be included into the state vector as an unknown state. In Equation (14) the mooring forces are considered as a unmodelled phenomena. In real applications, it would be difficult to directly measure the action of the moorings so they are included into the state vector as unknow states to be estimate. The ISWEC mooring system is designed to minimize the impact on the pitching motion of the device, appointed to the power conversion chain [31]. As demonstrated in Figure 7b, the mooring forces have a very slow dynamics such as the surge component and a constant load in heave and pitch directions compared to the WEF. In normal working conditions, snatch loads do not appear, and the mooring forces have mainly the behaviour as shown in Figure 7a. $F_m(t)$ is synthesised using the same method that was detailed for $F_w(t)$, using a harmonic model to describe its behaviour. For the sake of simplicity, the mooring forces are modelled with only one frequency for each component representing their main spectral nature.

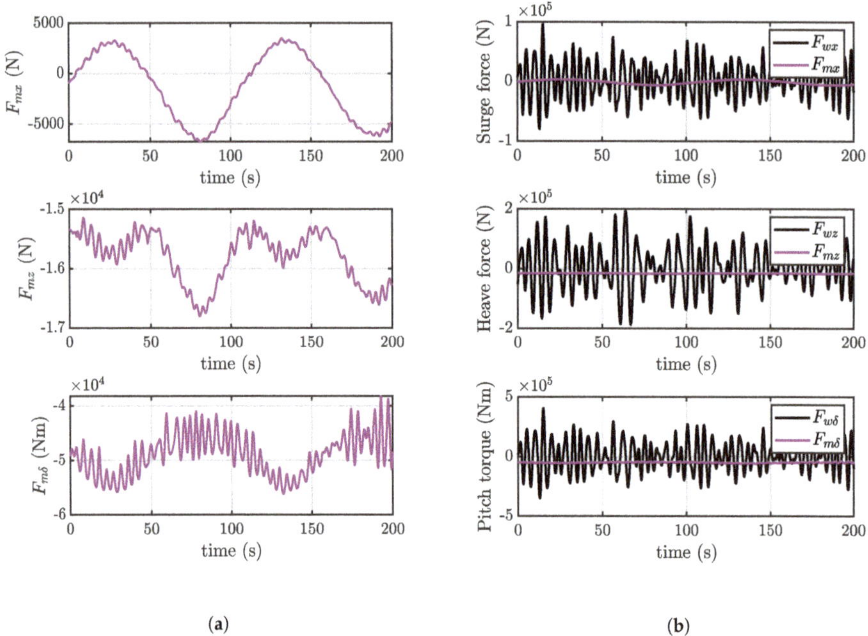

(a)         (b)

**Figure 7.** Mooring forces magnitude (**a**) and comparison with WEFs (**b**) (Wave Id 2). (**a**) Mooring forces; (**b**) WEFs and mooring forces comparison.

Under these assumptions the state vector $X(t)$ is augmented including the estimation of $F_w(t)$ and $F_m(t)$ and the System (14) is re-written:

$$\dot{X}_a(t) = A_a X_a(t) + B_a F_g(t)$$
$$Y(t) = C_a X_a(t) + D_a F_g(t) \tag{17}$$

where $X_a(t) \in \mathbb{R}^{n_F}$ is the augmented state vector defined as:

$$X_a(t) = \begin{bmatrix} X(t) & \hat{F}(t) & \dot{\hat{F}}(t) \end{bmatrix}^T \tag{18}$$

Here $n_F = n_S + 2n_D(n_W + 1)$ is the augmented state vector dimension. $\hat{F} \in \mathbb{R}^{n_D \times (n_W+1)}$ is the unknown force to be estimated:

$$\hat{F}(t) = \begin{bmatrix} \hat{F}_x(t) & \hat{F}_z(t) & \hat{F}_\delta(t) \end{bmatrix}^T \tag{19}$$

The $j$-th compoenent of the estimated force is given by the WEF harmonics and the mooring force as follows:

$$\hat{F}_j(t) = \begin{bmatrix} \hat{F}_{wj_1}(t) & \hat{F}_{wj_2}(t) & \cdots & \hat{F}_{wj_{n_W}}(t) & \hat{F}_{mj}(t) \end{bmatrix}^T \tag{20}$$

Therefore, the estimated $\hat{F}_w(t)$ is obtained by summing up all the harmonic contributions for each of its components:

$$\hat{F}_w(t) = \begin{bmatrix} \sum_{i=1}^{n_W} F_{wx_i}(t) & \sum_{i=1}^{n_W} F_{wz_i}(t) & \sum_{i=1}^{n_W} F_{w\delta_i}(t) \end{bmatrix}^T \tag{21}$$

The augmented matrices $A_a \in \mathbb{R}^{n_F \times n_F}$, $B_a \in \mathbb{R}^{n_F \times n_D}$, $C_a \in \mathbb{R}^{m_S \times n_F}$ and $D_a \in \mathbb{R}^{m_S \times n_D}$ are given by:

$$A_a = \begin{bmatrix} 0 & I & 0 & 0 & 0 \\ -KM^{-1} & -B_v M^{-1} & -C_r M^{-1} & M^{-1}N & 0 \\ 0 & B_r & A_r & 0 & 0 \\ 0 & 0 & 0 & 0 & I \\ 0 & 0 & 0 & -\Omega & 0 \end{bmatrix}, \quad B_a = \begin{bmatrix} 0 \\ M^{-1} \\ 0 \\ 0 \\ 0 \end{bmatrix}$$

$$C_a = \begin{bmatrix} I & 0 & 0 & 0 & 0 \\ 0 & I & 0 & 0 & 0 \\ -KM^{-1} & -B_v M^{-1} & -C_r M^{-1} & M^{-1}N & 0 \end{bmatrix}, \quad D_a = \begin{bmatrix} 0 \\ 0 \\ M^{-1} \end{bmatrix} \tag{22}$$

Again, $I$ and $0$ are identity and zero matrices according to the context. $N \in \mathbb{R}^{n_D \times (n_W+1)}$ is defined as:

$$N = \begin{bmatrix} I_{1,n_W+1} & 0 & 0 \\ 0 & I_{1,n_W+1} & 0 \\ 0 & 0 & I_{1,n_W+1} \end{bmatrix} \tag{23}$$

where $I_{1,n_W+1}$ is a $1 \times (n_W + 1)$ vector of ones. $\Omega \in \mathbb{R}^{3(n_W+1) \times 3(n_W+1)}$ is the diagonal matrix with the frequencies identified to approximate the unknown forces along the three DoFs:

$$\Omega = \begin{bmatrix} \Omega_x & 0 & 0 \\ 0 & \Omega_z & 0 \\ 0 & 0 & \Omega_\delta \end{bmatrix} \tag{24}$$

In Equation (24), $\Omega_x \in \mathbb{R}^{(n_W+1) \times (n_W+1)}$, $\Omega_z \in \mathbb{R}^{(n_W+1) \times (n_W+1)}$ and $\Omega_\delta \in \mathbb{R}^{(n_W+1) \times (n_W+1)}$ are diagonal matrices containing the frequencies for unknown force components:

$$\Omega_j = \begin{bmatrix} diag(\omega_{wj}) & 0 \\ 0 & \omega_{mj} \end{bmatrix} \tag{25}$$

where $\omega_{wj} \in \mathbb{R}^{n_W}$ stores the $n_W$ harmonics to model the WEF components. The number of frequencies in $\Omega$, have to be chosen in order to have a compromise between the accuracy of the estimation for each excitation force component and the computational effort. The matrices (25) include a term at low frequency to represent the contribution of the mooring forces for each DoF.

Let us consider the following linear time-invariant stochastic discrete model representing the discrete-time version of the augmented System (17):

$$\begin{aligned}
\hat{X}_{ad}(k+1) &= A_{ad}\hat{X}_{ad}(k) + B_{ad}F_g(k) + \Gamma w(k) \\
Y_d(k) &= C_{ad}\hat{X}_{ad}(k) + D_{ad}F_g(k) + v(k) \\
w(k) &\sim \mathcal{N}(0,Q) \\
v(k) &\sim \mathcal{N}(0,R)
\end{aligned} \tag{26}$$

where $\hat{X}_{ad}(k)$ represents the system estimated states, $F_g(k)$ is the known input and $Y_d(k)$ contains the measurements of the system dynamics. $w(k)$ and $v(k)$ are zero mean white noise sequences with known covariance, uncorrelated with each other and with the initial state of the system. $A_{ad}$, $B_{ad}$, $C_{ad}$ and $D_{ad}$ stand for the discretised versions of the matrices $A_a$, $B_a$, $C_a$ and $D_a$. $\Gamma$ is the weighting matrix for the process disturbances. $Q$ and $R$ are the covariance matrices of the process and measurements noise. The KF algorithm performs the estimation in the form of a feedback control: the filter estimates the process state at some time and then obtains feedback in the form of (noisy) measurements. As such, the equations for the KF fall into two groups: time-update equations and measurement-update equations [43]:

Time Update:

$$\begin{aligned}
P^-(k) &= A_{ad}P(k-1)A_{ad}^T + \Gamma Q\Gamma^T \\
\hat{X}_a^-(k) &= A_{ad}\hat{X}_a(k-1) + B_{ad}F_g(k-1)
\end{aligned} \tag{27}$$

Measurement Update:

$$\begin{aligned}
K(k) &= P_k^- C^T \left( C_{ad}P^-(k)C_{ad}^T + R \right)^{-1} \\
P(k) &= \left( I + K(k)C_{ad} \right) P^-(k) \\
\hat{X}_{ad}(k) &= \hat{X}_{ad}^-(k) + K(k)\left( Y(k) - C_{ad}\hat{X}_{ad}^-(k) - D_{ad}F_g(k) \right)
\end{aligned} \tag{28}$$

The time update equations can also be thought of as predictor equations, while the measurement update equations can be thought of as corrector equations. In this framework, the KF algorithm can be implemented to estimate the unknown WEF vector $\hat{F}_w(k)$ by measured system dynamics $Y_d(k)$ and known input $F_g(k)$ at any instant $k$.

## 4. Wave Excitation Force Estimation with Neural Network

Despite its simplicity and efficacy, the KF filter observer may suffer from several drawbacks: the non-linear effects emerging in sever sea-states conditions as well as the reliabiliy of the WEC model could negatively affect the estimation performances. For example, the slack-mooring of the ISWEC is modelled with a quasi-static approach (developed in-house). The main advantage consists in reducing the computational burden of the simulation at the expense of model accuracy. More accurate mooring models are presented in [44] considering two dynamic lumped-mass approaches (the open source MoorDyn [45] and the commercial OrcaFlex 11.0e [46]) where mooring actions are resolved coupling both the hydrodynamics and gyroscope model of the WEC in the MoorDyn and Orcaflex enviorments. When the system model is not reducible to a series of analytical equations or, even more, is based only on observed data the implementation KF model is not trivial. This argument can be extended to any other complex aspect of WEC modelling that cannot be analytically formulated with acceptable accuracy. In this context, artifical NNs represents powerful tools to map the non-linear relations from sets of input-output data. In NN models the parameters are tuned to fit the input-output data, without reference to the physical background and no information about the model architecture.

In general mathematical terms, WEFs acting on the ISWEC can be expressed as a non-linear function $f(\bullet)$ of the system known inputs and measurements as follows [14]:

$$\hat{F}_w(k) = f(F_g(k), \ldots, F_g(k - k_N + 1), Y_d(k), \ldots, Y_d(k - k_N + 1)) \tag{29}$$

Equation (29) shows that the estimation of the WEF at instant $k$ is, at least in principle, addressed combining a series of known system inputs and measurements collected from discrete time $k - k_N + 1$ to $k$ where $k_N$ represents the delay steps of the available data. More in detail, the WEF could be obtained from the set of WEC motion measurements from the ISWEC on-board sensors and the gyroscopic reaction. In [14] the same approach is considered including the measures provided by the IMU and encoders mounted on the ISWEC adding the velocity $\dot{z}_d(k)$ and position $z_d(k)$ obtained by numerical integration of the acceleration $\ddot{z}_d(k)$. These two inputs were included to improve the estimation accuracy. However, this measurements framework does not consider the effect of the sensor noise: in practice, accelerometer signals are often very noisy, hence velocity and position integration from acceleration are likely to be unreliable, resulting in unreliable estimations in practice. In this work, the NN is evaluated for all the configurations of Table 2 and the sensitiveness and roubustness analysis is extended to all the wave domain of the installation site. The input-output architecture of the NN is shown in Figure 8. The feedforward NN is composed of linked neurons arranged in three layers. The input layer collects a set of inputs $I_I$ multiplying them by a set of weights, assigned to the data on the basis of their relative importance to other inputs. The hidden neurons apply a non-linear activation function $\sigma$ to the weighted sum of their inputs. Then, the outputs of each hidden neuron are linearly combined by the output functions $\Sigma$ to produce the network outputs $O_O$. The use of delays in the input variables is considered to increase the reliability of the estimate. For a generic dynamic system, one way to consider its dynamic behaviour using static neurons is to employ past values of the inputs [47], resulting in good performances in term of estimation accuracy and robustness to different wave conditions as demonstrated in [14].

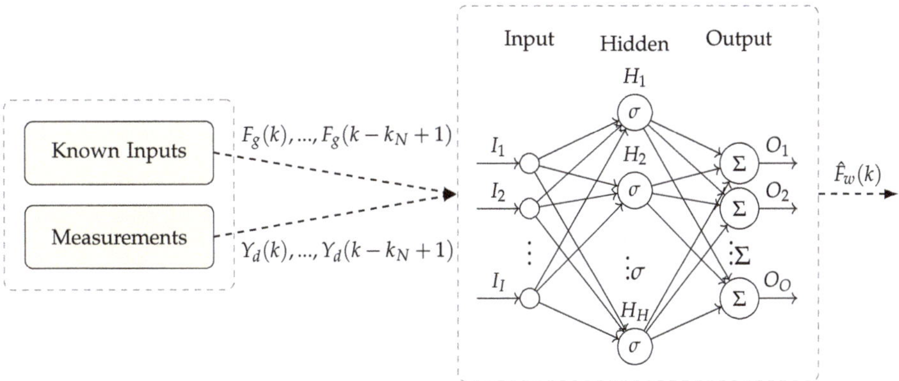


**Figure 8.** Neural Network architecture for ISWEC.

## 5. Kalman Filter and Neural Network Tuning

The matrix $\Omega$ containing the frequencies to model the WEFs as well as $Q$ and $R$ are tuned for the system under study. A sensitivity analysis has been performed to tune the number of frequencies. Then, through an iterative process the diagonal coefficients of $Q$ and $R$ are identified in order to have the best estimation performances for each measurement framework considered. Numerical experiments are conduced considering the four wave profiles of Table 1. The estimation performances are valuated considering the Goodness-of-Fit ($GoF$) proposed by [48]:

$$GoF_j = 1 - \frac{\sqrt{\sum_{k=1}^{T_s} \left(F_{wj}(k) - \hat{F}_{wj}(k)\right)^2}}{\sqrt{\sum_{k=1}^{T_s} F_{wj}(k)^2}} \tag{30}$$

In Equation (30), $F_{wj}(k)$ and $\hat{F}_{wj}(k)$ are the true and estimated WEF for the $j$-th DoF at discrete time instant $k$, respectively. $T_s$ is the total number of samples.

### 5.1. Kalman Filter Parameters

For the presented case, 1, 3, 6 and 9 frequencies are tested in order to find a compromise between the accuracy of the estimation process and the KF complexity. The interval is chosen between period 3 s and period 9 s, linearly spaced. Each point refers to the mean of each $GoF_j$ obtained from four tuning sea-states. The mooring components are modelled with only one period: 100 s for $F_{mx}(t)$ and 2000 s for $F_{mz}(t)$ and $F_{m\delta}(t)$, considered as low frequency contributions (Figure 7). The FM framework has been considered for this tuning process. Table 3 sets out the results obtained from each tuning wave gruped in respect to the number of frequencies $n_W$.

Table 3. $GoF$ results of the KF observer for each tuning wave for different number of frequencies $n_W$. The last row stores the mean values.

| Id | $n_W = 1$ | | | $n_W = 3$ | | | $n_W = 6$ | | | $n_W = 9$ | | |
|----|-------|-------|-------|-------|-------|-------|-------|-------|-------|-------|-------|-------|
|    | $GoF_x$ | $GoF_z$ | $GoF_\delta$ | $GoF_x$ | $GoF_z$ | $GoF_\delta$ | $GoF_x$ | $GoF_z$ | $GoF_\delta$ | $GoF_x$ | $GoF_z$ | $GoF_\delta$ |
| 1 | 0.665 | 0.640 | 0.675 | 0.901 | 0.956 | 0.898 | 0.917 | 0.962 | 0.894 | 0.917 | 0.965 | 0.899 |
| 2 | 0.747 | 0.846 | 0.706 | 0.931 | 0.968 | 0.919 | 0.942 | 0.986 | 0.919 | 0.943 | 0.987 | 0.925 |
| 3 | 0.771 | 0.899 | 0.642 | 0.911 | 0.974 | 0.927 | 0.924 | 0.988 | 0.927 | 0.925 | 0.988 | 0.930 |
| 4 | 0.785 | 0.933 | 0.537 | 0.914 | 0.970 | 0.925 | 0.930 | 0.990 | 0.927 | 0.930 | 0.991 | 0.928 |
|   | 0.755 | 0.824 | 0.685 | 0.917 | 0.962 | 0.913 | 0.928 | 0.981 | 0.917 | 0.928 | 0.982 | 0.921 |

Increasing the number of harmonics from $n_W = 1$ to $n_W = 3$ leads to a significant improvement of all $GoF$. On average, an advance of 0.162 is obtained for $GoF_x$, 0.138 for $GoF_z$ and 0.228 for $GoF_\delta$. Passing from $n_W = 3$ to $n_W = 6$ shows a slight increase of $GoF_x$ and $GoF_z$, enchancing the quality of the $\hat{F}_{wx}(t)$ and $\hat{F}_{wz}(t)$ estimation of 0.011 and 0.021, respectively. Further enlarge of $n_W$ results in a minimal or null upgrade of any $GoF$. In this regard, $n_W = 6$ is considered a good trade-off between accuracy and complexity of the observer.

For what concerns the covariance matrices, repeated simulations are conduced to tune the diagonal elements of $Q$ and $R$ for each mesurement framework. They are chosen to guarantee an accurate estimation of the unknown states without amplifying the noise level. Figure 9a–c compare the $Q$ coeffcients relative to the WEF components for each measurement framework. From the chart, it can be seen that the $Q$ coeffcients are reduced as the magnitude of the noise increases and are tuned in respect to the energy content of the WEF. In this way, the most relevant frequency components are amplified more than the others, resulting in a more accurate estimation. Figure 9d provides the $R$ coefficients, balanced in order to penalize the most inaccurate measurements (e.g., DGPS positions) giving less importance to the related signals. Data indicate that the $R$ coefficients grow as the noise magnitudes increase. Missing markers mean that the measurements are not available for that framework.

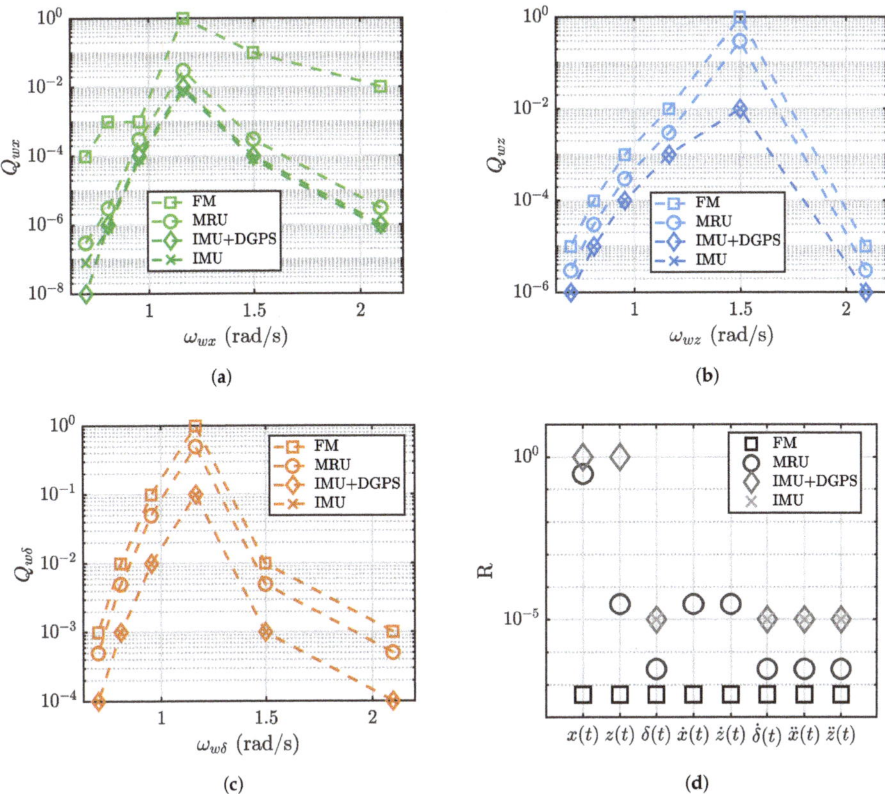

**Figure 9.** $Q$ and $R$ coefficients for each measurement framework and WEF compoenents: $Q$ values for $\hat{F}_{wx}$ (**a**), $Q$ values for $\hat{F}_{wz}$ (**b**), $Q$ values for $\hat{F}_{w\delta}$ (**c**) and $R$ values. (**a**) $Q$ coefficients of surge components; (**b**) $Q$ coefficients of heave components; (**c**) $Q$ coefficients of pitch components; (**d**) $R$ coefficients.

### 5.2. Neural Network Parameters

The NN training aims to obtain the weights and bias of the network evaluating the sensitivity of the model to both the delay steps $k_N$ and neurons $n_N$. The four wave profiles of Table 1 are used to obtain the training data for each measurement framework. The time-series of the WEFs are concatenated to generate a long training set applied to the ISWEC numerical model obtaining the WEC motion measurements. Then, all the time-series are normalized in the $[-1, 1]$ range to avoid problems due to different magnitude between input-output signals. Since the system is non-linear, the data normalization could negatively affect the trainign process if, for example, saturations of the PTO appear. To avoid saturations, the training data are chosen to cover the operating range of the ISWEC. However, in sever sea-states the WEC motion could overcome the normalization range causing estimation errors but, in such extreme conditions, the WEC is shut-down for safety purposes and no control is applied. In order to address over-fitting problem the data are randomly divided into three parts: 50% for training purposes, 30% for validation and 20% for performance evaluation. The performance function is the Mean Squared Error (MSE) normalized between $-1$ and $1$, ensuring that the relative accuracy of output elements of different magnitude are treated as equally important. Then, 1, 3, 5, and 7 delay steps are tested while the number of neurons is fixed to 10; 5, 10, 15 and 20 neurons are evaluated with 3 delay steps. Specific tuning processes are applied for each framework with the same number of delay steps and neurons, considering noisy signals. The FM framework has been considered for tuning the network hyperparameters.

The sensitivity analysis on the delay steps is reported in Table 4. The mean value of each $GoF$ reveals that increasing the number of delay steps from $k_N = 1$ to $k_N = 3$ results in a revelant increase of estimation accuracy, expecially for the $GoF_\delta$ that passes from 0.880 to 0.977. A similar behaviour is obtained for the $GoF_x$, where the mean estimation performance grows of 0.053 points. On the other hand, no relevant improvements are obtained for the $GoF_z$ as well as for the other DoFs employing 5 and 7 delay steps. In this regard, 3 unit delays are chosen for the NN architecture.

**Table 4.** $GoF$ results of the NN model for each tuning wave for different delay steps $k_N$. The number of neurons are fixed to 10. The last row stores the mean values.

| Id | $k_N = 1$ | | | $k_N = 3$ | | | $k_N = 5$ | | | $k_N = 7$ | | |
|---|---|---|---|---|---|---|---|---|---|---|---|---|
| | $GoF_x$ | $GoF_z$ | $GoF_\delta$ | $GoF_x$ | $GoF_z$ | $GoF_\delta$ | $GoF_x$ | $GoF_z$ | $GoF_\delta$ | $GoF_x$ | $GoF_z$ | $GoF_\delta$ |
| 1 | 0.900 | 0.935 | 0.839 | 0.963 | 0.938 | 0.966 | 0.969 | 0.939 | 0.966 | 0.978 | 0.951 | 0.973 |
| 2 | 0.922 | 0.967 | 0.891 | 0.977 | 0.970 | 0.980 | 0.980 | 0.972 | 0.978 | 0.985 | 0.978 | 0.978 |
| 3 | 0.927 | 0.969 | 0.889 | 0.979 | 0.976 | 0.980 | 0.980 | 0.977 | 0.978 | 0.985 | 0.982 | 0.982 |
| 4 | 0.933 | 0.968 | 0.903 | 0.979 | 0.978 | 0.981 | 0.978 | 0.977 | 0.978 | 0.985 | 0.985 | 0.982 |
| | 0.921 | 0.960 | 0.880 | 0.974 | 0.965 | 0.977 | 0.977 | 0.966 | 0.975 | 0.983 | 0.974 | 0.980 |

Table 5 reports the estimation results obtained for different values of neurons. 10 neurons represents the best choise ensuring the good estimation performances with acceptable network complexity. Surprisingly, increasing the number of neurons over 10 leads to a slight degradation of performances, suggesting that the learning algorithm is not able to converge properly during the learning process over-fitting noisy data.

**Table 5.** $GoF$ results of the NN model for each tuning wave for different neurons $n_N$. The delay steps are fixed to 3. The last row stores the mean values.

| Id | $n_N = 5$ | | | $n_N = 10$ | | | $n_N = 15$ | | | $n_N = 20$ | | |
|---|---|---|---|---|---|---|---|---|---|---|---|---|
| | $GoF_x$ | $GoF_z$ | $GoF_\delta$ | $GoF_x$ | $GoF_z$ | $GoF_\delta$ | $GoF_x$ | $GoF_z$ | $GoF_\delta$ | $GoF_x$ | $GoF_z$ | $GoF_\delta$ |
| 1 | 0.952 | 0.929 | 0.947 | 0.963 | 0.938 | 0.966 | 0.958 | 0.937 | 0.964 | 0.956 | 0.934 | 0.964 |
| 2 | 0.969 | 0.967 | 0.964 | 0.977 | 0.970 | 0.980 | 0.972 | 0.971 | 0.980 | 0.972 | 0.969 | 0.978 |
| 3 | 0.968 | 0.972 | 0.964 | 0.979 | 0.976 | 0.980 | 0.974 | 0.976 | 0.980 | 0.972 | 0.975 | 0.979 |
| 4 | 0.970 | 0.974 | 0.963 | 0.979 | 0.978 | 0.981 | 0.976 | 0.976 | 0.981 | 0.973 | 0.976 | 0.981 |
| | 0.965 | 0.961 | 0.959 | 0.974 | 0.965 | 0.977 | 0.970 | 0.965 | 0.976 | 0.970 | 0.964 | 0.975 |

## 6. Numerical Results and Discussions

In this section, several results are carried out for a comprehensive analysis of the KF and NN performances. An exhaustive evaluation in different wave conditions as well as a robustness evaluation to different measurements and plant inaccuracies is provided. Both the KF and NN are tested and compared extending the wave domain to the 35 waves represented in Figure 4. Each $GoF_j$ is weighted on the annual energy of the specific site in analysis:

$$\overline{GoF}_j = \frac{\sum_{v=1}^{V} GoF_j(v)E(v)}{\sum_{v=1}^{V} E(v)} \tag{31}$$

In Equation (31), $E(v)$ is the annual energy associated to $v$-th wave and $V$ the total number of waves considered. Then, the percentage difference between the KF and NN results is defined as follows:

$$\Delta\overline{GoF_j} = \frac{\overline{GoF_j}\big|_{NN} - \overline{GoF_j}\big|_{KF}}{\overline{GoF_j}\big|_{KF}} 100 \tag{32}$$

where $\overline{GoF_j}\big|_{KF}$ and $\overline{GoF_j}\big|_{NN}$ are the weighted $GoF$ obtained with KF and NN, respectively.

### 6.1. Influence of the Measurement Framework

The available measurements and their accuracy play an important role in the estimation process. The main question concerns the influence of the available signals and how missing data can affect the estimation performances. The three measurement configurations MRU, IMU+DGPS and IMU are compared considering noisless data. In order to improve readability, some of the bulky figures are placed in the Appendix A.

Focusing on the KF results, Figure A1 is quite revealing in several ways. Overrall, maximum estiamtion performances are found near the tuning waves (marked with black squares) except for the $GoF_z$ provided by the IMU framework. In particular, best performances are achieved in the range $[4, 8]$ seconds as the period interval is chosen between 3 s and 9 s. In a practical setup, an accurate analysis the dominant periods is crucial to achieve the best performances from the KF. Moreover, promising results are obtained in all the sea-states considered, suggesting a good versatility of the observer where the system is not tuned. The accuracy is almost the same for the force components in surge and pitch directions. Despite the configuration IMU+DGPS does not provide the $\dot{x}(t)$ signal, the IMU+DGPS equipment is able to estimate the $\hat{F}_{wx}(t)$ component with the same accuracy as the MRU one. The same argument can be applied considering the IMU results. In spite of the IMU does not provide both $x(t)$ and $\dot{x}(t)$, its accuracy of the pitch force estimation equals the MRU and IMU+DGPS one. Since surge and pitch DoFs are coupled due to the device geometry, the surge force can be succesfully estimated from the pitching motion of the device (available in all the frameworks). Morevoer, the effect of the damping is negligible and the hydrostatic stiffness is null in the surge direction, not affecting the estimate of $\hat{F}_{wx}(t)$. For what concern the heave component, there is a significant difference between the $GoF_z$ of the IMU and the other two frameworks. The absence of both $z(t)$ and $\dot{z}(t)$ leads a decrease of performances since, as illustrated in Figure 10, the KF is not able to estimate correctly the position $z(t)$ resulting in a loss of estimation accuracy.

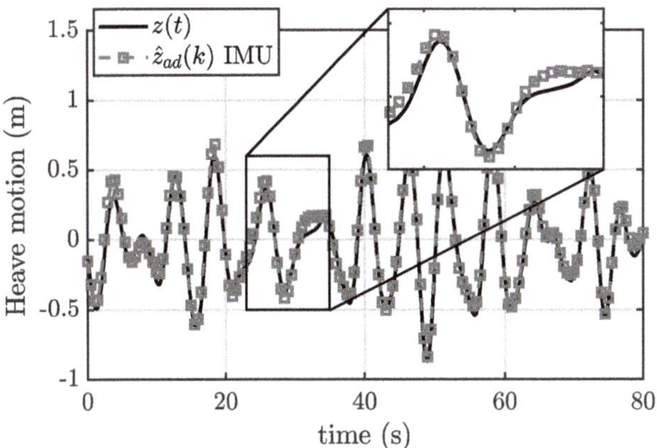


**Figure 10.** Heave motion compared with its estimation obtained with the IMU framework (Wave Id 2).

Considering the NN, Figure A2 demonstrates that the NN performance is maximized with a MRU measurement framework and near the tuning waves. In agreement with the KF, removing the absolute velocities $\dot{x}(t)$ and $\dot{z}(t)$ leads to a minimal decrease in estimation accuracy for all the DoFs. This finding is expected for both surge and pitch DoFs since the measurements that influence the $\hat{F}_{wx}(t)$ and $\hat{F}_{w\delta}(t)$

estimation ($\delta(t)$, $\dot{\delta}(t)$ and $\ddot{x}(t)$) are provided both in MRU and IMU+DGPS configurations. The lack of performance increase for the IMU framework, especially for the heave DoF where the goodness of estimation significantly reduces for all the waves, suggesting that the NN is not able to obtain an appropriate fit of the $F_{wz}(t)$ if a full measurement framework is not provided. On the other hand, $\delta(t)$, $\dot{\delta}(t)$ and $\ddot{x}(t)$ are adequate to succesfully estimate the $\hat{F}_{wx}(t)$ and $\hat{F}_{w\delta}(t)$ components, as demonstrated for the KF observer.

*6.2. Sensor Noise Effect*

Once the influence of the available measurements is evaluated, realistic data are applied both to the KF and NN to assess the effect of the sensor disturbances. Starting with the KF observer, Figure A3 highlights that the MRU framework gives almost the same precision compared to the noiseless case, especially for the surge and pitch excitation forces. MRU units are affected by minimal disturbances and offer an accurate measures of angular rates, orientations, accelerations, velocities and positions. The use of IMU+DGPS intruduces higher estimation errors due to the increase of noise. In particular, the positions are the most polluted due to a noise RMS equal to 0.5 m (Table 2) as depicted in Figure 11. The errors introduced by the DGPS are the same order of manitude of the heave motion $z(t)$ providing a significant decrease in performance of the $\hat{F}_{wz}(t)$ estimation. The performances of the IMU framework almost equalize the IMU+DGPS one for all the DoF. As demonstrated for the noiseless case, it is apparent that removing the DPGS noisy measurements does not affect the quality of the estimation. Having a good quality in $\delta(t)$, $\dot{\delta}(t)$ and $\ddot{x}(t)$ acquisitions allows to effectively handle the estimation of the coupled DoF since they mutually influence each other. For what concern the $GoF_z$ value, the error introduced by the DGPS forces to increase $R$ coefficient associated to $z(t)$ to such an extent that its measurements are not taken into account. This demonstrate why IMU+DGPS and IMU have similar performances for the heave component.

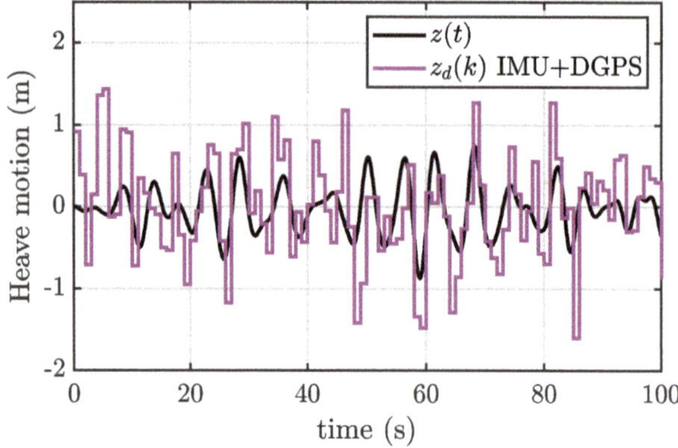

**Figure 11.** Heave motion compared with its DGPS acquired signal (Wave Id 2).

As shown in Figure A4, the introduction of noisy measurements negatively affects the estimation accuracy of the NN, expecially for the heave DoF. A relevant degradation of performance is shown for the IMU+DGPS and IMU framework estimating the wave exctitation force along the heave direction. As demostrated in Figure 11, the heave motion provided by the DGPS is heavily polluted by sensor noise making the NN unable to estimate the heave component with acceptable accuracy. In this context, the KF implementing the IMU+DGPS or IMU unit exploits the heave acceleration to estimate the $\hat{F}_{wz}(t)$ component with almost double precision than the NN.

## 6.3. Sensitivity to Plant Variations

In order to evaluate the effectiveness of KF and NN, the architectures are tested considering a typical case may happen in practice: variation of the physical properties of the floater. In fact, the WEC model may differ from its built version passing from the theoretical design to its construction in the shipyard and, even more, during its life cycle due to wear and biofouling. A further analysis is performed, testing the behavior of the frameworks modifying the mass matrix $M$ and stiffness matrix $K$. An iterative method is proposed varying both the mass and stiffness matrix in respect to the nominal value as follows:

$$M_c = \overline{M}M$$
$$K_c = \overline{K}K \tag{33}$$

where $\overline{M}$ and $\overline{K}$ are the correction coefficients and $M_c$ and $K_c$ the corrected values. $\overline{M}$ and $\overline{K}$ span from 0.85 to 1.15. At this point, the performances of the KF are shown in Figure 12 in term of $\overline{GoF}$. The decrease of performances is limited for the surge component. The variation of the mass matrix results in a decay of almost 0.1 of $\overline{GoF}_x$ for all the measurement frameworks. The stiffness matrix does not influence the estimation of the $\hat{F}_{wx}(t)$ since the hydrostatic stiffness is null in the $x$ direction. In this regard, no variation of $\overline{GoF}_x$ is detected in Figure 12d. The degradation of performance is quite sensitive to the mass and the stiffness variation for heave and pitch DoFs. As to the mass matrix, the $\overline{GoF}_{delta}$ in Figure 12c shows an higher decrease since pitch-surge modes are coupled and the estimation is influenced both by the diagonal and off-diagonal terms of $M$; as demonstrated in [14], the pitch force estimation is more influenced than the heave one because more terms of $M$ are modified at once. The $K$ variation results in a decay of almost 0.2 in both $\overline{GoF}_z$ and $\overline{GoF}_\delta$ for all the measurement frameworks as depicted in Figure 12e,f.

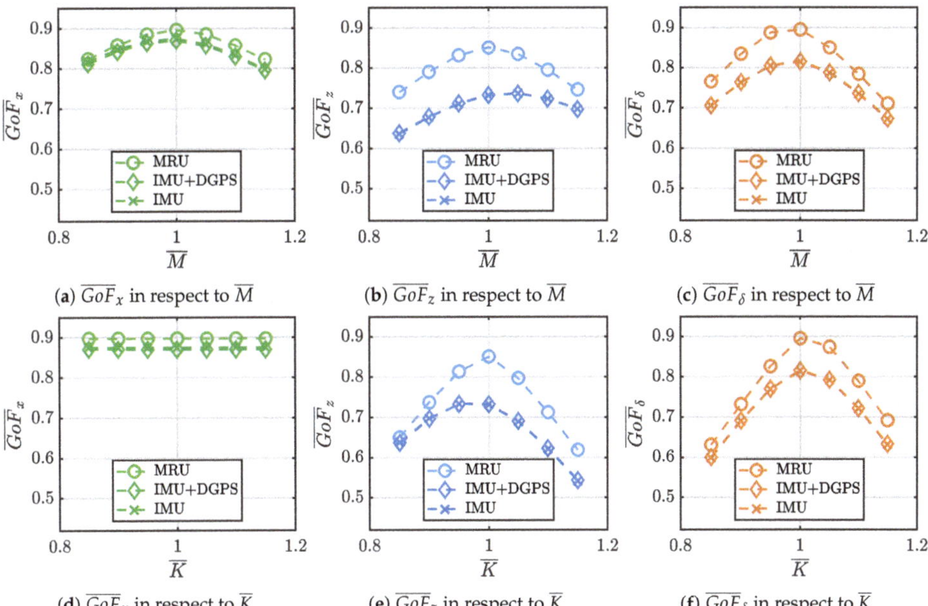

**Figure 12.** $\overline{GoF}$ results obtained with the KF varying $\overline{M}$ and $\overline{K}$. The first column refers to the $x$ DoF (a,d), the second to the $z$ DoF (b,e) and the third to the $\delta$ DoF (c,f).

Figure 13 illustrates the influence of $\overline{M}$ and $\overline{K}$ for the NN framework.

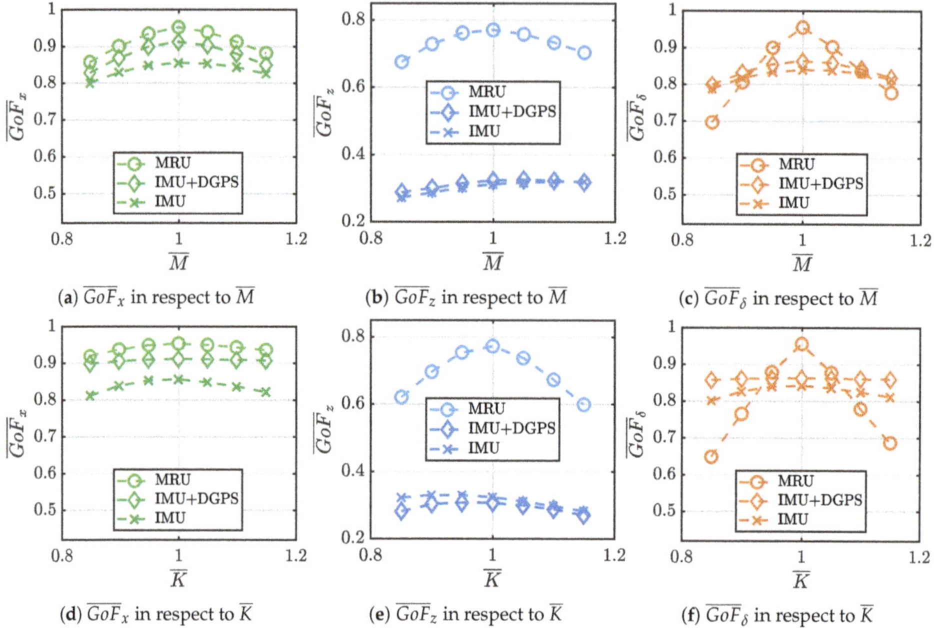

(a) $\overline{GoF}_x$ in respect to $\overline{M}$     (b) $\overline{GoF}_z$ in respect to $\overline{M}$     (c) $\overline{GoF}_\delta$ in respect to $\overline{M}$

(d) $\overline{GoF}_x$ in respect to $\overline{K}$     (e) $\overline{GoF}_z$ in respect to $\overline{K}$     (f) $\overline{GoF}_\delta$ in respect to $\overline{K}$

**Figure 13.** $\overline{GoF}$ results obtained with the NN varying $\overline{M}$ and $\overline{K}$. The first column refers to the $x$ DoF (**a,d**), the second to the $z$ DoF (**b,e**) and the third to the $\delta$ DoF (**c,f**).

A mean decrease of 0.1 is shown for all the measurement frameworks in presence of mass matrix variations, except for a high reduction of $GoF_\delta$ with the MRU configuration. The NN is merely a non-linear function that maps a serie of input-output arguments regardless of any phisical background. The training algorithm considers all the measurements available from the MRU to train the NN and a variation of the mass matrix influences all the measurements data provided to the network leading to a lower estimation accuracy. On the other hand, IMU+DGPS and IMU consider few measurements and appear to be less sensitive to plant variations. In particular, the $GoF_\delta$ suffers of a decrease up to 0.3 for a $\overline{K}$ of 0.85 and almost 0.25 for a $\overline{K}$ equal to 1.15, suggesting again that the MRU framework is less robust in presence of plant variations with the NN model. Likewise, a stiffness matrix variation negatively affects the estimation accuracy of $\hat{F}_{wx}$ despite its estimation is influenced only by $\delta(t)$, $\dot{\delta}(t)$ and $\ddot{x}(t)$ acquisitions as demostrated for the KF observer.

## 6.4. Comparison and Summary

The weighted GoFs and percentage differences of the KF and NN results are summarized in Table 6. The apex * means that no disturbances are considered. Starting from noiseless results, the $\overline{GoF}_x$ and $\overline{GoF}_\delta$ values confirm that the measurements of an IMU* unit are sufficient to estimate the surge and pitch components with acceptable accuracy. In detail, excluding absolute displacements and velocity measurements the $\overline{GoF}_x\big|_{KF}$ decreases from 0.910 to 0.893, the $\overline{GoF}_\delta\big|_{KF}$ from 0.901 to 0.899, the $\overline{GoF}_x\big|_{NN}$ from 0.951 to 0.931 and the $\overline{GoF}_\delta\big|_{KF}$ from 0.961 to 0.940. $\overline{GoF}_z$ follows the same behaviour, especially from the IMU+DGPS* to IMU* framework where a decrease of 0.229 and 0.397 appear for the KF and NN, respectively. Comparing the two techniques, the performance of the NN model surpasses the KF perfromance for surge and pitch DoFs, especially for the pitch DoF where an increment of 6.65%, 6.31% and 5.73% is achieved. However, the estimation of the heave force is not reliable with the NN, where a difference of $-28.5\%$ is obtained. Interestingly, the NN is able to handle noisy data better than the KF for both surge and pitch DoF, except for the IMU case. Specifically, a minimal difference of 2.73% is obtained with the IMU framework in surge direction. The NN always overcomes the $\overline{GoF_\delta}$ of the KF with a peak of 6.70% with a MRU unit. On the other hand, poor performances are carried out for the heave DoF with the NN, obtaining a weighted GoF almost equal to 0.3 for both IMU+DGPS and IMU.

**Table 6.** $\overline{GoF}$ and $\Delta\overline{GoF}$ results of the KF observer and NN model for each measurement frameworks. The measurements are considered with and without noise.

| Framework | Noise | Kalman Filter | | | Neural Network | | | | | |
|---|---|---|---|---|---|---|---|---|---|---|
| | | $\overline{GoF_x}$ | $\overline{GoF_z}$ | $\overline{GoF_\delta}$ | $\overline{GoF_x}$ | $\overline{GoF_z}$ | $\overline{GoF_\delta}$ | $\Delta\overline{GoF_x}$ (%) | $\Delta\overline{GoF_z}$ (%) | $\Delta\overline{GoF_\delta}$ (%) |
| MRU* | × | 0.910 | 0.971 | 0.901 | 0.951 | 0.941 | 0.961 | 4.50 | −3.08 | 6.65 |
| IMU+DGPS* | × | 0.910 | 0.967 | 0.902 | 0.950 | 0.939 | 0.959 | 4.39 | −2.89 | 6.31 |
| IMU* | × | 0.893 | 0.758 | 0.889 | 0.913 | 0.542 | 0.940 | 2.23 | −28.5 | 5.73 |
| MRU | ✓ | 0.898 | 0.873 | 0.895 | 0.950 | 0.771 | 0.955 | 5.79 | −11.6 | 6.70 |
| IMU+DGPS | ✓ | 0.875 | 0.744 | 0.824 | 0.912 | 0.327 | 0.863 | 4.22 | −56.1 | 4.73 |
| IMU | ✓ | 0.878 | 0.744 | 0.819 | 0.854 | 0.322 | 0.839 | −2.73 | −56.7 | 1.82 |

Surprisingly, the sensitivity of the KF in presence of plant inaccuracies is evident compared to the NN. In particular, Figure 14 highlight a relevant decrease in performance for $\hat{F}_{w\delta}(t)$ estimation with a $\overline{M}$ and $\overline{K}$ variation when the KF is employed (IMU framework). Using the KF, the results show a decrease of performances up to 0.18 points varying $\overline{M}$ and 0.20 varying $\overline{K}$. In contrast, the NN guarantee a weighted $\overline{GoF_\delta}$ always greater than 0.77 suggesting low sensitivity when plant variations appear. In all likelihood, the sensitivity of the KF could be diminished tuning the matrix $Q$ in order to give less importance to the model in favor of measurements. However, more research on this topic needs to be undertaken before the association between $Q$ and KF sensitivity is more clearly understood.

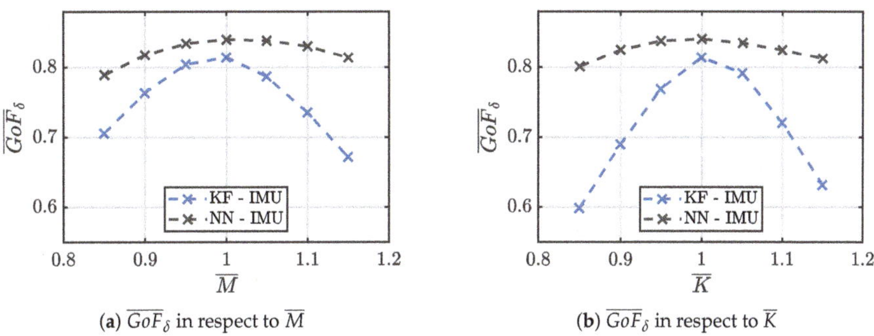

(a) $\overline{GoF_\delta}$ in respect to $\overline{M}$ (b) $\overline{GoF_\delta}$ in respect to $\overline{K}$

**Figure 14.** $\overline{GoF_\delta}$ comparison varying $\overline{M}$ (**a**) and $\overline{K}$ (**b**) for the IMU framworks.

The real ISWEC system is equipped with the MTi unit presented in Section 2.2.7. Unlike the MRU, the IMU unit does not provide positioning data and the measurement noise are one order of magnitude higher than the MRU ones. Although more information are available with the MRU, the IMU framework provide acceptable accuracy for both surge and pitch wave forces, directly involved in the power extraction. In detail, in case of noisy measurements and KF observer, the weighted accuracy approaches 0.88 and 0.82 for $\overline{GoF_x}$ and $\overline{GoF_\delta}$, respectively; with the NN the accuracy is about 0.85 and 0.84 for $\overline{GoF_x}$ and $\overline{GoF_\delta}$, respectively. Moreover, the MRU units are (usually) one order of magnitude more expensive than the IMU ones, and the precision provided does not justify the little increase of performances, expecially in the KF case. Despite the low performances obtained for the WEF in heave direction (not directly involted in the power extraction), the IMU sensor is considered reliable for the estimation.

## 7. Conclusions

The purpose of this work is to estimate the WEFs of a non-linear WEC employing a KF observer and a NN model. This study proposes a methodology for tuning both estimators and compares them for a wide range of sea-states in presence of noise disturbances and plant variations. Four different measurement frameworks are proposed, one ideal (full measurements available without noise) and three real frameworks composed by three different sensors commercially available. The main aim is to assess the estimation performances in term of GoF for different sensor equipments in all the operating conditions of the ISWEC device. The non-linear 3-DoF model of the ISWEC is considered as the plant of reference. A linear 3-DoF hydrodynamic model is used in the KF assuming linear wave theory and linear viscous damping along the pitch DoF. Moreover, the filter model assumes no viscous damping in both surge and heave directions and the mooring forces are considered as an unknow state to be estimated. The key of the KF is to approximate the expression of the WEFs as a linear superposition of finite harmonic components with variable amplitudes and fixed frequencies. Hence, it is possible to include the excitation forces into the state vector of the system model and perform an unknown state estimation. This method restricts the bandwidth of the estimated disturbance as it can only estimate at the specified discrete frequencies. This makes it more robust to other external disturbances such as unmodelled hydrodynamic forces outside from the frequency range considered. The feedforward NN is designed according to the hydrodynamic equation of the ISWEC. The WEFs are expressed as a function of the system dynamics at current and past time instants taking into account the dynamic memory of the plant using static neurons with no feedback data.

First, the KF parameters are tuned according to the frequency range, stating the WEF frequencies and the mooring frequencies. The wave energy matrix of the sea-site of interest has been considered to identify the principal wave periods of the incoming sea-states to decouple them from the mooring actions. Moreover, according to the noise magnitudes of the sensors, matrices $Q$ and $R$ have been balanced to obtain the best GoF from each measurement framework. The NN has been tuned with the same intent, chosing both delay steps and number of neurons guaranteeing the best compromise between network complexity and estimation accuracy. Second, numerical simulations are performed to investigate the influence of the measurement framework and sensors accuracy. Overrall, it is demonstrated that the 3-DoF KF performs well when applied to a non-linear WEC model. The KF shows best performances with the MRU* and IMU+DGPS* frameworks, expecially in surge and pitch directions where a $\overline{GoF}_x$ and $\overline{GoF}_\delta$ greater than 0.9 are guaranteed. The IMU* framework gives the worst performances along the heave direction performing a $\overline{GoF}_z$ almost equal to 0.75. The same argument applies for the NN, where the estimation performances are maximized in surge and pitch. Adding noisy measurements results in an acceptable decrease of accuracy for $\overline{GoF}_x$ and $\overline{GoF}_\delta$. In detail, the comparison shows that the estimation accuracies of NN and KF are approximately the same and a $\overline{GoF}_x$ and $\overline{GoF}_\delta$ greater than 0.83 is obtained. However, both the KF and NN are considered not reliable for the WEF estimation along the heave DoF if DGPS and IMU are employed. Then, intersting results are obtained comparing the estimation performances under plant variations. Contrary to expectations, the KF is affected by plant variations more than the NN. Despite the KF can handle inaccuracies of the numerical model tuning on the $Q$ matrix, the NN shows good performances when these inaccuracies become relevant. Further work is required to establish the underlying cause of this outcome, acting on the $Q$ matrix to improve the mean performances of the KF observer. Sections 6.1–6.3 demonstrated the good reliability of the IMU framework for the WEF estimation in surge and pitch directions for both KF and NN. However, it is evident how the estimation of the heave component is affected by the absence of the heave motion in the KF and both heave motion and velocity in NN. Despite this outcome, the use of the IMU unit is encourageed since the $\hat{F}_{wz}(t)$ is not directly involved in the power extraction of the ISWEC.

In conclusion, the main advantage of the model-based approach is that in presence of a real plant, it is possible to tune the observer on a small number of waves to obtain accurate estimation performances for a large number of sea states. The main strength of the KFHO is to consider only the frequency bandwidth specified; the knowlegde of the spectral properties of the signal to estimate allow to exclude all the undesired components and disturbance from the estimation (e.g., Mooring forces). On the other hand, a model-free non-linear approach should be more suitable to model complex hydrodynamic phenomena when their analytical expression are not available. Future work will approach the problem of the WEF estimation using more non-linear approaches (e.g., Recurrent Neural Networks and Extended Kalman Filter). The non-linear approach is expected to be more accurate in presence of strong non-linearity (e.g., PTO saturations and non-linear mooring models) in respect to a linear model-based observer. The aim will be elaborate on the estimation results between a model-based and model-free approach in term of estimation performances for a broad range of sea states. The best estimation approach will be used for the implementation of the MPC strategy on ISWEC.

**Author Contributions:** Conceptualization, M.B., A.H., S.A.S., P.D., G.B., G.M. and A.P.; methodology, M.B., A.H., S.A.S., P.D., G.B., G.M. and A.P.; software, M.B. and A.H.; validation, M.B., A.H., S.A.S. and P.D.; formal analysis, M.B., A.H., S.A.S., P.D. and G.B.; investigation, M.B. and A.H.; resources, G.B., G.M. and A.P.; data curation, M.B. and A.H.; writing—original draft preparation, M.B. and A.H.; writing—review and editing, M.B., A.H., S.A.S., P.D. and G.B.; visualization, M.B., A.H., S.A.S. and G.B.; supervision, G.M. and A.P.; project administration, G.M. and A.P.; funding acquisition, G.B. and G.M. All authors have read and agreed to the published version of the manuscript.

**Funding:** This research received no external funding.

**Conflicts of Interest:** The authors declare no conflict of interest.

## Appendix A Figures

*Appendix A.1. GoF with Noiseless Data and KF Observer*

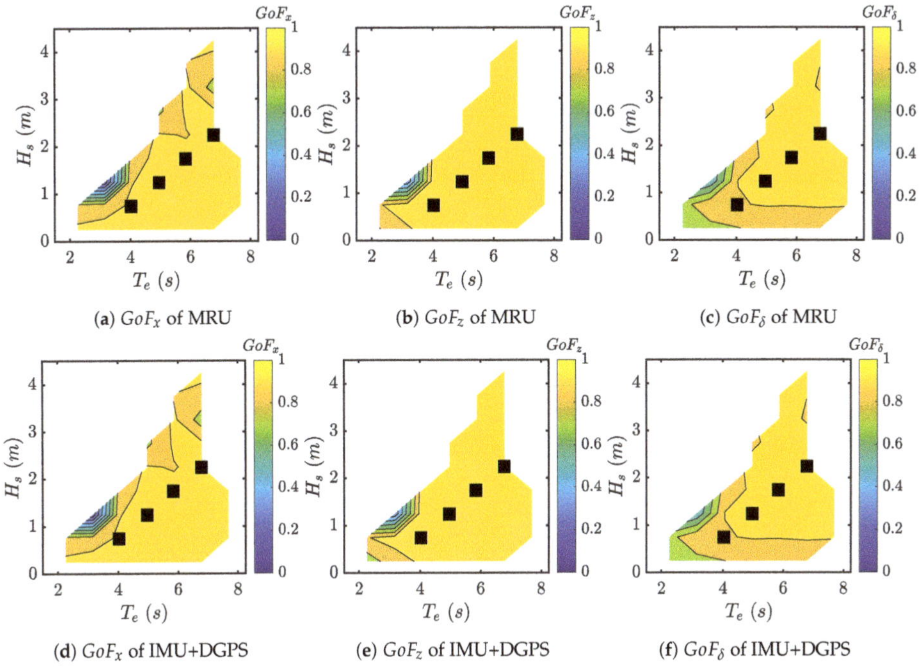

(**a**) $GoF_x$ of MRU      (**b**) $GoF_z$ of MRU      (**c**) $GoF_\delta$ of MRU

(**d**) $GoF_x$ of IMU+DGPS      (**e**) $GoF_z$ of IMU+DGPS      (**f**) $GoF_\delta$ of IMU+DGPS

**Figure A1.** *Cont.*

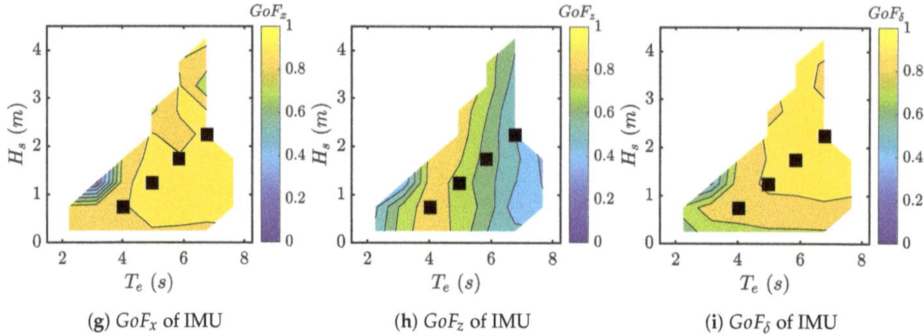

**(g)** $GoF_x$ of IMU

**(h)** $GoF_z$ of IMU

**(i)** $GoF_\delta$ of IMU

**Figure A1.** *GoF* results with noisless measurements for each measurement framework obtained with the KF. The first row refers to the MRU framework (**a–c**), the second to the IMU+DGPS framework (**d–f**), the thrid to the IMU framework (**g–i**).

*Appendix A.2. GoF with Noiseless Data and NN Model*

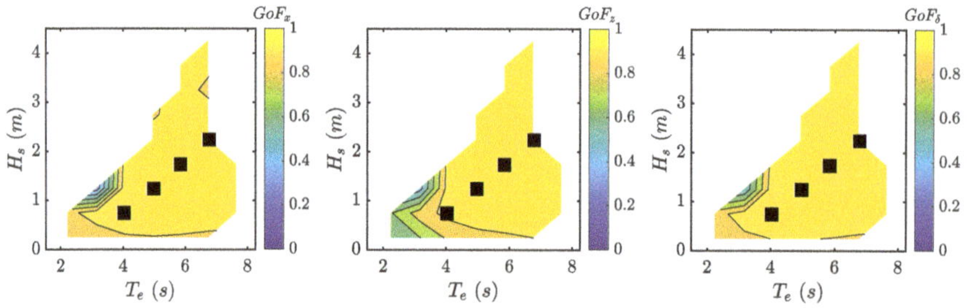

**(a)** $GoF_x$ of MRU

**(b)** $GoF_z$ of MRU

**(c)** $GoF_\delta$ of MRU

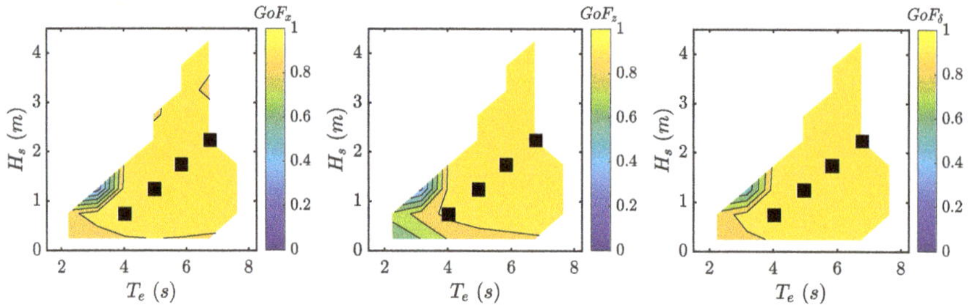

**(d)** $GoF_x$ of IMU+DGPS

**(e)** $GoF_z$ of IMU+DGPS

**(f)** $GoF_\delta$ of IMU+DGPS

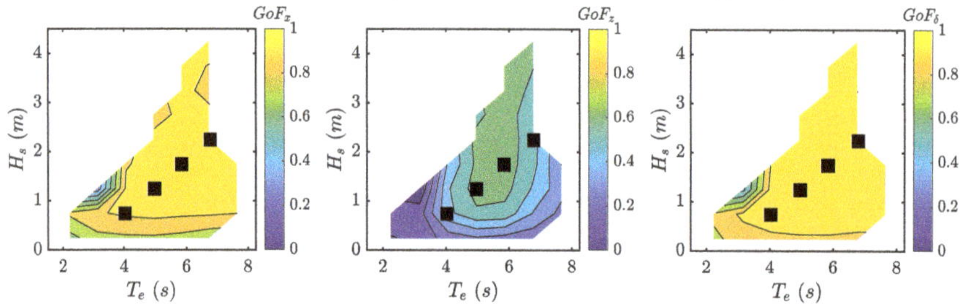

**(g)** $GoF_x$ of IMU

**(h)** $GoF_z$ of IMU

**(i)** $GoF_\delta$ of IMU

**Figure A2.** *GoF* results with noisless measurements for each measurement framework obtained with the NN. The first row refers to the MRU framework (**a–c**), the second to the IMU+DGPS framework (**d–f**), the thrid to the IMU framework (**b–i**).

*Appendix A.3. GoF with Noisy Data and KF Observer*

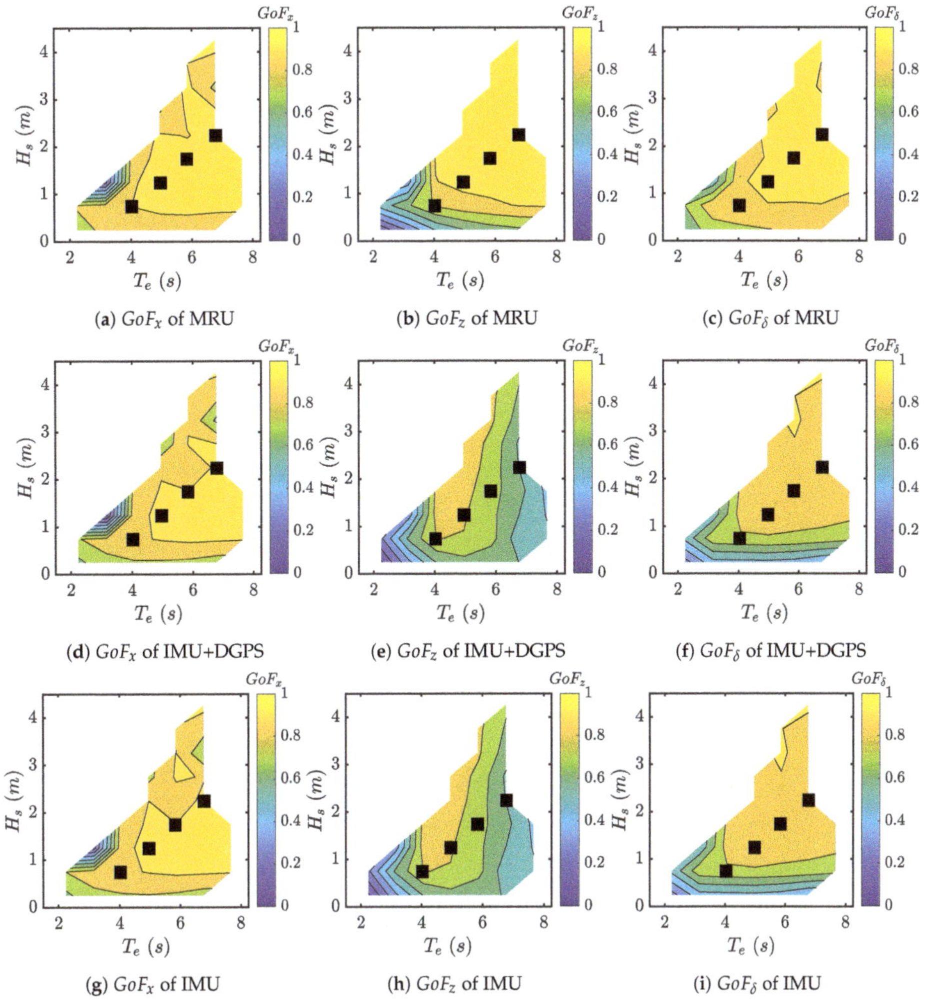

**(a)** $GoF_x$ of MRU

**(b)** $GoF_z$ of MRU

**(c)** $GoF_\delta$ of MRU

**(d)** $GoF_x$ of IMU+DGPS

**(e)** $GoF_z$ of IMU+DGPS

**(f)** $GoF_\delta$ of IMU+DGPS

**(g)** $GoF_x$ of IMU

**(h)** $GoF_z$ of IMU

**(i)** $GoF_\delta$ of IMU

**Figure A3.** *GoF* results with noisy measurements for each measurement framework obtained with the KF. The first row refers to the MRU framework (**a–c**), the second to the IMU+DGPS framework (**d–f**), the thrid to the IMU framework (**g–i**).

*Appendix A.4. GoF with Noisy Data and NN Model*

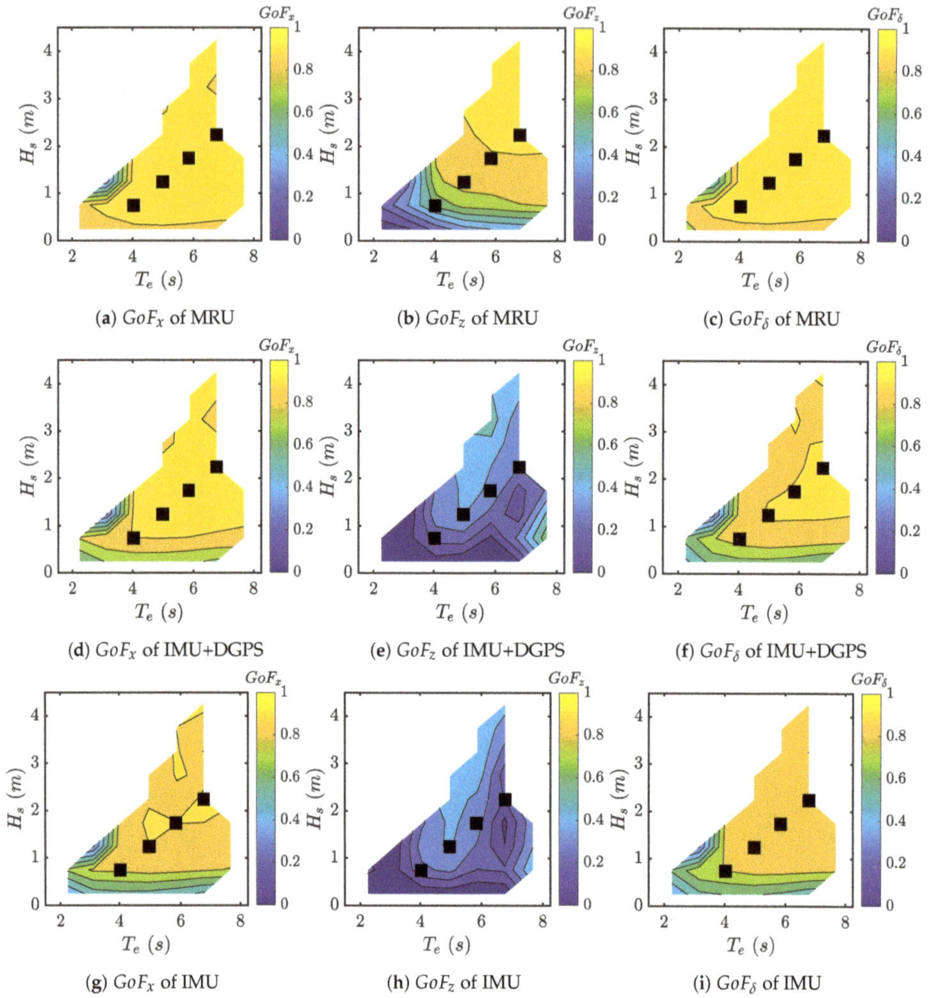

**Figure A4.** *GoF* results with noisy measurements for each measurement framework obtained with the NN. The first row refers to the MRU framework (**a–c**), the second to the IMU+DGPS framework (**d–f**), the thrid to the IMU framework (**g–i**).

## References

1. Faedo, N.; Olaya, S.; Ringwood, J.V. Optimal control, MPC and MPC-like algorithms for wave energy systems: An overview. *IFAC J. Syst. Control* **2017**, *1*, 37–56. [CrossRef]
2. Pena-Sanchez, Y.; Windt, C.; Davidson, J.; Ringwood, J.V. A Critical Comparison of Excitation Force Estimators for Wave-Energy Devices. *IEEE Trans. Control Syst. Technol.* **2019**, 1–13. [CrossRef]
3. Ling, B.A.; Batten, B.A. Real Time Estimation and Prediction of Wave Excitation Forces on a Heaving Body. In Proceedings of the International Conference on Offshore Mechanics and Arctic Engineering, St. John's, NL, Canada, 31 May–5 June 2015.
4. Garcia-Abril, M.; Paparella, F.; Ringwood, J.V. Excitation force estimation and forecasting for wave energy applications. *IFAC Pap.* **2017**, *50*, 14692–14697. [CrossRef]

5. Pena-Sanchez, Y.; Garcia-Abril, M.; Paparella, F.; Ringwood, J.V. Estimation and Forecasting of Excitation Force for Arrays of Wave Energy Devices. *IEEE Trans. Sustain. Energy* **2018**, *9*, 1672–1680. [CrossRef]

6. Nguyen, H.N.; Tona, P. Wave Excitation Force Estimation for Wave Energy Converters of the Point-Absorber Type. *IEEE Trans. Control Syst. Technol.* **2018**, *26*, 2173–2181. [CrossRef]

7. Abdelrahman, M.; Patton, R.; Guo, B.; Lan, J. Estimation of wave excitation force for wave energy converters. In Proceedings of the 2016 3rd Conference on Control and Fault-Tolerant Systems (SysTol), Barcelona, Spain, 7–9 September 2016; pp. 654–659.

8. Abdelkhalik, O.; Zou, S.; Robinett, R.; Bacelli, G.; Wilson, D. Estimation of excitation forces for wave energy converters control using pressure measurements. *Int. J. Control* **2017**, *90*, 1793–1805. [CrossRef]

9. Hillis, A.J.; Brask, A.; Whitlam, C. Real-time wave excitation force estimation for an experimental multi-DOF WEC. *Ocean Eng.* **2020**, *213*, 107788. [CrossRef]

10. Li, L.; Gao, Z.; Yuan, Z.M. On the sensitivity and uncertainty of wave energy conversion with an artificial neural-network-based controller. *Ocean Eng.* **2019**. [CrossRef]

11. Desouky, M.A.A.; Abdelkhalik, O. Wave prediction using wave rider position measurements and NARX network in wave energy conversion. *Appl. Ocean Res.* **2019**, *82*, 10–21. [CrossRef]

12. Genuardi, L.; Bracco, G.; Sirigu, S.; Bonfanti, M.; Paduano, B.; Dafnakis, P.; Mattiazzo, G. An application of model predictive control logic to inertial sea wave energy converter. *Adv. Mech. Mach. Sci.* **2019**. [CrossRef]

13. Sirigu, S.A.; Bracco, G.; Bonfanti, M.; Dafnakis, P.; Mattiazzo, G. On-board sea state estimation method validation based on measured floater motion. In Proceedings of the 11th IFAC Conference on Control Applications in Marine Systems, Robotics, and Vehicles CAMS 2018, Opatija, Croatia, 10–12 September 2018.

14. Bonfanti, M.; Carapellese, F.; Sirigu, S.A.; Bracco, G.; Mattiazzo, G. Excitation Forces Estimation for Non-linear Wave Energy Converters: A Neural Network Approach. Unpublished work.

15. Vissio, G. ISWEC toward the Sea-Development, Optimization and Testing of the Device Control Architecture. Ph.D. Thesis, Politecnico di Torino, Torino, Italy, 2018.

16. Bracco, G.; Casassa, M.; Giorcelli, E.; Giorgi, G.; Martini, M.; Mattiazzo, G.; Passione, B.; Raffero, M.; Vissio, G. Application of sub-optimal control techniques to a gyroscopic Wave Energy Converter. *Renew. Energies Offshore* **2014**, 265–269. [CrossRef]

17. Bonfanti, M.; Bracco, G.; Dafnakis, P.; Giorcelli, E.; Passione, B.; Pozzi, N.; Sirigu, S.; Mattiazzo, G. Application of a passive control technique to the ISWEC: Experimental tests on a 1:8 test rig. In Proceedings of the NAV International Conference on Ship and Shipping Research, Trieste, Italy, 20–22 June 2018.

18. Bracco, G.; Giorcelli, E.; Giorgi, G.; Mattiazzo, G.; Passione, B.; Raffero, M.; Vissio, G. Performance assessment of the full scale ISWEC system. In Proceedings of the 2015 IEEE International Conference on Industrial Technology (ICIT), Seville, Spain, 17–19 March 2015; pp. 2499–2505.

19. Bracco, G.; Cagninei, A.; Giorcelli, E.; Mattiazzo, G.; Poggi, D.; Raffero, M. Experimental validation of the ISWEC wave to PTO model. *Ocean. Eng.* **2016**, *120*, 40–51. [CrossRef]

20. Cagninei, A.; Raffero, M.; Bracco, G.; Giorcelli, E.; Mattiazzo, G.; Poggi, D. Productivity analysis of the full scale inertial sea wave energy converter prototype: A test case in Pantelleria Island. *J. Renew. Sustain. Energy* **2015**, *7*, 61703. [CrossRef]

21. Raffero, M.; Martini, M.; Passione, B.; Mattiazzo, G.; Giorcelli, E.; Bracco, G. Stochastic control of inertial sea wave energy converter. *Sci. World J.* **2015**, *2015*. [CrossRef]

22. Wendt, F.; Nielsen, K.; Yu, Y.H.; Bingham, H.; Eskilsson, C.; Kramer, M.; Babarit, A.; Bunnik, T.; Costello, R.; Crowley, S.; et al. Ocean energy systemswave energy modelling task: Modelling, verification and validation ofwave energy converters. *J. Mar. Sci. Eng.* **2019**, *7*, 379. [CrossRef]

23. Ransley, E.; Yan, S.; Brown, S.; Hann, M.; Graham, D.; Windt, C.; Schmitt, P.; Davidson, J.; Ringwood, J.; Musiedlak, P.H.; et al. A blind comparative study of focused wave interactions with floating structures (CCP-WSI blind test series 3). *Int. J. Offshore Polar Eng.* **2020**, *30*, 1–10. [CrossRef]

24. Faltinsen, O.M. *Sea Loads on Ships and Offshore Structures*; Cambridge University Press: Cambridge, UK, 1993.

25. Cummins, W.E. *The Impulse Response Function and Ship Motions*; Technical Report 1661; Department of the Navy: Port Hueneme, CA, USA, 1962.

26. Perez, T.; Fossen, T.I. Joint identification of infinite-frequency added mass and fluid-memory models of marine structures. *Model. Identif. Control* **2008**, *29*, 93–102. [CrossRef]

27. Pérez, T.; Fossen, T.I. Time-vs. frequency-domain Identification of parametric radiation force models for marine structures at zero speed. *Model. Identif. Control* **2008**, *29*, 1–19. [CrossRef]

28. Fontana, M.; Casalone, P.; Sirigu, S.A.; Giorgi, G. Viscous Damping Identification for a Wave Energy Converter Using CFD-URANS Simulations. *J. Mar. Sci. Eng.* **2020**, *8*, 355. [CrossRef]

29. Pozzi, N.; Bracco, G.; Passione, B.; Sirigu, S.A.; Mattiazzo, G. PeWEC: Experimental validation of wave to PTO numerical model. *Ocean. Eng.* **2018**, *167*, 114–129. [CrossRef]

30. Newman, J.N. Second-order, slowly-varying Forces on Vessels in Irregular Waves. In *International Symposium on the Dynamics of Marine Vehicles and Structures in Waves*; IME: London, UK, 1974; pp. 182–186.

31. Sirigu, S.A.; Bonfanti, M.; Begovic, E.; Bertorello, C.; Dafnakis, P.; Giorgi, G.; Bracco, G.; Mattiazzo, G. Experimental investigation of the mooring system of a wave energy converter in operating and extreme wave conditions. *J. Mar. Sci. Eng.* **2020**, *8*, 180. [CrossRef]

32. Pozzi, N.; Bonfanti, M.; Mattiazzo, G. Mathematical Modeling and Scaling of the Friction Losses of a Mechanical Gyroscope. *Int. J. Appl. Mech.* **2018**, *10*, 1–21. [CrossRef]

33. Ochi, M.K. *Ocean Waves: The Stochastic Approach*; Cambridge Ocean Technology Series; Cambridge University Press: Cambridge, UK, 1998.

34. Mei, C.; Stiassnie, M.; Yue, D. *Theory and Applications of Ocean Surface Waves*; Technion-Israel Institute of Technology: Haifa, Israel, 2005; Volume 23, p. 503.

35. Merigaud, A. A Harmonic Balance Framework for the Numerical Simulation of Non-Linear Wave Energy Converter Models in Random Seas. Ph.D. Thesis, National University of Ireland Maynooth, Kildare, Ireland, 2018.

36. Sirigu, S.A. Development of A Resonance-Tunable Wave Energy Converter. Ph.D. Thesis, Politecnico di Torino, Torino, Italy, 2019.

37. Hasselmann, K.; Barnett, T.; Bouws, E.; Carlson, H.; Cartwright, D.; Enke, K.; Ewing, J.; Gienapp, H.; Hasselmann, D.; Kruseman, P.; et al. Measurements of wind-wave growth and swell decay during the Joint North Sea Wave Project (JONSWAP). *Deut. Hydrogr. Z.* **1973**, *8*, 1–95.

38. MTi User Manual. 2020. Available online: https://www.xsens.com/hubfs/Downloads/usermanual/MTi_usermanual.pdf (accessed on 20 October 2020)

39. *ECN 413, ECN 425, ERN 487, Product Information ECN*; HEIDENHAIN: Traunreut, Germany, 2017.

40. NI cRIO-9030 User Manual. 2020. Available online: https://www.ni.com/pdf/manuals/376260a_02.pdf (accessed on 20 October 2020)

41. Khaleghi, B.; Khamis, A.; Karray, F.O.; Razavi, S.N. Multisensor data fusion: A review of the state-of-the-art. *Inf. Fusion* **2013**, *14*, 28–44. [CrossRef]

42. Fung, M.L.; Chen, M.Z.; Chen, Y.H. Sensor fusion: A review of methods and applications. In Proceedings of the 29th Chinese Control and Decision Conference, Chongqing, China, 28–30 May 2017; pp. 3853–3860.

43. Crassidis, J.L.; Junkins, J.L. *Optimal Estimation of Dynamic Systems (Chapman & Hall/CRC Applied Mathematics & Nonlinear Science)*, 2nd ed.; Chapman & Hall/CRC: Boca Raton, FL, USA, 2011.

44. Paduano, B.; Giorgi, G.; Gomes, R.P.; Pasta, E.; Henriques, J.C.; Gato, L.M.; Mattiazzo, G. Experimental validation and comparison of numerical models for the mooring system of a floating wave energy converter. *J. Mar. Sci. Eng.* **2020**, *8*, 565. [CrossRef]

45. Hall, M. MoorDyn User's Guide. 2015. Available online: http://www.matt-hall.ca/ (accessed on 20 October 2020).

46. Orcina-Ltd. OrcaFlex Software. 2014. Available online: http://orcina.com/ (accessed on 20 October 2020).

47. Ablameyko, S.; Goras, L.; Gori, M.; Piuri, V. *Neural Networks for Instrumentation, Measurement and Related Industrial Applications*; IOS Press: Amsterdam, The Netherlands, 2003.

48. Laurent, Q. Estimation and Prediction of Wave Input and System States Based on Local Hydropressure and Machinery Response Measurements. Ph.D. Thesis, KTH, Optimization and Systems Theory, Stockholm, Sweden, 2016. Available online: http://urn.kb.se/resolve?urn=urn:nbn:se:kth:diva-191995 (accessed on 20 October 2020).

Journal of
*Marine Science and Engineering*

*Article*

# Towards Real-Time Reinforcement Learning Control of a Wave Energy Converter

Enrico Anderlini [1,*], Salman Husain [2], Gordon G. Parker [2], Mohammad Abusara [3] and Giles Thomas [1]

[1] Department of Mechanical Engineering, University College London, London WC1E 6BT, UK; giles.thomas@ucl.ac.uk
[2] Department of Mechanical Engineering—Engineering Mechanics, Michigan Technological University, Houghton, MI 49931, USA; shusain@mtu.edu (S.H.); ggparker@mtu.edu (G.G.P.)
[3] College of Engineering, Mathematics and Physical Sciences, University of Exeter, Penryn Campus, Penryn, Cornwall TR10 9FE, UK; M.Abusara@exeter.ac.uk
* Correspondence: e.anderlini@ucl.ac.uk

Received: 30 September 2020; Accepted: 23 October 2020; Published: 28 October 2020

**Abstract:** The levellised cost of energy of wave energy converters (WECs) is not competitive with fossil fuel-powered stations yet. To improve the feasibility of wave energy, it is necessary to develop effective control strategies that maximise energy absorption in mild sea states, whilst limiting motions in high waves. Due to their model-based nature, state-of-the-art control schemes struggle to deal with model uncertainties, adapt to changes in the system dynamics with time, and provide real-time centralised control for large arrays of WECs. Here, an alternative solution is introduced to address these challenges, applying deep reinforcement learning (DRL) to the control of WECs for the first time. A DRL agent is initialised from data collected in multiple sea states under linear model predictive control in a linear simulation environment. The agent outperforms model predictive control for high wave heights and periods, but suffers close to the resonant period of the WEC. The computational cost at deployment time of DRL is also much lower by diverting the computational effort from deployment time to training. This provides confidence in the application of DRL to large arrays of WECs, enabling economies of scale. Additionally, model-free reinforcement learning can autonomously adapt to changes in the system dynamics, enabling fault-tolerant control.

**Keywords:** wave energy converter; control; reinforcement learning; deep reinforcement learning; deep learning; adaptive control

## 1. Introduction

Ocean wave energy is a type of renewable energy with the potential to contribute significantly to the future energy mix. Despite an estimated global resource of 146 TWh/yr [1], the wave energy industry is still in its infancy. In 2014, there were less than 10 MW of installed capacity worldwide due to the high levellised cost of energy (LCoE) of approximately EUR 330–630/MWh [1]. The main operational challenge is the maximisation of energy extraction in the common, low-energetic sea states, whilst ensuring the survival of the wave energy converters (WECs) in storms [2]. Two major contributors to the lowering of the LCoE to EUR 150/MWh by 2030 are expected to be the achievement of economies of scale and the development of effective control strategies [3].

Over the past decade, model predictive control (MPC) has attracted much research interest, as it can offer improved performance over the control strategies developed in the 1970s and 1980s, based on hydrodynamic principles. Assuming knowledge of the wave excitation force, MPC computes the control action, typically the force applied by the power take-off system (PTO), that results in optimal energy absorption over a future time horizon using a model of the WEC dynamics. The controller applies only the first value of the PTO force, recomputing the optimal control action at the next time horizon. The iterative procedure enables the controller to reduce the negative impact of inaccuracies in the prediction of the future excitation force and modelling errors. Additionally, the MPC framework enables the inclusion of constraints on both the control action and the system dynamics. A good review of MPC for WEC control can be found in [4]. Li and Belmont first proposed a fully convex implementation, which trades off the energy absorption, the energy consumed by the actuator and safe operation [5]. An even more efficient implementation cast in a quadratic programming form has been proposed by Zhong and Yeung [6]. Fundamentally, the convex form enables the strategy to be generalised to the control of multiple WECs in real-time [7,8]. However, linear MPC relies on a linear model of the WEC dynamics. In energetic waves, nonlinearities in the static and dynamic Froude–Krylov forces (i.e., the hydrostatic restoring and wave incidence forces) and viscous drag effects can become significant [9]. Although nonlinear MPC strategies have been proposed [10,11] and even tested experimentally [12], achieving a successful real-time, centralised control implementation for multiple WECs is expected to be challenging [13].

As described in [14,15], alternative strategies have been developed for the control of WECs. Some, like simple-and-effective control [16], present similar performance to MPC at a much lower computational cost. Alternative solutions based on machine learning have been recently considered thanks to the advancements in the field of artificial intelligence. Most commonly, the neural networks are used to provide a data-based, nonlinear model of the system dynamics, i.e., for system identification. After training, the identified model is coupled with standard strategies used for the control of WECs, e.g., resistive or damping control in [17], reactive or impedance-matching in [18] and latching control in [19,20]. On the one hand, some studies have proposed the use of neural networks to find the optimal parameters for impedance-matching control on a time-averaged basis [21], thus being readily applicable to the centralised control of multiple WECs [22]. On the other hand, other works have focused on real-time control [19,20,23], exploiting the capability of neural networks to handle the predicted wave elevation over a future time horizon, similar to MPC. The main advantage of machine learning models for the system identification of WECs is that the same method can be used for different WEC technologies and is potentially adaptive to changes in the system dynamics, e.g., to subsystem failures or biofouling.

A promising solution to developing an optimal, real-time nonlinear controller for WECs inclusive of constraints on both the state and action is to cast the problem in a dynamic programming framework, similarly to MPC for WEC problems. In non-linear dynamic programming solutions, a neural network is used as a critic to approximate the time-dependent optimal cost value expressed as a Hamilton–Jacobi–Bellman equation. Numerical studies have shown the effectiveness and robustness of this approach for the control of a single WEC [24–26]. In particular, dynamic programming, also classified as model-based reinforcement learning (RL), is much more data-efficient than model-free RL [27]. Using a machine learning model trained on the collected data, such as Gaussian processes [27] or neural networks [24–26], enables the controller to plan off-line, thus significantly speeding up the learning of a suitable control policy even from a small set of samples. Conversely, model-free RL methods which learn from direct interactions with the environment require a much larger number of samples (in order of $10^8$ as opposed to $10^4$ for complex control tasks [28]). For this reason, to date model-free RL has been applied only to the time-averaged resistive and reactive control of WECs with discrete PTO damping and stiffness coefficients [29–31], with lower level controllers necessary to ensure constraints abidance [32]. However, model-free RL schemes are known to find the optimal control policy, even for real-time applications and very complex systems [28].

This paper introduces the world-first deep reinforcement learning (DRL) control method for WECs. The novel approach enables the real-time, nonlinear optimal control of WECs based on model-free RL. Deep learning allows the method to treat continuous control input and output features efficiently at deployment time.

To avoid unpredictable behaviour during the initial learning stage, WECs are expected to be controlled with model-based, robust methods once first deployed in the future. For a real-time implementation on complex WECs, these are likely to rely on linear models considering current technologies. After sufficient data samples are collected, the controller can move to the proposed data-driven model, whose computational cost at deployment does not increase if nonlinearities are present and can adapt to changes in the system dynamics or noncritical faults if retrained regularly. Hence, first of all, the WEC is operated in a range of representative sea states under the convex MPC proposed in [6,8]. Data samples are collected from 15-minute-long wave traces in each sea state. Subsequently, the dataset is used to train a deep neural network (DNN), defined as a neural network with more than one hidden layer according to [33], which mimics the controller behaviour. The DNN thus corresponds to the actor of an actor-critic RL strategy. The actor will be then used to initialise a model-free RL controller, as in [28]. The agent will then be further trained to optimise its behaviour as in [34].

In this article, the analysis is limited to the simulation of a standard spherical point absorber constrained to heaving motions [9]. The simulation environment is currently based on a linear model as presented in Section 2. The new DRL-based control method for WECs is described in Section 3. Finally, the performance of the trained actor is assessed directly against the original linear MPC developed in [6] in Section 4, with conclusions drawn in Section 5.

## 2. Linear Model of a Heaving Point Absorber

A heaving point absorber is shown schematically in Figure 1. Assuming linear potential wave theory, the equation of motion of a heaving point absorber can be expressed in the time domain as [35]

$$m\ddot{z}(t) = f_e(t) + f_r(t) + f_h(t) + f_m(t) + f_{PTO}(t), \tag{1}$$

where $t$ indicates time, $z$ the heave displacement, $f_e$ the wave excitation force inclusive of wave incidence and diffraction effects (or dynamic Froude–Krylov and scattering forces), $f_r$ the wave radiation force, $f_h$ the hydrostatic restoring force (or static Froude–Krylov), $f_m$ the mooring force, $f_{PTO}$ the PTO or control force, and $m$ is the mass of the buoy. In this study, the mooring forces are ignored for simplicity.

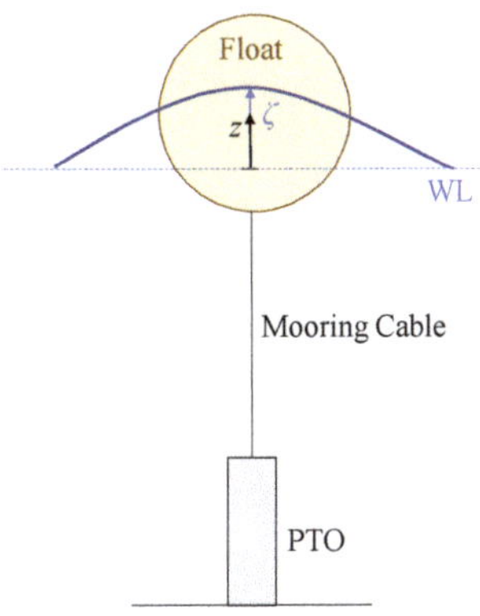


**Figure 1.** Heaving point absorber.

The linear excitation wave force can be obtained using the excitation impulse response function $H(t)$ as

$$f_e(t) = \int_{-\infty}^{\infty} H(t-\tau)\zeta(t)d\tau, \tag{2}$$

where $\zeta$ is the wave elevation. Similarly, the linear hydrostatic restoring force is

$$f_h(t) = -\rho g A_w z(t), \tag{3}$$

where $\rho$ is the water density, $g$ the gravitational acceleration and $A_w$ the waterplane area. Using Cummins' equation [36], the radiation force can be expressed as

$$f_r(t) = -A(\infty)\ddot{z}(t) - \int_{-\infty}^{t} K(t-\tau)\dot{z}(\tau)d\tau, \tag{4}$$

where $A(\infty)$ is the heave added mass at infinite wave frequency and $K$ is the radiation impulse response function. The convolution integral in (4) causes significant challenges for control tasks. Hence, it is common practice to approximate the convolution integral with a state-space model to improve the computational performance and ensure controllability. Here, the approach based on moment matching proposed in [37] is followed. Hence, (4) is reformulated as

$$\dot{x}_{ss}(t) = A_{ss}x_{ss} + B_{ss}\dot{z}, \tag{5a}$$

$$\int_{-\infty}^{t} K(t-\tau)\dot{z}(\tau)d\tau \approx C_{ss}x_{ss} + D_{ss}\dot{z}. \tag{5b}$$

Substituting (3)–(5) into (1) allows the equation of motion of the heaving point absorber to be expressed in state space form:

$$\dot{x}(t) = Ax(t) + B_u u(t) + B_v v(t), \tag{6a}$$

$$y = Cx, \text{ where} \tag{6b}$$

$$x = \begin{bmatrix} z & w & x_{ss}^T \end{bmatrix}^T, \tag{6c}$$

$$u = f_{PTO}, \tag{6d}$$

$$v = f_e, \tag{6e}$$

$$B_u = B_v = \begin{bmatrix} 0 & M^{-1} & 0^T \end{bmatrix}^T, \tag{6f}$$

$$A = \begin{bmatrix} 0 & 1 & 0 \\ -M^{-1}C & 0 & -M^{-1}C_{ss} \\ 0 & B_{ss} & A_{ss} \end{bmatrix}, \tag{6g}$$

$$C = \begin{bmatrix} 1 & 1 & 0^T \end{bmatrix}^T, \tag{6h}$$

with $M = m + A(\infty)$. The net useful energy that can be absorbed from the waves between times $t_0$ and $t_f$ is given by

$$E = -\int_{t_0}^{t_f} f_{PTO}(t)\dot{z}(t)dt. \tag{7}$$

### 3. Real-Time Reinforcement Learning Control of a Wave Energy Converter

RL is a decision-making framework in which an agent learns a desired behaviour, or policy $\pi$, from direct interactions with the environment [38]. As shown in Figure 2, at each time step, the agent is in a state $s$ and takes and action $a$, thus landing in a new state $s'$ while receiving a reward $r$. A Markov decision process is used to model the action selection depending on the value function

$Q(s, a)$, which represents an estimate of the future reward. By interacting with the environment for a long time, the agent learns an optimal policy, which maximises the total expected reward.

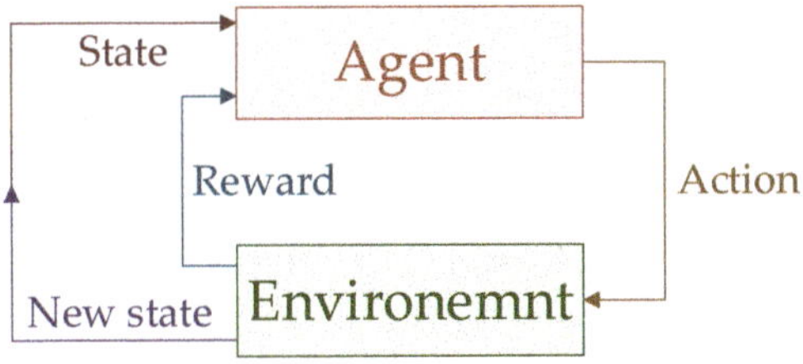


**Figure 2.** Block diagram of reinforcement learning (RL) control.

*3.1. Problem Formulation*

As a decision-making framework, RL is typically used to train an agent, or system, to perform a task that is particularly challenging to express in a standard control setting, e.g., walking for a legged robot. These tasks are usually described as episodic, i.e., the experience can be subdivided into discrete trials whose end is determined by either success in achieving the desired task, e.g., the robot has correctly made a walking step, or failure, e.g., the robot has fallen over and needs to start again. However, WEC control is clearly continuous, which will require a reformulation of the RL schemes as shown in [29–32].

In a feedforward configuration, the control of WECs is dependent on the wave excitation force and its predicted value over a future time horizon. Here, a simple state space is selected, which includes only the WEC displacement and velocity and capture the wave excitation information from the wave elevation and its rate. Therefore, the RL state space for a heaving point absorber is defined as

$$s = \begin{bmatrix} z & \dot{z} & \zeta & \dot{\zeta} \end{bmatrix}^T. \tag{8}$$

The action space is identical to the control input $u$:

$$a = f_{\text{PTO}}. \tag{9}$$

Although RL originated with the treatment of discrete actions, as for instance shown in [29–32], successful strategies with continuous action spaces have been recently proposed [34,39,40]. With either solution, constraints on the action can be easily imposed, so that $|a| < f_{\max}$.

Specifying an appropriate reward function is fundamental to have the agent learn the desired behaviour. Note that in RL, the optimisation problem is typically cast as a maximisation rather than a minimisation. Taking inspiration from [24–26], the reward function can be defined as

$$r = -f_{\text{PTO}}\dot{z} - w_u f_{\text{PTO}}^2 - w_z p_z, \tag{10}$$

where the weights $w_u$ and $w_z$ can be used to tune the penalty on the control action and heave displacement, respectively. Whilst a constraint can be placed on the PTO force, as it coincides with the control action, it is not possible to impose proper limits on the heave displacement. Therefore, a discontinuous function is used to determine the penalty term $p_z$ to produce an aggressive controller:

$$p_z = \begin{cases} 0 & \text{if } |z| \leq z_{\max}, & (11\text{a}) \\ 1 & \text{if } |z| > z_{\max}, & (11\text{b}) \end{cases}$$

where $z_{\max}$ defines the displacement limit.

Another difference between the RL and MPC frameworks consists of the way the information on the future incoming waves is treated, i.e., the prediction step. In feedforward MPC, an external method, e.g., autoregressive techniques and the excitation impulse response function [41], is used to predict the incoming wave force and the information is included in the cost function to select the control action. In the RL framework, the agent learns an optimal policy for the maximisation of the total reward, which is a function of the current reward as well as discounted future rewards deriving from following either the current or the optimal policy. This means that the reward function should be specified for the current time step rather than include information from predicted future time steps. The prediction step is embedded within the RL system in a probabilistic setting.

### 3.2. RL Real-Time WEC Control Framework

Although trust region policy optimisation is used in [28] for a model-free controller initialised with samples obtained from an MPC controller, here actor-critic strategies are considered for the real-time control of a WEC. An example of a successful scheme with continuous state and action spaces, which are necessary for improved control performance, is soft actor-critic (SAC) [34]. In particular, SAC [34], which is the most advanced actor-critic DRL algorithm at the time of writing, is selected here for the control framework for the point absorber.

As shown in Figure 3, the controller would be split into an actor and a critic. The function of the critic is to evaluate the policy, thus updating the action-value function, which is a measure of the total discounted reward, using the samples collected from observations of the environment. The discounted reward estimated by the critic is then fed to the actor. Using the estimated action-value function, the actor selects an action based on the current state, directly interacting with the environment. The policy is then improved by learning from the collected observations. As SAC is an off-line, off-policy algorithm, the critic and actor can be updated using batches of data samples, known as experience replay buffer.

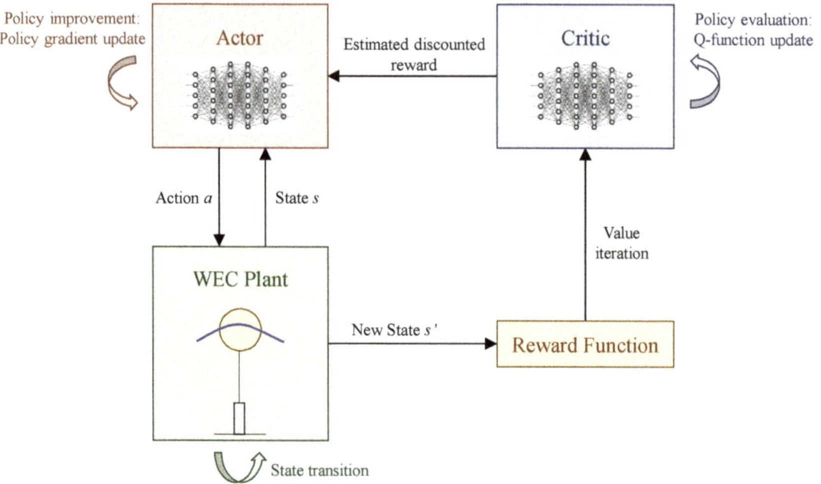


**Figure 3.** Diagram of the soft actor-critic (SAC) algorithm for the real-time control of a wave energy converter (WEC).

The agent seeks to maximise not only the environment's expected reward, but also the policy's entropy. The concept of entropy ensures that the agent selects random actions to explore the environment through a parameter $\alpha$ defined as the entropy temperature. The parameter is automatically adjusted with gradient descent to ensure sufficient exploration at the start of learning, and subsequently a greater emphasis on the maximisation of the expected reward. A DNN is used to model the mean of the log of the standard deviation of the policy. For the policy improvement step, the policy distribution is updated towards the softmax distribution for the current Q function by minimizing the Kullback–Leibler divergence.

In SAC, two DNNs are used to approximate the critic's policy evaluation to mitigate positive bias in the policy evaluation step. The DNNs are trained off-line using batches of data collected by the actor during deployment. The minimum value of the two soft Q-functions is used in the gradient descent during training, which has been found to significantly speed up convergence. Additionally, target networks that are obtained as an exponentially moving average of the soft Q-function weights are used to smooth out the effects of noise in the sampled data.

The SAC algorithm is summarised in Algorithm 1. For a full explanation, the reader is referred to [34].

---

**Algorithm 1:** SAC algorithm taken from [34].

---

**Result:** Optimised actor and critic DNNs
initialise policy, two soft Q and two target soft Q DNNs;
initialise experience replay buffer with MPC samples;
**for** *each episode* **do**
    **for** *each step* **do**
        sample actions from the policy;
        sample transition from the environment;
        store the transition in the replay buffer;
    **end**
    **for** *each gradient update step* **do**
        update the soft Q DNN weights;
        update the policy DNN weights;
        adjust the entropy temperature;
        update the target DNN weights;
    **end**
**end**

---

As compared with the RL solutions presented in [29–32], once trained the new DRL implementation can be implemented in real time, with a control time step similar to the one used by other control methods, e.g., MPC.

## 4. Results and Discussion

### 4.1. Case Study

A spherical point absorber as shown in Figure 1 is selected as a case study for the development of the real-time RL WEC control scheme. The spherical buoy represents a standard case study based on the Wavestar prototype WEC, which has also been used in [9,37,42] among other studies. Additionally, the simple geometry enables the inclusion of computationally efficient nonlinear Froude–Kryolv and viscous forces in the future. The properties of the point absorber in the simulation environment can be found in Table 1. The hydrodynamic coefficients have been computed in the panel-code WAMIT. However, the same matrices as in [37] have been used for the state-space approximation of the radiation convolution integral. The problem has been programmed in the Python/Pytorch framework for the SAC controller.

In addition, the robust and computationally efficient MPC strategy described in [6,8] is selected to initialise the training and benchmark the results of the DRL scheme. The method is implemented in the MATLAB/Simulink framework using the quadratic-programming solver quadprog, after discretising the matrix equation in (6) with a zero-order hold. Similarly to [6,8], the future wave elevation is assumed here to be known exactly, as prediction methods with 90% accuracy up to 10 s into the future have been proposed [41,43].

For both control methods, a first-order Euler scheme is used for the time integration of the simulations with a time step of 0.01 s.

**Table 1.** Properties of the spherical point absorber in the simulation environment.

| Property | Value |
|---|---|
| Buoy diameter [m] | 5 |
| Buoy mass [kg] | 32.725 |
| Buoy resonant period [s] | 3.17 |
| Water depth [m] | $\infty$ |
| $\rho$ [kg/m$^3$] | 1000 |
| $g$ [m/s$^2$] | 9.81 |

Typical ocean waves have an energy wave period approximately ranging from 5 s to 20 s [44]. Hence, it is clear that the selected point absorber will need significant control effort to extract energy from realistic ocean waves, since its resonant period is lower, as shown in Table 1. In this work, the peak wave period is considered to range from 4 s to 10 s, which is expected to be realistic for the small point absorber. Additionally, as the simulation environment is based here on a linear model, only small wave amplitudes up to 1 m are analysed. As a result, no constraints on either the buoy displacement or the PTO force are set on the MPC controller. The penalty on the slew rate is expected to be sufficient for the achievement of a suitable WEC response, by setting $r_{MPC} = 10^{-5}$ to ensure convexity according to [8].

Zhong and Yeung [6] have shown that, in regular waves, the mean absorbed power does not increase with time horizon duration after the horizon is one wave period long. Hence, here we set $H = 10$ s, since it corresponds to the longest wave period that is analysed and corresponds to realistic prediction timeframes [41,43]. The control time step length is set to $\delta t = 0.2$ s.

A maximum PTO force $f_{max} = 10^5$ N and displacement $z_{max} = 2.5$ m are selected. Additionally, the weights of (10) are set to $w_u = 10^{-5}$ and $w_z = 10^6$. The hyperparameters used for the SAC agent are the same as in [34] and are reported in Table 2 for greater clarity.

**Table 2.** Hyperparameters of the SAC agent.

| Parameter | Value |
|---|---|
| optimizer | Adam |
| learning rate | $3 \times 10^{-4}$ |
| discount factor | 0.99 |
| replay buffer size | $10^6$ |
| number of hidden layers (all networks) | 2 |
| number of hidden units per layer | 256 |
| number of samples per minibatch | 256 |
| entropy target | $-1$ |
| activation function | ReLU |
| target smoothing coefficient | 0.005 |
| target update interval | 2 |
| gradient steps | 1 |

*4.2. Results in Irregular Waves*

To generate sufficient data samples for the training of the actor DNN, 28 wave traces of irregular waves lasting 15 min each are produced, with the significant wave height ranging from 0.5 m to 2 m in steps of 0.5 m and the peak wave period from 4 s to 10 s in steps of 1 s. A Bretschneider spectrum is used [44]. The controller is started only after 100 s from the start of the wave trace to avoid numerical instabilities during the initial transient. The wave trace is logged after an additional 50 s for 900 s.

### 4.2.1. Training

The sampled data is used to initialise the experience replay buffer of the SAC agent. Each episode consists of a randomly initialised wave trace whose significant wave height and peak wave period are randomly selected in the 1.5–2 m and 6–8 s range, respectively. The wave trace lasts for 200 s and is initialised with no control force for the first 100 s to avoid numerical instabilities. Hence, each episode lasts a total of 2001 steps for the selected control time step of 0.2 s. The same control time step is selected for the DRL controller. To ensure the robustness of the algorithm, the agent is trained with five different seed values to the random number generator.

As can be seen in Figure 4, after the initialisation with the samples collected by the MPC controller, the agent learns a policy to maximise the expected reward after approximately 50 episodes (or approximately $10^5$ steps). Note that in Figure 4, the total reward per episode is highly dependent on the randomly selected significant wave height and peak period; hence, large variations are possible even after training, due to the different level of energy in the waves.

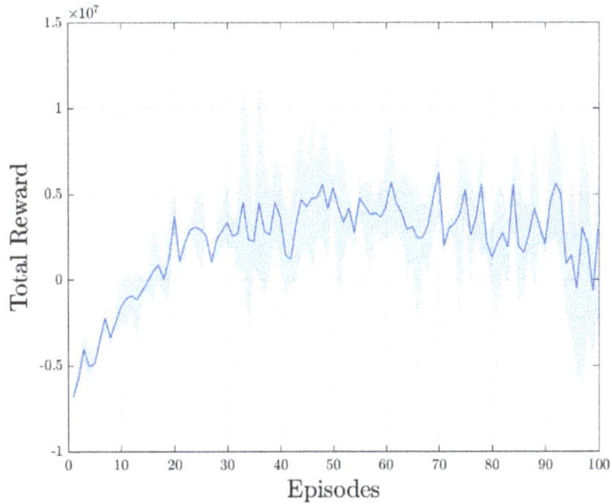

**Figure 4.** Total reward per episode during training.

The convergence time corresponds to approximately 5000 s of experience in addition to the previous 25,200 s with MPC control for a total of approximately 8.4 h. This is a really short time over the life time of the WEC and provides confidence in the controller being able to deliver adaptive control in practical implementations. However, the negative absorbed powers shown during the first episodes are highly worrying. At the start of learning, the agent is preferring random actions to ensure exploration. However, during exploration the device, and in particular the PTO, may fail. Therefore, in the future, a fixed entropy temperature may reduce exploration at the start and thus its associated risks if the controller is already initialised with data from a robust controller. This solution is however likely to slow down the training time.

### 4.2.2. Comparison between SAC and MPC

To assess the performance of the DRL control, MPC and SAC are tested in unseen waves. The traces have a Bretschneider spectrum in the same range of significant wave height and peak wave period, but different seed numbers to the random number generator from the training set. They last 1050 s, with the controller initialised after 100 s and the averaging to compute the mean power started after a further 50 s.

The mean useful or net absorbed power is shown in the dashed lines in Figure 5 for MPC. The reactive power, $P_{rea}$, is defined as the power transferred from the PTO to the point absorber, whilst the active or resistive power, $P_{act}$, is the power transferred from the absorber to the PTO. The net useful power is thus $P_u = P_{act} - P_{rea}$. For the MPC, the extracted power at higher wave periods is curtailed by the penalty on the slew rate. In Figure 6, the ratio of the reactive and active power for the MPC can be seen in the dashed lines. Reactive power is primarily used to speed the WEC up in shorter waves, i.e., when the wave period is shorter than the resonant period, whereas passive damping can be used to slow the device down for longer wave periods. The steady increase in the ratio of the reactive and active power for higher wave periods for the MPC is thus unexpected. The greater control effort the further from the resonance period (3.17 s for the point absorber) is however visible also in [6] and can be explained with the decrease of the absolute value of the active power for longer waves. Furthermore, a comparison with the case studies in [6,8] shows that the selected value of $r_{MPC}$ is providing a stronger influence in this example.

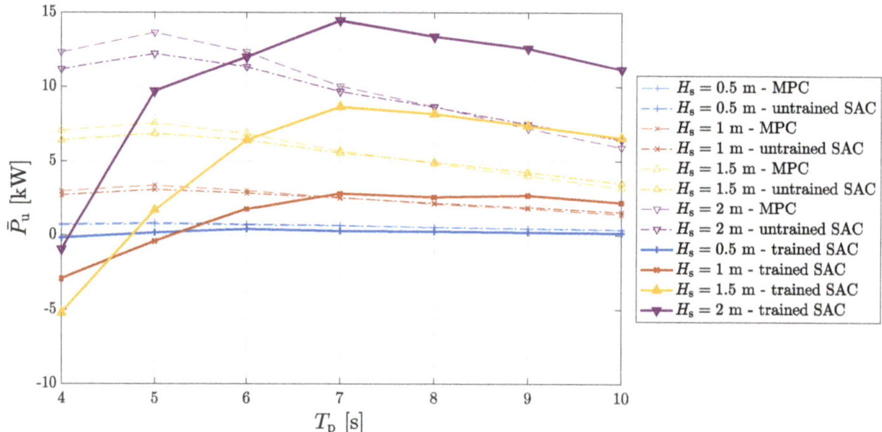

**Figure 5.** Variation with peak wave period of the mean useful power absorbed by the heaving sphere for the considered range of significant wave height.

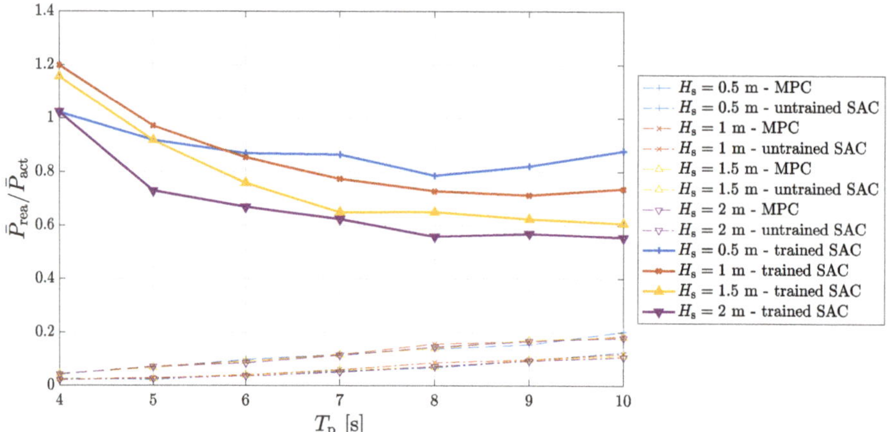

**Figure 6.** Variation with peak wave period of the ratio of the mean reactive and active power absorbed by the heaving sphere for the considered range of significant wave height.

Figures 5 and 6 also display the performance of the SAC agent before and after training. The dot-dash lines correspond to the SAC agent before training, with the entropy set to zero. In this case, the agent replicates closely the MPC behaviour, even though there are differences in the actual time-domain response. After training, the SAC agent (shown with thick continuous lines) improves the energy absorption for the higher peak wave period values ($T_p > 6$ s) and the higher significant wave height values ($T_p > 1$ m). These ranges correspond with the ranges used during training and show poor generalisation ability. The negative mean absorbed power values for the lower periods close to the WEC's resonant period are particularly worrying. From Figure 6, it is clear that the main cause for this behaviour is the aggressive policy that the SAC agents selects. The large flows of reactive power are useful for periods smaller than the resonant period, i.e., in shorter waves, but unhelpful close to the resonant period or for long wave periods, where damping is more useful to slow the WEC down. The problem may be caused by the low resonant period of the point absorber. A larger device whose resonant period is within the typical ocean waves period range should be selected in the future to assess the behaviour of the controller for both short and long waves.

In Figure 7, the magnitude of the maximum heave displacement and PTO force can be seen. Although the SAC algorithm presents higher displacements than MPC, the maximum value of 2.5 m is not exceeded. This hints at the efficacy of the discontinuous penalty term in (10). However, designing a method to guarantee the constraint handling for the displacement is critical for the DRL controller to find an industrial application in the future. The aggressive behaviour of the SAC agent is further underlined in Figure 7b, where the peak PTO force is hit in all sea states.

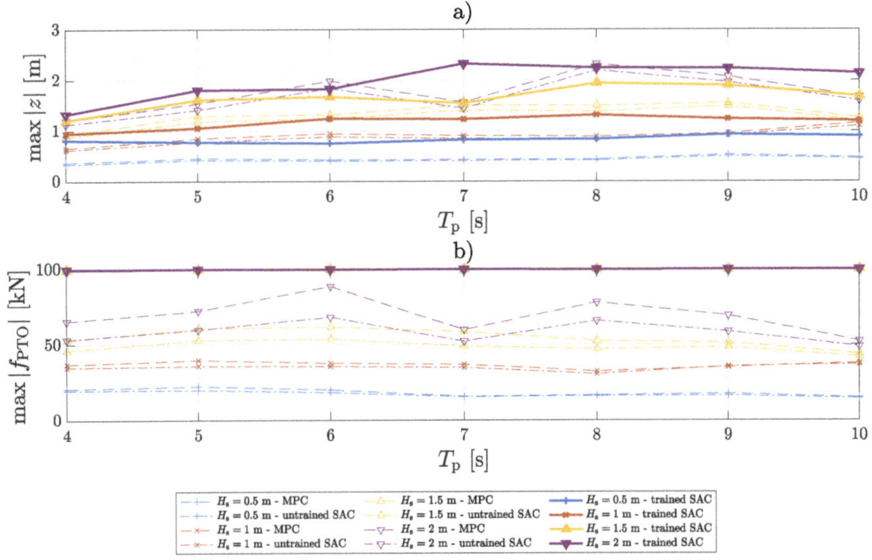

**Figure 7.** Variation with peak wave period of the maximum absolute displacement (**a**) and PTO force (**b**) for the heaving sphere for the considered range of significant wave height.

The response of the MPC and SAC algorithms in the time domain can be seen in Figure 8. The figure shows an extract of one of the simulations used to verify the performance of the SAC scheme against MPC ($H_s = 2$ m and $T_p = 6$ s). From the figure, it is clear that the SAC converges onto an aggressive bang-bang control policy.

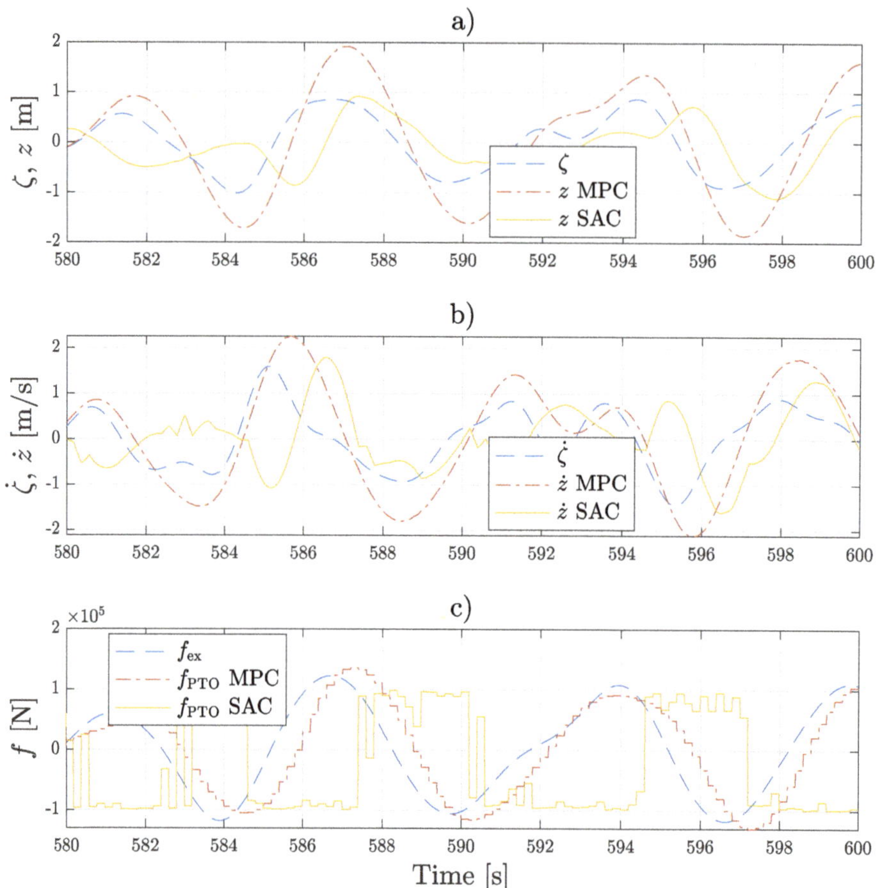

**Figure 8.** Time variation of the wave elevation and float heave displacement (**a**), wave and heave velocity (**b**) and wave excitation and PTO force (**c**) for the simulation of the WEC in an irregular wave trace with $H_s = 2$ m and $T_p = 6$ s.

The aggressiveness of the controller can be reduced by increasing the penalty on the PTO force (through $w_u$). The poor ability of the SAC agent to generalise to unseen wave conditions is problematic and symptomatic of a possibly over-simplistic selection of the state-space. In [34], the information from a number of past time steps is captured in the state-space to ensure convergence for the control of a walking robot. A similar approach will be needed for the control of a WEC to capture the oscillatory nature of gravity waves, similar to MPC. Furthermore, it is clear that the experience replay buffer should include data samples from a broad range of sea states, in particular with regards to period both below and above the resonance period of the device. Currently, the memory buffer is updated with new samples by removing the oldest sample if the memory is full. This technique will be changed by binning the data by wave period and height and ensuring a minimum number of samples per bin.

Table 3 shows the computational time required to train the SAC controller (over 100 episodes) and to run a simulation of the WEC lasting 1050 s using the MPC and trained SAC schemes. The mean from the 28 simulations employed to compare the two strategies is used. Note that the first 150 s are needed to initialise the WEC dynamics and power averaging and that the control time step is 0.2 s for both algorithms. Hence, there are 4501 control time steps per simulation, leading to the values for the computational time per time step shown in Table 3. The simulations are run on a laptop with an Intel 5, 2.3 GHz, dual-core processor and 16 GB RAM.

**Table 3.** Training time (if applicable), mean total simulation time and time per control time step for the simulations used to verify and compare the MPC and SAC control algorithms.

| Scheme | Training Time [s] | Total Simulation Time [s] | Time Per Control Time Step [s] |
|---|---|---|---|
| MPC | - | 11.798 | $2.6 \times 10^{-3}$ |
| SAC | 869 | 2.607 | $5.8 \times 10^{-4}$ |

As can be seen in Table 3, the SAC algorithm requires approximately 15 min to train over 100 episodes for analysed point absorber. The large computational time prevents an on-line application, although training can happen regularly off-line in practice, once significantly large new batches of data are collected. Conversely, once trained, the computational effort is 40 times lower than the simulation control time step, thus enabling a real-time implementation with ease. Additionally, the computational effort associated with SAC is one order of magnitude smaller than for linear MPC. In fact, the computational time per control time step shown in Table 3 is overly conservative for SAC, as it includes the overhead from the dynamic simulation in Python. Conversely, the linear MPC code is implemented in a very efficient MALTAB/Simulink script with C-coded S-functions. Therefore, a gain in performance as high as one additional order of magnitude is expected from compiled solutions if the system has to be implemented on an actual WEC [45].

For reproducibility, the results of the SAC algorithm in the unseen test wave traces can be accessed on Github[1], including the wave elevation, vertical velocity and excitation force.

## 5. Conclusions

In this article, an established convex MPC has been used to generate observations for a heaving point absorber in a range of irregular waves in a linear simulation environment. The samples have been used to initialise a DRL agent, which learns an optimal policy from direct interactions with with the environment for the maximisation of the energy absorption. By being off-line and off-policy, the SAC algorithm enables the training to be decoupled from deployment, thus shifting the computational effort on the training. This is a fundamental trait, as the control of large groups of WECs in the future to achieve economies of scale is reliant on having an effective, real-time centralised strategy.

The DRL control improves the energy absorption of the point absorber over convex MPC for wave periods higher than the resonant period of the device, whilst meeting the displacement and force constraints. This is achieved by adopting a more aggressive policy with higher slew rate. However, poorer performance is shown for lower wave height and period values. These problems will be addressed by reformulating the state-space, updating the reward function and the sampling of data for the experience replay. Additionally, the exploration will be reduced from the start of training to prevent the controller from taking actions that damage the PTO. Furthermore, the DRL controller will be tested in a simulation environment inclusive of nonlinear effects, e.g., nonlinear static and dynamic Froude–Krylov forces as in [42] and viscous drag. A sensitivity analysis will be run to assess the impact of modelling errors on the performance of the DRL and MPC algorithms.

**Author Contributions:** Conceptualization, E.A.; methodology, E.A.; software, E.A. and S.H. (hydrodynamics); validation, E.A.; formal analysis, E.A.; investigation, E.A.; resources, E.A. and G.G.P.; data curation, E.A. and S.H.; writing—original draft preparation, E.A.; writing—review and editing, G.G.P., M.A. and G.T.; visualization, E.A. All authors have read and agreed to the published version of the manuscript.

**Funding:** This research received no external funding.

---

[1]  https://github.com/enricoande/Results-for-jmse-967614

*J. Mar. Sci. Eng.* **2020**, *8*, 845


**Acknowledgments:** The authors would like to acknowledge the help provided by Giuseppe Giorgi in identifying the Special Issue.

**Conflicts of Interest:** The authors declare no conflict of interest.

## Abbreviations

The following abbreviations are used in this manuscript:

| | |
|---|---|
| DNN | Deep Neural Network |
| DRL | Deep Reinforcement Learning |
| LCoE | Levellised Cost of Energy |
| MPC | Model Predictive Control |
| PTO | Power Take-Off |
| RL | Reinforcement Learning |
| SAC | Soft Actor-Critic |
| WEC | Wave Energy Converter |

## References

1. Kempener, R.; Neumann, F. *Wave Energy: Technology Brief 4*; International Renewable Energy Agency Technical Report; International Renewable Energy Agency: Abu Dhabi, UAE, 2014.

2. Sgurr Control; Quoceant. *Control Requirements for Wave Energy Converters Landscaping Study: Final Report*; Technical report; Wave Energy Scotland: Inverness, Scotland, 2016.

3. Luis Villate, J.; Ruiz-Minguela, P.; Berque, J.; Pirttimaa, L.; Cagney, D.; Cochrane, C.; Jeffrey, H. *Strategic Research and Innovation Agenda for Ocean Energy*; Technical report; ETIPOCEAN: Bruxelles, Belgium, 2020.

4. Faedo, N.; Olaya, S.; Ringwood, J.V. Optimal control, MPC and MPC-like algorithms for wave energy systems: An overview. *IFAC J. Syst. Control.* **2017**. [CrossRef]

5. Li, G.; Belmont, M.R. Model predictive control of sea wave energy converters—Part I: A convex approach for the case of a single device. *Renew. Energy* **2014**, *69*, 453–463. [CrossRef]

6. Zhong, Q.; Yeung, R.W. An efficient convex formulation for model predictive control on wave-energy converters. In Proceedings of the 36th International Conference on Ocean, Offshore and Arctic Engineering, Trondheim, Norway, 25–30 June 2017. [CrossRef]

7. Li, G.; Belmont, M.R. Model predictive control of sea wave energy converters—Part II: The case of an array of devices. *Renew. Energy* **2014**, *68*, 540–549. [CrossRef]

8. Zhong, Q.; Yeung, R.W. Model-Predictive Control Strategy for an Array of Wave-Energy Converters. *J. Mar. Sci. Appl.* **2019**, *18*, 26–37. [CrossRef]

9. Giorgi, G.; Ringwood, J.V. Nonlinear Froude-Krylov and viscous drag representations for wave energy converters in the computation/fidelity continuum. *Ocean Eng.* **2017**, *141*, 164–175. [CrossRef]

10. Richter, M.; Magana, M.E.; Sawodny, O.; Brekken, T.K.A. Nonlinear Model Predictive Control of a Point Absorber Wave Energy Converter. *IEEE Trans. Sustain. Energy* **2013**, *4*, 118–126. [CrossRef]

11. Li, G. Nonlinear model predictive control of a wave energy converter based on differential flatness parameterisation. *Int. J. Control* **2017**, *90*, 68–77. [CrossRef]

12. Son, D.; Yeung, R.W. Optimizing ocean-wave energy extraction of a dual coaxial-cylinder WEC using nonlinear model predictive control. *Appl. Energy* **2017**, *187*, 746–757. [CrossRef]

13. Oetinger, D.; Magaña, M.E.; Sawodny, O. Centralised model predictive controller design for wave energy converter arrays. *IET Renew. Power Gener.* **2015**, *9*, 142–153. [CrossRef]

14. Ringwood, J.V.; Bacelli, G.; Fusco, F. Energy-Maximizing Control of Wave-Energy Converters: The Development of Control System Technology to Optimize Their Operation. *IEEE Control Syst. Mag.* **2014**, *34*, 30–55. [CrossRef]

15. Korde, U.A.; Ringwood, J.V. *Hydrodynamic Control of Wave Energy Devices*; Cambridge University Press: Cambridge, UK, 2016.

16. Fusco, F.; Ringwood, J.V. A simple and effective real-time controller for wave energy converters. *IEEE Trans. Sustain. Energy* **2013**, *4*, 21–30. [CrossRef]

17. Gaspar, J.F.; Kamarlouei, M.; Sinha, A.; Xu, H.; Calvário, M.; Faÿ, F.X.; Robles, E.; Soares, C.G. Speed control of oil-hydraulic power take-off system for oscillating body type wave energy converters. *Renew. Energy* **2016**, *97*, 769–783. [CrossRef]

18. Valério, D.; Mendes, M.J.G.C.; Beirão, P.; Sá da Costa, J. Identification and control of the AWS using neural network models. *Appl. Ocean Res.* **2008**, *30*, 178–188, [CrossRef]

19. Li, L.; Yuan, Z.; Gao, Y. Maximization of energy absorption for a wave energy converter using the deep machine learning. *Energy* **2018**, *165*, 340–349. [CrossRef]

20. Li, L.; Gao, Z.; Yuan, Z.M. On the sensitivity and uncertainty of wave energy conversion with an artificial neural-network-based controller. *Ocean Eng.* **2019**, *183*, 282–293. [CrossRef]

21. Anderlini, E.; Forehand, D.I.; Bannon, E.; Abusara, M. Reactive control of a wave energy converter using artificial neural networks. *Int. J. Mar. Energy* **2017**, *19*, 207–220. [CrossRef]

22. Thomas, S.; Giassi, M.; Eriksson, M.; Göteman, M.; Isberg, J.; Ransley, E.; Hann, M.; Engström, J. A Model Free Control Based on Machine Learning for Energy Converters in an Array. *Big Data Cogn. Comput.* **2018**, *2*, 36. [CrossRef]

23. Tri, N.M.; Truong, D.Q.; Thinh, D.H.; Binh, P.C.; Dung, D.T.; Lee, S.; Park, H.G.; Ahn, K.K. A novel control method to maximize the energy-harvesting capability of an adjustable slope angle wave energy converter. *Renew. Energy* **2016**, *97*, 518–531. [CrossRef]

24. Na, J.; Li, G.; Wang, B.; Herrmann, G.; Zhan, S. Robust Optimal Control of Wave Energy Converters Based on Adaptive Dynamic Programming. *IEEE Trans. Sustain. Energy* **2019**, *10*, 961–970. [CrossRef]

25. Na, J.; Wang, B.; Li, G.; Zhan, S.; He, W. Nonlinear constrained optimal control of wave energy converters with adaptive dynamic programming. *IEEE Trans. Ind. Electron.* **2019**, *66*, 7904–7915. [CrossRef]

26. Zhan, S.; Na, J.; Li, G. Nonlinear Noncausal Optimal Control of Wave Energy Converters via Approximate Dynamic Programming. *IEEE Trans. Ind. Infor.* **2019**, *15*, 6070–6079. [CrossRef]

27. Kamthe, S.; Deisenroth, M.P. Data-Efficient Reinforcement Learning with Probabilistic Model Predictive Control. In Proceedings of the Machine Learning Research, Lanzarote, Spain, 9–11 April 2018; Volume 84, pp. 1701–1710.

28. Nagabandi, A.; Kahn, G.; Fearing, R.S.; Levine, S. Neural Network Dynamics for Model-Based Deep Reinforcement Learning with Model-Free Fine-Tuning. *arXiv* **2017**, arXiv:1708.02596v2.

29. Anderlini, E.; Forehand, D.I.M.; Stansell, P.; Xiao, Q.; Abusara, M. Control of a Point Absorber using Reinforcement Learning. *IEEE Trans. Sustain. Energy* **2016**, *7*, 1681–1690. [CrossRef]

30. Anderlini, E.; Forehand, D.I.; Bannon, E.; Abusara, M. Control of a Realistic Wave Energy Converter Model Using Least-Squares Policy Iteration. *IEEE Trans. Sustain. Energy* **2017**, *8*, 1618–1628. [CrossRef]

31. Anderlini, E.; Forehand, D.I.; Bannon, E.; Xiao, Q.; Abusara, M. Reactive control of a two-body point absorber using reinforcement learning. *Ocean Eng.* **2018**, *148*, 650–658. [CrossRef]

32. Anderlini, E.; Forehand, D.I.; Bannon, E.; Abusara, M. Constraints Implementation in the Application of Reinforcement Learning to the Reactive Control of a Point Absorber. In Proceedings of the 36th International Conference on Ocean, Offshore and Arctic Engineering, Trondheim, Norway, 25–30 June 2017. [CrossRef]

33. LeCun, Y.; Bengio, Y.; Hinton, G. Deep learning. *Nature* **2015**, *521*, 436–444. [CrossRef]

34. Haarnoja, T.; Zhou, A.; Hartikainen, K.; Tucker, G.; Ha, S.; Tan, J.; Kumar, V.; Zhu, H.; Gupta, A.; Abbeel, P.; et al. Soft Actor-Critic Algorithms and Applications. *arXiv* **2018**, arXiv:1812.05905.

35. Falnes, J. *Ocean Waves and Oscillating Systems*, paperback ed.; Cambridge University Press: Cambridge, UK, 2005. [CrossRef]

36. Cummins, W.E. The impulse response function and ship motions. *Schiffstechnik* **1962**, *47*, 101–109.

37. Faedo, N.; Peña-Sanchez, Y.; Ringwood, J.V. Finite-order hydrodynamic model determination for wave energy applications using moment-matching. *Ocean Eng.* **2018**, *163*, 251–263. [CrossRef]

38. Sutton, R.S.; Barto, A.G. *Reinforcement Learning*, hardcover ed.; MIT Press: Cambridge, MA, USA, 1998; p. 344.

39. Lillicrap, T.P.; Hunt, J.J.; Pritzel, A.; Heess, N.; Erez, T.; Tassa, Y.; Silver, D.; Wierstra, D. Continuous control with deep reinforcement learning. In Proceedings of the International Conference on Learning Representations, San Juan, PR, USA, 2–4 May 2016. [CrossRef]

40. Haarnoja, T.; Zhou, A.; Abbeel, P.; Levine, S. Soft Actor-Critic: Off-Policy Maximum Entropy Deep Reinforcement Learning with a Stochastic Actor. In Proceedings of the 35th International Conference on Machine Learning, Stockholm, Sweden, 10–15 July 2018.

41. Fusco, F.; Ringwood, J. Short-Term Wave Forecasting for time-domain Control of Wave Energy Converters. *IEEE Trans. Sustain. Energy* **2010**, *1*, 99–106. [CrossRef]
42. Giorgi, G.; Ringwood, J.V. Computationally efficient nonlinear Froude–Krylov force calculations for heaving axisymmetric wave energy point absorbers. *J. Ocean. Eng. Mar. Energy* **2017**, *3*, 21–33. [CrossRef]
43. Paparella, F.; Monk, K.; Winands, V.; Lopes, M.F.; Conley, D.; Ringwood, J.V. Up-wave and autoregressive methods for short-term wave forecasting for an oscillating water column. *IEEE Trans. Sustain. Energy* **2015**, *6*, 171–178. [CrossRef]
44. Holthuijsen, L.H. *Waves in Oceanic and Coastal Waters*; Cambridge University Press: Cambridge, UK, 2007. [CrossRef]
45. Fourment, M.; Gillings, M.R. A comparison of common programming languages used in bioinformatics. *BMC Bioinform.* **2008**, *9*, 82. [CrossRef] [PubMed]

Journal of
*Marine Science and Engineering*

*Article*

# Spectral Control of Wave Energy Converters with Non-Ideal Power Take-off Systems

Alexis Mérigaud * and Paolino Tona

IFP Energies Nouvelles, 1-4 Avenue du Bois Préau, 92852 Rueil-Malmaison, France; paolino.tona@ifpen.fr
* Correspondence: alexis.merigaud@ifpen.fr

Received: 7 October 2020; Accepted: 21 October 2020; Published: 28 October 2020

**Abstract:** Spectral control is an accurate and computationally efficient approach to power-maximising control of wave energy converters (WECs). This work investigates spectral control calculations with explicit derivative computation, applied to WECs with non-ideal power take-off (PTO) systems characterised by an efficiency factor smaller than unity. To ensure the computational efficiency of the spectral control approach, it is proposed in this work to approximate the discontinuous efficiency function by means of a smooth function. A non-ideal efficiency function implies that the cost function is non-quadratic, which requires a slight generalisation of the derivative-based spectral control approach, initially introduced for quadratic cost functions. This generalisation is derived here in some detail given its practical interest. Two application case studies are considered: the Wavestar scale model, employed for the WEC control competition (WECCCOMP), and the 3rd reference model (RM3) two-body heaving point absorber. In both cases, with the approximate efficiency function, the spectral approach calculates WEC trajectory and control force solutions, for which the mean electrical power is shown to lie within a few percent of the true optimal electrical power. Regarding the effect of a non-ideal PTO efficiency upon achievable power production, and concerning heaving point-absorbers, the results obtained are significantly less pessimistic than those of previous studies: the power achieved lies within 80–95% of that obtained by simply applying the efficiency factor to the optimal power with ideal PTO.

**Keywords:** wave energy converters; spectral control; optimal control; power take-off

---

## 1. Introduction

Power-maximising control is a promising path to improve the economic competitiveness of WECs [1]. Various control methods have been proposed, of which the most basic are simple parametric control strategies, whereby one or two scalar control parameters are adjusted on a sea state-by-sea state basis. Recent years have witnessed a growing interest for model predictive control (MPC) strategies [2], whereby an optimisation problem is solved in real-time to update a prescribed control input, or a reference trajectory to be followed by the WEC. MPC strategies take profit from wave excitation predictions. They can, in theory, handle nonlinear WEC models, through the use of nonlinear optimisation algorithms, as well as constraints on the WEC dynamics, control force or instantaneous power.

First introduced in [3] for WEC control applications, spectral and pseudo-spectral MPC methods discretise the wave inputs and optimisation variables by means of basis functions, typically Fourier [4,5] (when a periodic wave input is considered), or Chebyshev-like [6] (to handle directly the non-periodic wave signal, seen within the receding-horizon control window). In terms of signal approximation (i.e., when it comes to approximate a given function by means of a finite number of "spectral" basis functions), spectral techniques are said to have *superlinear convergence* properties [7], which means that the approximation error decreases more than linearly with the number of basis functions employed, provided that the function approximated is continuous (with a geometric convergence rate for functions with a finite degree of smoothness, and an exponential convergence rate for infinitely-smooth functions such as linear functions). As a consequence, spectral methods require a small number of coefficients to describe the control inputs or WEC dynamics with a prescribed accuracy, thus reducing the number of optimisation variables, and the associated computation cost. Spectral methods can thus handle relatively long control horizons. In addition, the "aggressiveness" of the control solution, i.e., how abruptly the solution oscillates, can be tuned, by adjusting the number of basis functions.

In [5], a computationally efficient spectral control formulation for nonlinear, 1-DoF WEC models is introduced in which, using the dynamical equation, the control variable is eliminated in the expression of the objective function, thus resulting in a problem where the only variables to optimise are those, corresponding to the WEC motion. The approach in [5], combined with the explicit computation of the first- and second-order derivatives of the objective function, is shown to be orders of magnitude faster than usual spectral control formulations, whereby the dynamical equations are considered as a set of equality constraints. Although the proposed approach is based on Fourier basis functions, it has been successfully implemented in a receding-horizon set-up [8,9]. Also note that combining spectral control (for optimal reference generation) with tracking control entails a certain degree of robustness to modelling errors, in contrast to direct computation of the optimal control force, which requires accurate modelling of all the WEC dynamics. Indeed, the controller presented in [9] is shown to be little affected by tracking loop imperfection as well as by estimation and forecast errors.

In all references previously mentioned, like in the vast majority of the wave energy control literature, only hydrodynamic power absorption is considered. From an optimal control perspective, the objective function, here denoted as $f$, is generally the raw mechanical power absorbed by the PTO system [2]. For a WEC with one mechanical degree of freedom, the raw absorbed power is written as follows,

$$f(x,u) = -\frac{1}{T}\int_{t=0}^{T} \dot{x}u\,dt \tag{1}$$

where $T$ is the length of the control horizon, $x$ denotes the WEC generalised position and $u$ the PTO force (which is the control input).

However, in practical applications, the PTO system is not ideal, so that the electrical power, $P_e$, is not equal to the absorbed mechanical power, $P_a = -u\dot{x}$, as highlighted by a number of authors [10–17]. More specifically, when power is mechanically absorbed from the waves, the electrical power $P_e$, actually available at the end of the mechanical-to-electrical conversion stage, is smaller than the absorbed power $P_a$, because of the losses occurring in the conversion stage ($0 \leq P_e \leq P_a$). Conversely, if a reactive control strategy is adopted [1], at times some power must be returned from the electromechanical system back into the ocean ($P_a \leq 0$). At those instants, because of the losses in the PTO system, the electrical power provided by the grid to the PTO system must be greater than $|P_a|$, so that $P_e \leq P_a \leq 0$. PTO losses should be taken into account when designing the WEC control strategy [10–17]. In fact, a reactive control strategy, obtained without modelling losses in the objective function, can yield negative average generated power when applied to the true system, where losses occur (see, for example, in [13]).

Therefore, it is of high practical interest to extend the methodology proposed in [5] to more general objective functions of the form

$$f(x,u) = \frac{1}{T}\int_{t=0}^{T} \alpha(x,\dot{x},u)\,dt \tag{2}$$

where $\alpha$ is some scalar function of $x$, $\dot{x}$ and $u$ (while the control calculations presented in [5] can only deal with cost functions in the form $-u\dot{x}$, quadratic in the problem variables). $P_e$, being modelled as a function of $P_a = -u\dot{x}$, is a special case of the formulation in Equation (2). Such a formulation can also include, for example, quadratic terms introduced to penalise the WEC excursions or control effort.

This work, like that in [5], is concerned with the calculation of optimal control solutions in an "off-line" fashion (as opposed to receding-horizon implementations such as in [6,8,9]), taking into account the totality of a relatively long, simulated irregular wave input signal comprising a few tens of wave periods. Therefore, the insight provided, regarding the effect of a non-ideal PTO system onto optimal control results, is reasonably general, and remains free of the many effects, which interact when considering a particular receding-horizon implementation [9]. Furthermore, the optimal control solution, obtained off-line, and the corresponding generated power, provide a reference to assess the performance of real-time, receding-horizon control algorithms (which cannot, in theory, outperform the optimal solution obtained off-line). The present study is restricted to WECs with one degree of freedom.

The rest of this paper is organised as follows.

- Section 2 revisits the method presented in [5], for an objective function of the form of Equation (2).
- Section 3 introduces the simple non-ideal PTO model retained in this work, and two numerical case studies: the Wavestar model used in the WECCCOMP challenge [18], and the RM3 two-body heaving point absorber [19].
- For both case studies, numerical results are shown and discussed in Section 4;
- Finally, conclusions are presented in Section 5.

## 2. An Efficient Spectral Control Method for a 1-DoF, Non-Linear WEC Model

### 2.1. WEC Dynamical Model

Consider a WEC, modelled with one degree of freedom, denoted $x$. Newton's second law, which describes the WEC motion, typically takes the following mathematical form (see in [5], and detailed justifications in Chapter 4 of the work in [20]):

$$\mathcal{L}\{x\} + n(x, \dot{x}) = u + e \tag{3}$$

where the dependence in the time variable $t$ has been omitted to alleviate notations, and

- $\mathcal{L}$ represents a linear integro-differential operator, containing some of, or all the terms of the well-known Cummins' equation [21]:

$$\mathcal{L}\{x\}(t) := (I + I_\infty)\ddot{x}(t) + C\dot{x}(t) + \int_{\tau=0}^{\infty} k_r(\tau)\dot{x}(t - \tau)d\tau + Sx(t) \tag{4}$$

where $I$ represents the system inertia; $I_\infty$ and $k_r$ are the radiation infinite-frequency added inertia and damping convolution kernel, respectively; $C$ models some linear damping term (for example, to take friction effects into account); and $S$ is the hydrostatic restoring force (resulting from the balance between gravity and Archimede's force);
- $n$ represents some nonlinear function, with the aim of modelling nonlinear dynamical effects, not accounted for by a linear modelling approach. If $n$ includes effects, usually modelled linearly in Cummins' Equation (4), then the corresponding linear coefficients in Equation (4) can be assumed equal to zero;
- $e(t)$ is the wave excitation force; and
- $u(t)$ is the control force, exerted by the PTO system onto the WEC's moving part.

The operator $\mathcal{L}$ is represented by a frequency-response mapping $z(\omega) = S - \omega^2(I + A_r(\omega)) - j\omega(C + B_r(\omega))$, where the radiation frequency-dependent added mass $A_r(\omega)$ and damping $B_r(\omega)$ are related to $I_\infty$ and $k_r$ through Ogilvie's relation [22].

### 2.2. Spectral Control Problem Formulation

Consider an optimal control problem of the following form,

$$\begin{cases} \max\limits_{x,u} f(x,u) \\ \text{s.t. } \mathcal{L}\{x\} + n(x,\dot{x}) = u + e \end{cases} \tag{5}$$

where $f(x,u) := \frac{1}{T}\int_{t=0}^{T}\alpha(x,\dot{x},u)dt$; $T$ is the control time horizon; the equality constraint expresses the fact that the dynamical Equation (3) must be satisfied at every instant; and the function $\alpha$, as mentioned in the introduction, represents the "instantaneous" objective function. Given the particular form of the dynamical equation, $u$ can be expressed as a function of the other variables, so that Problem (5) can be recast as follows,

$$\max\limits_{x} f(x) \tag{6}$$

where $f(x) = \frac{1}{T}\int_{t=0}^{T}\alpha(x,\dot{x},\mathcal{L}\{x\} + n(x,\dot{x}) - e)dt$, while the variable $u$, as well as the equality constraint, are eliminated, so that only the device trajectory is now optimised.

Assume a periodic, polychromatic excitation signal with period $T$, of the form

$$e(t) = \frac{1}{\sqrt{2}}\hat{e}_1 + \sum_{n=1}^{N}\hat{e}_{2n}\cos(\omega_n t) + \hat{e}_{2n+1}\sin(\omega_n t) \tag{7}$$

where the frequencies $\omega_n$ are defined harmonically with $\Delta\omega = \frac{2\pi}{T}$ and $\forall n \in \{1...N\}, \omega_n = n\Delta\omega$. In the following, consistently with a periodic excitation signal, solutions $x$ are searched among periodic signals with period $T$, and the WEC motion is assumed of the form

$$x(t) = \frac{1}{\sqrt{2}}\hat{x}_1 + \sum_{n=1}^{N}\hat{x}_{2n}\cos(\omega_n t) + \hat{x}_{2n+1}\sin(\omega_n t) \tag{8}$$

Note that, in Equations (7) and (8), the number $N$ of harmonics being formally identical does not imply any loss of generality: for example, the number of harmonics in $x$ can be considered larger than that in $e$, by setting to zero higher frequency components of $e$. In the extreme case of a zero-mean, monochromatic wave input, all components $\hat{e}_n$, except for $\hat{e}_2$ and $\hat{e}_3$, are zero, while the solution $x$ can have an arbitrarily large number of components $N$.

The next step consists of discretising the integral in objective function, using a set of $M$ equally-spaced points $t_m$, spanning the interval $[0;T]$, with typically $M \geq 2N + 1$, so that the maximisation problem becomes

$$\max \tilde{f}(x) := \sum_{m=1}^{M}\alpha(t_m) \tag{9}$$

where, for the sake of conciseness, $\alpha(t_m)$ denotes the value of $\alpha$ when its arguments are evaluated at time $t_m$.

Define the matrix $\Phi \in \mathbb{R}^{M\times(2N+1)}$ as

$$\forall i \in \{1...M\}, j \in \{1...N\}, \begin{cases} \Phi_{i,1} = \frac{1}{\sqrt{2}} \\ \Phi_{i,2j} = \cos(\omega_j t_i) \\ \Phi_{i,2j+1} = \sin(\omega_j t_i) \end{cases} \tag{10}$$

and the matrix

$$\Omega = \begin{pmatrix} \Omega_0 & \cdots & 0 \\ \vdots & \ddots & \vdots \\ 0 & \cdots & \Omega_N \end{pmatrix} \tag{11}$$

where $\Omega_0 = 0$ and, $\forall n \geq 1$, $\Omega_n = \begin{pmatrix} 0 & \omega_n \\ -\omega_n & 0 \end{pmatrix}$.

Define vectors $\hat{\mathbf{x}} := \begin{pmatrix} \hat{x}_1 \cdots \hat{x}_N \end{pmatrix}^{\mathsf{T}}$, $\hat{\mathbf{e}} := \begin{pmatrix} \hat{e}_1 \cdots \hat{e}_N \end{pmatrix}^{\mathsf{T}}$, $\mathbf{x} := \begin{pmatrix} x(t_1) \cdots x(t_M) \end{pmatrix}^{\mathsf{T}}$, $\dot{\mathbf{x}} := \begin{pmatrix} \dot{x}(t_1) \cdots \dot{x}(t_M) \end{pmatrix}^{\mathsf{T}}$ and $\mathbf{e} := \begin{pmatrix} e(t_1) \cdots e(t_M) \end{pmatrix}^{\mathsf{T}}$. Then, it is easy to see that $\mathbf{x} = \Phi\hat{\mathbf{x}}$, $\dot{\mathbf{x}} = \Phi\Omega\hat{\mathbf{x}}$ and $\mathbf{e} = \Phi\hat{\mathbf{e}}$.

Furthermore, the linear terms of $\mathcal{L}$ evaluated at times $t_1...t_M$, $l := \begin{pmatrix} \mathcal{L}\{x\}(t_1) & \cdots & \mathcal{L}\{x\}(t_M) \end{pmatrix}^{\mathsf{T}}$, can be calculated as $l = \mathbf{L}\hat{\mathbf{x}}$, where

$$\mathbf{L} = \begin{pmatrix} L_0 & \cdots & 0 \\ \vdots & \ddots & \vdots \\ 0 & \cdots & L_N \end{pmatrix} \tag{12}$$

with blocks $L_0 = \Re\{z(0)\}$ and $\forall n \in \{1...N\}$, $L_n = \begin{pmatrix} \Re\{z(\omega_n)\} & \Im\{z(\omega_n)\} \\ -\Im\{z(\omega_n)\} & \Re\{z(\omega_n)\} \end{pmatrix}$.

Finally, the objective function can be expressed as a function of $\hat{\mathbf{x}}$, as follows,

$$\tilde{f}(\hat{\mathbf{x}}) = \mathbf{1}_{M\times 1}^{\mathsf{T}} \alpha(\Phi\hat{\mathbf{x}}, \Phi\Omega\hat{\mathbf{x}}, \Phi\mathbf{L}\hat{\mathbf{x}} + n(\Phi\hat{\mathbf{x}}, \Phi\Omega\hat{\mathbf{x}}) - \Phi\hat{\mathbf{e}}) \tag{13}$$

where $\mathbf{1}_{M\times 1}$ is the unit vector of size $M \times 1$ and, with a slight abuse of notation, the functions $\alpha$ and $n$ are applied component-wise to the vectors used as arguments, i.e., for example, $g(\mathbf{v}) = \begin{pmatrix} g(\mathbf{v}_1)...g(\mathbf{v}_M) \end{pmatrix}^{\mathsf{T}}$.

Note that Equation (13) differs from the objective function in [5], as $\alpha$ is, in general, not a quadratic function of its arguments. The calculations which follow constitute a useful generalisation of the results in [5].

### 2.3. Explicit Calculation of the Objective Function Derivatives

In this section, it is assumed that $\alpha$ and $n$ are continuous in all their arguments, and admit second-order derivatives. Under such an assumption, the first- and second-order derivatives of $\tilde{f}$, with respect to the components of $\hat{\mathbf{x}}$, can be put to good use within a gradient-based optimisation algorithm [5]. In the following, it is shown how to calculate explicitly those derivatives. The calculations do not present any particular difficulty, but they are not detailed here for the sake of brevity.

A useful convention is first introduced. Given a function $y$ defined on $[0; T]$, the notation $D_y$ represents the $M \times M$ diagonal matrix with diagonal entries $y(t_1), ...y(t_M)$. Then, the vector of 1st-order derivatives is given as

$$\frac{\partial \tilde{f}}{\partial \hat{\mathbf{x}}} = \mathbf{1}_{M\times 1}^{\mathsf{T}} \left[ D_{v_1}\Phi + D_{v_2}\Phi\Omega + D_{v_3}\Phi\mathbf{L} \right] \tag{14}$$

with

$$\begin{cases} v_1 = \frac{\partial \alpha}{\partial x} + \frac{\partial \alpha}{\partial u}\frac{\partial n}{\partial x} \\ v_2 = \frac{\partial \alpha}{\partial \dot{x}} + \frac{\partial \alpha}{\partial u}\frac{\partial n}{\partial \dot{x}} \\ v_3 = \frac{\partial \alpha}{\partial u} \end{cases}$$

The matrix of second-order derivatives is given as

$$\begin{aligned} \frac{\partial^2 \tilde{f}}{\partial \hat{\mathbf{x}}^2} = {}& \Phi^{\mathsf{T}} D_{w_1}\Phi + \Phi^{\mathsf{T}} D_{w_2}\Phi\Omega + \Omega^{\mathsf{T}}\Phi^{\mathsf{T}} D_{w_2}\Phi \\ & + \Omega^{\mathsf{T}}\Phi^{\mathsf{T}} D_{w_3}\Phi\Omega + \Phi^{\mathsf{T}} D_{w_4}\Phi\mathbf{L} + \mathbf{L}^{\mathsf{T}}\Phi^{\mathsf{T}} D_{w_4}\Phi \\ & + \Omega^{\mathsf{T}}\Phi^{\mathsf{T}} D_{w_5}\Phi\mathbf{L} + \mathbf{L}^{\mathsf{T}}\Phi^{\mathsf{T}} D_{w_5}\Phi\Omega \\ & + \mathbf{L}^{\mathsf{T}}\Phi^{\mathsf{T}} D_{w_6}\Phi\mathbf{L} \end{aligned} \tag{15}$$

with:

$$\begin{cases} w_1 = \frac{\partial^2 \alpha}{\partial x^2} + 2\frac{\partial^2 \alpha}{\partial x \partial u}\frac{\partial n}{\partial x} + \frac{\partial^2 \alpha}{\partial u^2}\left(\frac{\partial n}{\partial x}\right)^2 + \frac{\partial \alpha}{\partial u}\frac{\partial^2 n}{\partial x^2} \\ w_2 = \frac{\partial^2 \alpha}{\partial \dot{x} \partial x} + \frac{\partial^2 \alpha}{\partial u^2}\frac{\partial n}{\partial x}\frac{\partial n}{\partial \dot{x}} + \frac{\partial^2 \alpha}{\partial \dot{x} \partial u}\frac{\partial n}{\partial x} + \frac{\partial^2 \alpha}{\partial x \partial u}\frac{\partial n}{\partial \dot{x}} + \frac{\partial \alpha}{\partial u}\frac{\partial^2 n}{\partial x \partial \dot{x}} \\ w_3 = \frac{\partial^2 \alpha}{\partial \dot{x}^2} + 2\frac{\partial^2 \alpha}{\partial \dot{x} \partial u}\frac{\partial n}{\partial \dot{x}} + \frac{\partial^2 \alpha}{\partial u^2}\left(\frac{\partial n}{\partial \dot{x}}\right)^2 + \frac{\partial \alpha}{\partial u}\frac{\partial^2 n}{\partial \dot{x}^2} \\ w_4 = \frac{\partial^2 \alpha}{\partial u \partial x} + \frac{\partial^2 \alpha}{\partial u^2}\frac{\partial n}{\partial x} \\ w_5 = \frac{\partial^2 \alpha}{\partial u \partial \dot{x}} + \frac{\partial^2 \alpha}{\partial u^2}\frac{\partial n}{\partial \dot{x}} \\ w_6 = \frac{\partial^2 \alpha}{\partial u^2} \end{cases}$$

In practice, at each iteration of the optimisation algorithm, the objective function $\tilde{\alpha}$, the nonlinear terms $n$, and their 1st- and 2nd-order derivatives, are evaluated along the trajectory given by $\mathbf{x} = \Phi\hat{\mathbf{x}}$. $\tilde{f}$ is evaluated; $v_1 ... v_3$ are built in order to calculate $\frac{\partial \tilde{f}}{\partial \hat{x}}$ as in Equation (14), and $w_1 ... w_6$ are built in order to calculate $\frac{\partial^2 \tilde{f}}{\partial \hat{x}^2}$ as in Equation (15).

### 2.4. Inequality Constraints

The reader may have noticed that inequality constraints have not been introduced in the problem formulation. Yet, operational limitations, typically on the WEC position or velocity, on the control force or on instantaneous power, are essential for the safe operation of the device.

In particular, constraints which are nonlinear with respect to $x$ (e.g., force or power constraints) are computationally challenging. How their 1st- and 2nd-order derivatives can be explicitly computed, is detailed in [5]. The more general objective function, introduced in the present paper in Equation (5), does not result in any change in the constraint-related calculations, which are therefore not reproduced here.

## 3. Numerical Case Studies

### 3.1. Non-Ideal Power Take-Off Systems

The effect of a non-ideal PTO efficiency upon WEC control is specifically addressed in a number of studies [10–15,18]. In most of them [11,12,14–18], the non-ideal PTO system is modelled by means of a non-unity efficiency factor $\mu$ verifying $0 \leq \mu \leq 1$, so that the instantaneous electric power is expressed as follows.

$$P_e = \begin{cases} -\mu u\dot{x} & \text{if } -u\dot{x} \geq 0 \\ -\frac{1}{\mu}u\dot{x} & \text{if } -u\dot{x} < 0 \end{cases} \tag{16}$$

The simple model (16) is also retained in the present study, for the two WEC models described in Sections 3.2 and 3.3.

Reactive control techniques (which imply that, at times, $P_a$ is negative) are essential to achieve optimal or near-optimal hydrodynamic power absorption, when the incoming wave frequency is away from the system's resonant frequency. In the case of monochromatic waves, a simple PI control law, of the form $u = -b\dot{x} - kx$, can achieve optimal power absorption, following the principle of impedance matching [1]. In irregular waves, a PI control law also presents significant improvements in absorbed power, with respect to a passive law of the form $u = -b\dot{x}$, although the PI controller is then sub-optimal. A salient point of studies [11,12] is that reactive control is particularly affected by a non-ideal PTO efficiency. In particular, it appears that the smaller $\mu$ is, the less reactive control should be performed (i.e., $|k|$ should be smaller).

Overall, all studies [10–15,17,18] are relatively pessimistic, regarding the mean electrical power $\overline{P}_e$ which can be achieved with a non-ideal PTO system: a 10% decrease in efficiency results in a drop in $\overline{P}_e$ of more than 50%, even when control parameters are tuned taking the PTO efficiency into account[1].

Define $h(P_a)$ the efficiency function, so that $h(P_a) = \frac{P_e}{P_a}$. Consistently with Equation (16), $h$ is equal to $\mu$ for positive $P_a$, and $1/\mu$ for negative $P_a$, as illustrated in Figure 1. Such a discontinuous function is undesirable for the efficient implementation of the spectral control approach, detailed in Section 2, and which requires the explicit calculation of the 1st- and 2nd-order derivatives of the objective function. Therefore, in this work, $h(P_a)$ is approximated by the following function,

$$h_\kappa(P_a) = A \tanh(\kappa P_a) + B \tag{17}$$

where $A = \frac{1}{2}(\mu - 1/\mu)$, $B = \frac{1}{2}(\mu + 1\mu)$ and $\kappa > 0$ is a real parameter, which governs the accuracy of the approximation, as illustrated in Figure 1. Note that in [16], the use of a smoothed approximation of the function (16) is also advocated, in order to avoid issues in gradient-based optimisation for the considered optimal control problems. However, the sensitivity of the results to smoothing has not been investigated.

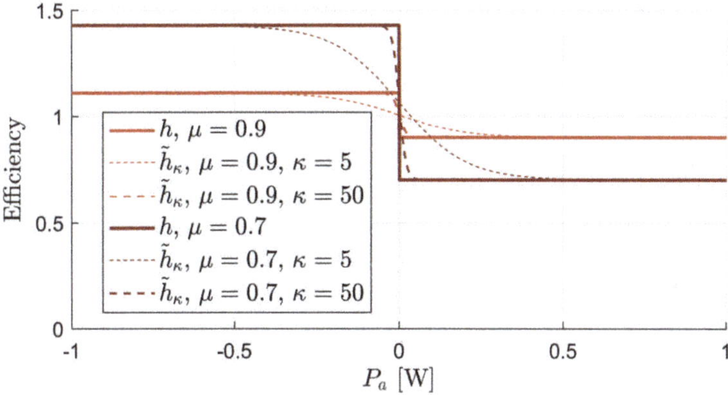

**Figure 1.** Efficiency function $h$ for $\mu = 0.9$ and $\mu = 0.7$, and their approximations $h_\kappa$.

If the *exact* efficiency is considered, the function $\alpha$ in the optimisation problem (5) is $\alpha(\dot{x}, u) := -u\dot{x}\, h(-u\dot{x})$. However, in practice, the problem is solved using the approximate function $\alpha_\kappa(\dot{x}, u) := -u\dot{x}\, h_\kappa(-u\dot{x})$. For all $(\dot{x}, u)$, $\alpha(\dot{x}, u) \leq \alpha_\kappa(\dot{x}, u)$. Therefore, the optimal mean electric power, $\overline{P}^*_{e,\kappa}$, obtained with $\alpha_\kappa$, is an upper bound to the optimal exact electric power, $\overline{P}^*_e$, which would be obtained if Problem (1) could be solved using the exact efficiency. Furthermore, for all $\kappa_1 \leq \kappa_2$, for all $(\dot{x}, u)$, $\alpha_{\kappa_2}(\dot{x}, u) \leq \alpha_{\kappa_1}(\dot{x}, u)$, which implies that $\overline{P}^*_{e,\kappa}$ decreases monotonically with $\kappa$, and has $\overline{P}^*_e$ as a limit when $\kappa \to \infty$.

$\overline{P}^*_e$ can also be bounded from below: consider the solution $x^*_\kappa, u^*_\kappa$, obtained with the approximate objective function. $x^*_\kappa, u^*_\kappa$ is necessarily sub-optimal, with respect to the exact objective function. Therefore, the exact mean electric power, effectively achieved with $(x^*_\kappa, u^*_\kappa)$, constitutes a lower bound to $\overline{P}^*_e$, and is calculated as follows.

$$\overline{P}^\dagger_{e,\kappa} = \frac{1}{T} \int_{t=0}^{T} \alpha(\dot{x}^*_\kappa, u^*_\kappa)\,dt \tag{18}$$

---

1 It should be noted, however, that the control strategies adopted, and the PTO efficiency models, differ across studies, so that accurate comparisons are difficult.

Therefore the following inequality holds.

$$\overline{P}^{\dagger}_{e,\kappa} \leq \overline{P}^{*}_{e} \leq \overline{P}^{*}_{e,\kappa} \tag{19}$$

Even if the true optimal electrical power $\overline{P}^{*}_{e}$ cannot be calculated for the exact efficiency, by setting $\kappa$ to a large enough value, it can be ensured that the electric power $\overline{P}^{\dagger}_{e,\kappa}$, effectively achieved with the solution found, lies within a prescribed percentage of $\overline{P}^{*}_{e}$.

Finally, let $(x^{*}, u^{*})|_{\mu}$ be the optimal solution for efficiency $\mu < 1$, and $\overline{P}^{*}_{e}|_{\mu}$ (resp. $\overline{P}_{a}|_{\mu}$) the mean electrical (resp. absorbed) power with the solution $(x^{*}, u^{*})|_{\mu}$. It is easy to show[2] that $\overline{P}^{*}_{e}|_{\mu} < \mu \overline{P}_{a}|_{\mu}$. Define $\overline{P}_{id}$ as the optimal power for an ideal PTO (absorbed and electric power are then identical). The solution $(x^{*}, u^{*})|_{\mu}$ is sub-optimal with respect to an ideal PTO, thus $\overline{P}_{a}|_{\mu} < \overline{P}^{*}_{id}$. Therefore, the following simple inequality holds.

$$\overline{P}^{*}_{e}|_{\mu} < \mu \overline{P}^{*}_{id} \tag{20}$$

Equation (20) means that the optimal mean electric power, for a non-ideal PTO efficiency, cannot exceed the optimal mean electric power for an ideal PTO efficiency, multiplied by the efficiency factor. In the particular case, where the optimal solution, for an ideal PTO efficiency, is purely passive, then that solution is also the optimal solution for a non-ideal PTO efficiency $\mu < 1$, and the inequality becomes an equality: $P^{*}_{e}|_{\mu} = \mu P^{*}_{e}|_{id}$. This can happen if the incoming wave frequency coincides with the WEC natural frequency, so that no reactive power is needed to achieve optimal control.

### 3.2. The WECCCOMP Model

The Wavestar device is a point absorber WEC, primarily operating in heave, rigidly connected to a rotation arm. The power is captured through the arm rotation around a fixed axis. In the WEC control competition (WECCCOMP), a scale model of the Wavestar device is considered. The linearised WEC dynamics can be modelled by expressing Newton's second law around the rotation axis, which results in a dynamical equation, of the form of Equation (3), where the nonlinear terms $n$ are set to zero. The variable $x$, in this case, corresponds to the angular displacement of the rotation arm. All the linear parameters of Equation (3) are derived from the linear model, provided by the competition organisers [18]. Lower and upper limits on the PTO torque ($\pm 10$ N.m) and angular position ($\pm 24°$) are considered as constraints. Six JONSWAP wave spectra [23], proposed by the organisers [18], and with characteristics listed in Table 1, are used to generate random excitation torque time series, with duration $T = 40$s, at least equal to 20 typical wave periods for the sea states considered.

In each random realisation, the optimal solution $\hat{x}^{*}_{\kappa}|_{\mu}$, and power $\overline{P}^{*}_{e,\kappa}|_{\mu}$, with $\mu = 1, 0.9, \ldots, 0.5$, are calculated using the *approximate* efficiency function $h_{\kappa}$. Then, the *exact* power $\overline{P}^{\dagger}_{e,\kappa}|_{\mu}$, obtained with $\hat{x}^{*}_{\kappa}|_{\mu}$, is also calculated.

For each pair of sea states (1–4, 2–5, 3–6), the appropriate range of values for $\kappa$ differs (when typical values for absorbed power are larger, a smaller $\kappa$ yields a satisfactory approximation). For each pair of sea states, $\kappa$ is gradually increased, until bounds given by Equation (19) are sufficiently close. Suitable values for $\kappa$ are indicated in Table 1.

**Table 1.** List of WECCCOMP JONSWAP sea states.

| ID | $(H_{m_0}, T_p, \gamma)$ | ID | $(H_{m_0}, T_p, \gamma)$ | $\kappa$ |
|----|--------------------------|----|--------------------------|----------|
| 1 | $(0.0208, 0.988, 1)$ | 4 | $(0.0208, 0.988, 3.3)$ | 100 |
| 2 | $(0.0625, 1.412, 1)$ | 5 | $(0.0625, 1.412, 3.3)$ | 5 |
| 3 | $(0.1042, 1.836, 1)$ | 6 | $(0.1042, 1.836, 3.3)$ | 2 |

---

[2]  To see this, decompose $\overline{P}_{a}$ into its passive and reactive components.

### 3.3. The RM3 Device

The RM3 device is a self-reacting, two-body point absorber, exploiting the relative heave motion of the two bodies [19]. The RM3 consists of a cylindrical float (B1) which vertically oscillates with waves, relative to a vertical column spar buoy (B2) damped by a subsurface reaction plate. Within the scope of linear hydrodynamics, the heaving mode of motion is decoupled from the other degrees of freedom, and therefore B1 and B2 may be modelled in terms of heave only. Assuming harmonic signals and adopting complex notations (omitting the frequency dependence for conciseness), the position in heave of B1 and B2, $x_1$ and $x_2$, can be described by the following two equations [24],

$$\begin{cases} z_1 \hat{x}_1 + z_{21} \hat{x}_2 = \hat{e}_1 + \hat{u} \\ z_{12} \hat{x}_1 + z_2 \hat{x}_2 = \hat{e}_2 - \hat{u} \end{cases} \tag{21}$$

where

- the complex mechanical impedance $z_i$ is built as $z_i(\omega) = S_i - \omega^2(I_i + A_{r,i}(\omega)) - j\omega(C_i + B_{r,i}(\omega))$ from the hydrodynamic coefficients of Body $i$,
- the coupling terms $z_{ij}(\omega) = z_{ji}(\omega) = -\omega^2 A_{r,i\to j}(\omega) - j\omega B_{r,i\to j}(\omega)$ are due to the radiation interaction between B1 and B2,
- $\hat{e}_i$ denotes the wave excitation force onto Body $i$ and
- the PTO system applies a force $u$ from B2 onto B1.

By rearranging Equation (21), it is easy to obtain a new formulation in terms of the relative motion[3], $x := x_1 - x_2$, sufficient to solve the control problem (see in [24] for more detail):

$$z_{eq}\hat{x} = \hat{e}_{eq} + \hat{u} \tag{22}$$

where $z_{eq} = \frac{z_2 z_1 - z_{12}^2}{z_0}$, $\hat{e}_{eq} = \frac{z_2 + z_{12}}{z_0}\hat{e}_1 + \frac{z_1 + z_{12}}{z_0}\hat{e}_2$ and $z_0 = z_1 + z_2 + 2z_{12}$. The proposed spectral method can thus be applied, using the equivalent impedance $z_{eq}$ to determine the entries of the matrix **L** as in Section 2.2.

Control calculations are carried out in four JONSWAP [23] sea states, inspired by the scatter diagram of the device's design location, given in [19], with $H_{m_0} = 2$m, $\gamma = 3$, and $T_p = 6, 8, 10$ and 12 s. The relative heave $x$ is subject to the constraint $-2m \leq x \leq 2m$ [19]. The wave signal duration is set to $T = 300$ s (more than 25 typical wave periods in all sea states). A value $\kappa$ of the order of $10^{-6}$ allows for accurate optimal power estimates[4].

## 4. Numerical Results

All constrained control results were obtained using the interior-point algorithm implemented in the Matlab[5] *fmincon* function. For the WECCCOMP model, the cut-off frequency for the control solution is set to 4 Hz (as the signal duration is $T = 40$s, the order $N$ of the approximation is therefore $N = 4 \times 40 = 160$). The number $M$ of collocation points is set to $M = 3N + 1$. For the RM3 device, the cut-off frequency for the control solution is set to 0.5 Hz (as the signal duration is $T = 300$ s, the order $N$ of the approximation is therefore $N = 0.5 \times 300 = 150$). The number $M$ of collocation points is set to $M = 4N + 1$.

---

[3]    For the motion to be fully determined, another equation is necessary, for example in terms of $y := x_1 + x_2$, but this does not have any influence in terms of power absorption.

[4]    The reader may have noted that this value is very different from those, which were found appropriate for the WECCCOMP model. This is because, given the form of Equation (17), the appropriate value for $\kappa$ depends on the typical magnitude of $P_a$: in the RM3 case, which is a full-scale model, $P_a$ takes much larger values than in the WECCCOMP case.

[5]    www.matlab.com.

### 4.1. The WECCCOMP Model

Figure 2 shows control results obtained with the WECCCOMP numerical model (see Section 3.2), in the six sea states of the competition, and values of $\mu$ between 1 and 0.5. In each sea state, power values are normalised by $\overline{P}_{id}^*$, the optimal power obtained with $\mu = 1$. $\overline{P}_{id}^*$ essentially represents the optimal absorbed mechanical power; it is a well-known theoretical limit in the wave energy literature [1], which can be calculated as $\overline{P}_{id}^* = \int_{\omega=0}^{\infty} \frac{-\omega |\hat{e}(\omega)|^2}{2\Im\{z(\omega)\}} d\omega$. As per Equation (19), $\overline{P}_{e,\kappa}^\dagger|_\mu$ and $\overline{P}_{e,\kappa}^*|_\mu$ together provide bounds to $\overline{P}_e^*|_\mu$, the optimal power for the exact efficiency function.

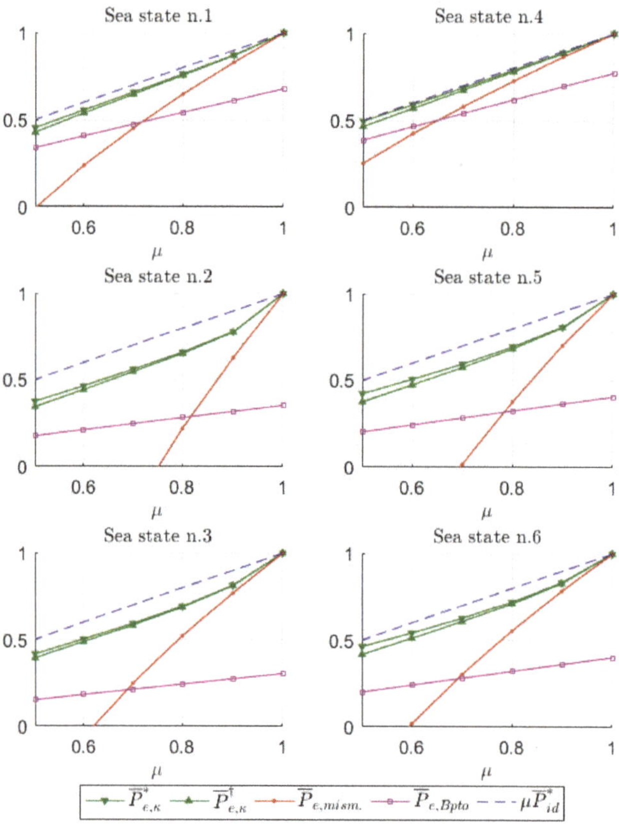


**Figure 2.** Normalised mean electric power for the model Wavestar device in the six WECCCOMP sea states, over a range of efficiency parameter values.

Also shown in Figure 2, the dotted lines labelled "$\mu\overline{P}_{id}^*$" represent the upper bound of Equation (20), i.e., a hypothetical scenario where $\overline{P}_{id}^*$ would "only" be affected by a factor of $\mu$. $\overline{P}_{mism.}$ is the actual electric power which would be obtained, if the solution $\hat{x}_{id}^*$ to the ideal control problem were followed by a WEC with imperfect PTO efficiency $\mu \leq 1$; in other words, $\overline{P}_{mism.}$ shows the effect of a *mismatch* between the nominal efficiency ($\mu = 1$) and the actual efficiency ($1 > \mu \geq 0.5$). Finally, $\overline{P}_{Bpto}$ represents the power obtained using a simple linear PTO damping, with coefficient $B_{pto}$ optimised for each sea state.

The importance of taking into account the non-ideal PTO efficiency in control calculations is evidenced by $\overline{P}_{mism.}$, which can even take negative values, similarly to the results of [13]. In contrast, the proposed spectral control approach, with approximate efficiency function, provides a solution for which the mean electric power, $\overline{P}^{\dagger}_{e,\kappa}|_{\mu}$, lies within a few percents of the true optimal power value. Unlike other studies [11–14], the controller's performance, optimised taking $\mu$ into account, is reasonably close to the "best-case scenario" $\mu\overline{P}^{*}_{id}$ (in fact, almost identical in sea states 1 and 4). Results also clearly evidence the interest of the proposed spectral control approach, with respect to an optimised linear damper, even in sea states 1 and 4, where the waves are close to the WEC resonant frequency.

Figure 3 shows the effect of a non-ideal PTO efficiency, upon the WEC optimal trajectory (in this case, represented by the WEC angular velocity $\dot{\theta}$, see Section 3.2), control torque, and instantaneous power, in Sea State 3. In the non-ideal case, $\mu$ is set to 0.7 as in the WECCCOMP.

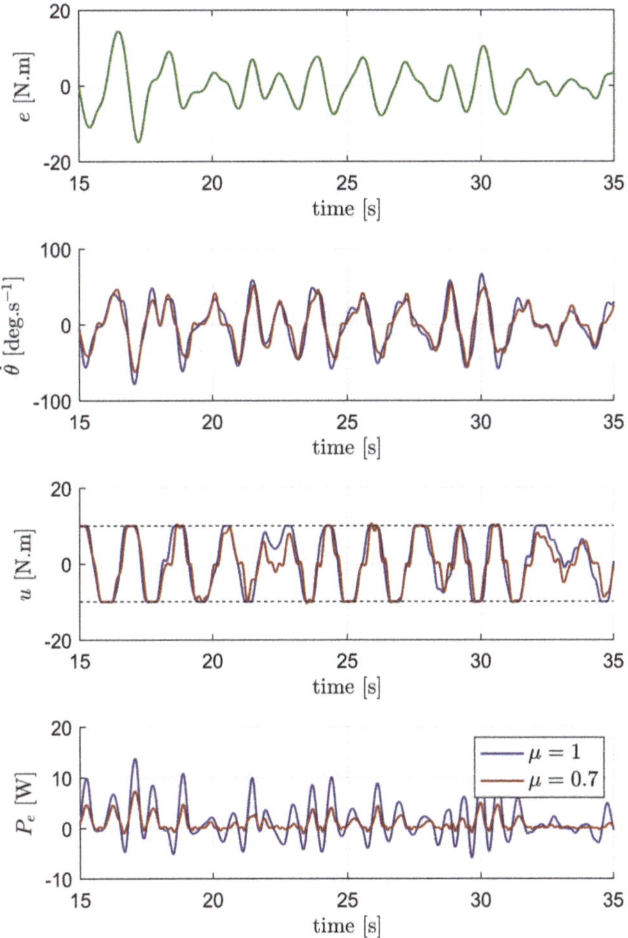

**Figure 3.** Excitation torque, optimal angular velocity, control torque and instantaneous electric power, with $\mu = 1$ and $\mu = 0.7$.

Although the two optimal velocities and control inputs look qualitatively similar[6], the optimal electric power is fundamentally modified by the non-ideal PTO efficiency. As could be expected, very little reactive power must be employed with $\mu = 0.7$; furthermore, the peaks in instantaneous electric power are reduced by a factor of at least 2 (while the average electric power still exceeds 60% of the ideal one).

### 4.2. The RM3 Device

Results for the RM3 device (see Section 3.3) are illustrated in a way similar to those of Section 4.1, in Figures 4 and 5. In Figure 4, the two bounds $\overline{P}^{*}_{e,\kappa}|_{\mu}$ and $\overline{P}^{\dagger}_{e,\kappa}|_{\mu}$ are close enough to indicate that the solution obtained is nearly optimal.

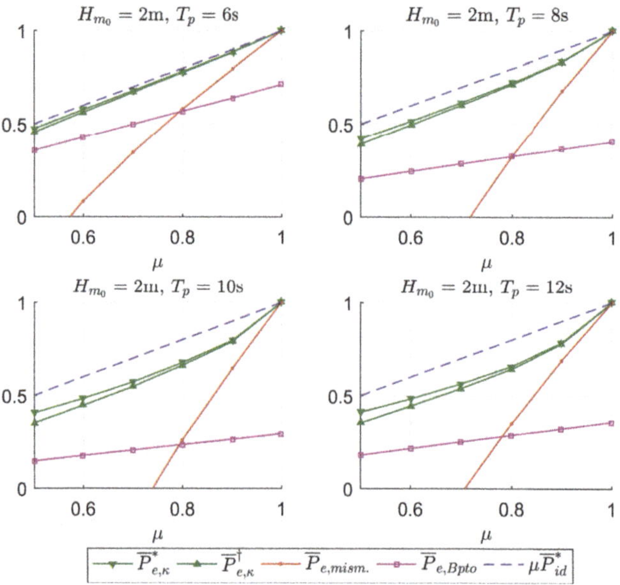

**Figure 4.** Normalised mean electric power for the RM3 device in four JONSWAP sea states with $\gamma = 3$, over a range of efficiency parameter values.

The RM3 resonant period of the device is of the order of 4.5s, which is shorter than typical waves encountered at the device's design location[7]. As a consequence, from Figure 4, it can be seen that for shorter $T_p$, less reactive power is used to achieve optimal power absorption, thus making the simple linear damper perform reasonably well with respect to the reactive control strategy. In contrast, when waves are away from the WEC resonant frequencies, more reactive power must be employed, and the control strategy is significantly affected by the non-ideal efficiency; see, in particular, $\overline{P}_{e,mism.}$. However, the efficiency-aware spectral control strongly mitigates such adverse effects, bringing $\overline{P}^{\dagger}_{e,\kappa}$ close to the hypothetical best-case scenario $\mu\overline{P}^{*}_{id}$.

Finally, similarly to Figure 3, the effect of a non-ideal efficiency upon the control strategy is examined more closely in Figure 5, for the JONSWAP sea state with $T_p = 8$ s. While the ideal PTO

---

6  In particular, in both cases the optimal velocity is in phase with the excitation torque, which is a classical WEC optimal control result [1]

7  This is typically the case for heaving point absorbers: their dimensions give them a relatively short resonant period, but reactive control can artificially make them resonate in longer wave periods.

yields large reactive power flows, taking into account a non-ideal efficiency $\mu = 0.7$ eliminates most of the reactive power, and spectacularly reduces the peak power amplitude. With the non-ideal PTO efficiency, the optimal relative heave velocity $\dot{x}$ (see Section 3.3) and control force $u$ exhibit visible higher-frequency oscillations. Finally, note that the relative position constraint is successfully implemented within the spectral control strategy, which would not be possible with a simpler control approach, such as a PI controller.

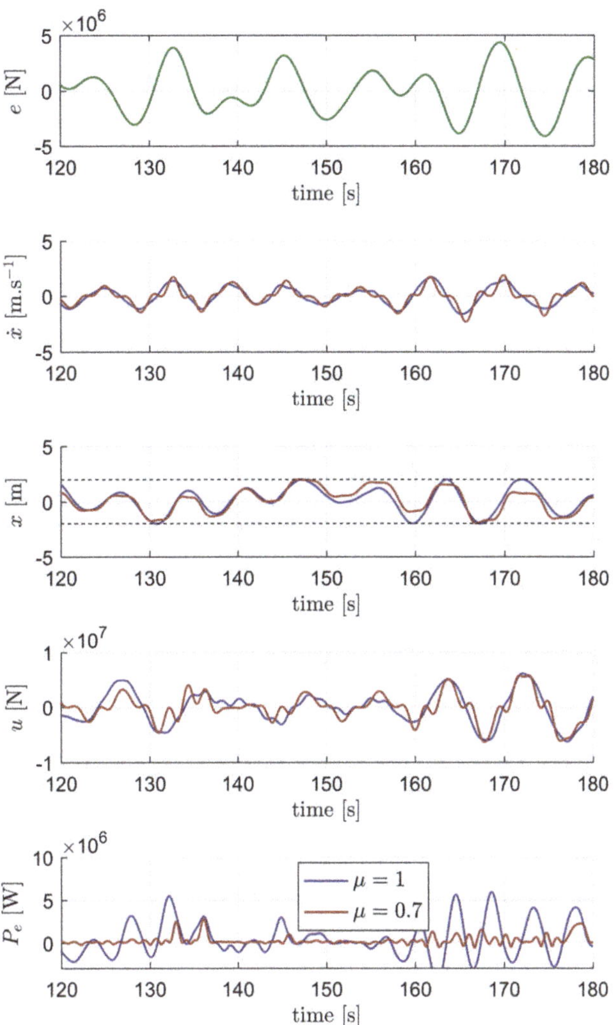

**Figure 5.** Excitation force, optimal relative velocity, relative position, control force, and instantaneous electric power, with $\mu = 1$ and $\mu = 0.7$, $T_p = 8$ s.

## 5. Conclusions

This work shows how the spectral control approach in [5] can be extended to handle more general objective functions, of the form outlined in Equation (2). The efficiency of the proposed technique allows optimal control results to be calculated at once in long wave signals, comprising more than

20 wave periods. Such calculations permit a theoretical analysis of WEC optimal control results, under a variety of modelling and control choices. Furthermore, as in [9], they can provide a robust point of comparison to assess the performance of more practical control approaches, typically implemented in a receding-horizon configuration. In fact, the authors have used the optimal spectral control approach proposed in this work, to assess the receding-horizon MPC solution, proposed by their institution for the WECCCOMP. This comparison will be the subject of a future paper.

A particularly important application of the proposed spectral method consists of WEC control with non-ideal PTO efficiency. In this study, a discontinuous efficiency function, given in Equation (16), is considered. A smooth, hyperbolic tangent approximation for the efficiency function allows the calculation of a near-optimal solution, which also provides upper and lower bounds to the exact optimal electric power.

In view of the results of Figures 2 and 4, concerning two different WECs representative of heaving point absorbers (both devices have resonant periods shorter than their design waves), a non-ideal PTO efficiency may not be as detrimental as suggested in previous studies [11,12,14,15], if the control strategy is able to take the non-ideal efficiency into account, which is the case for the proposed optimal spectral approach. In fact, with realistic efficiency values in the range of $0.7 \leq \mu \leq 0.9$ [11], it does not seem unreasonable to approximate the optimal electric power achievable with $\mu \overline{P}_{id}^{*}$, minus 5 to 15% (reminding that $\mu \overline{P}_{id}^{*}$ is merely a theoretical limit, that no control strategy can exceed if the PTO has an efficiency factor of $\mu$).

The fact that a non-ideal PTO efficiency necessitates a control strategy with less reactive power, already evidenced in previous studies [11,12], is clearly corroborated by the results of Figures 3 and 5. This is to be expected, irrespective to the WEC technology and the sea states considered. However, the extent of power production degradation when a control strategy neglects the non-ideal PTO characteristics will be, in general, system- and wave climate-dependent. Note, also, that the results of this study depend, to some extent, on the specific model adopted for the PTO efficiency. Other efficiency functions could be investigated, such as those in [10,13], which model the fact that the PTO efficiency drops for small load factors.

The computational speed of the proposed spectral control technique is not a central aspect of the present work; however, it was found that optimal solutions, for irregular wave signals with more than 5 wave periods, could be obtained in less than 1 second, with a Matlab implementation running on a 2.60 GHz, 4-core Intel® processor. A real-time, receding-horizon implementation could be considered in future work, either by employing windowing functions to make the receding-horizon control problem periodic, in the spirit of the works in [8,9], or through the use of non-periodic spectral basis functions [6].

**Author Contributions:** Conceptualising, A.M.; software, A.M.; writing—original draft preparation, A.M.; writing—review and editing, A.M. and P.T.; supervision, P.T.; project administration, P.T. All authors have read and agreed to the published version of the manuscript.

**Funding:** This research received no external funding.



## References

1. Ringwood, J.V.; Bacelli, G.; Fusco, F. Energy-maximizing control of wave-energy converters: The development of control system technology to optimize their operation. *IEEE Control Syst. Mag.* **2014**, *34*, 30–55.
2. Faedo, N.; Olaya, S.; Ringwood, J.V. Optimal control, MPC and MPC-like algorithms for wave energy systems: An overview. *IFAC J. Syst. Control* **2017**, *1*, 37–56. [CrossRef]
3. Bacelli, G.; Ringwood, J.V.; Gilloteaux, J.C. A control system for a self-reacting point absorber wave energy converter subject to constraints. *IFAC Proc. Vol.* **2011**, *44*, 11387–11392. [CrossRef]
4. Bacelli, G.; Ringwood, J.V. Nonlinear optimal wave energy converter control with application to a flap-type device. *IFAC Proc. Vol.* **2014**, *47*, 7696–7701. [CrossRef]

5.  Mérigaud, A.; Ringwood, J.V. Improving the computational performance of nonlinear pseudospectral control of wave energy converters. *IEEE Trans. Sustain. Energy* **2017**, *9*, 1419–1426. [CrossRef]

6.  Genest, R.; Ringwood, J.V. Receding horizon pseudospectral control for energy maximization with application to wave energy devices. *IEEE Trans. Control Syst. Technol.* **2016**, *25*, 29–38. [CrossRef]

7.  Boyd, J.P. *Chebyshev and Fourier Spectral Methods*; Courier Corporation: North Chelmsford, MA, USA, 2001.

8.  Auger, C.; Merigaud, A.P.L.; Ringwood, J.V. Receding-horizon pseudo-spectral control of wave energy converters using periodic basis functions. *IEEE Trans. Sustain. Energy* **2018**, *10*, 1644–1652. [CrossRef]

9.  Mérigaud, A.; Ringwood, J.V. Towards realistic nonlinear receding-horizon spectral control of wave energy converters. *Control Eng. Pract.* **2018**, *81*, 145–161.

10. Tedeschi, E.; Carraro, M.; Molinas, M.; Mattavelli, P. Effect of control strategies and power take-off efficiency on the power capture from sea waves. *IEEE Trans. Energy Convers.* **2011**, *26*, 1088–1098. [CrossRef]

11. Genest, R.; Bonnefoy, F.; Clément, A.H.; Babarit, A. Effect of non-ideal power take-off on the energy absorption of a reactively controlled one degree of freedom wave energy converter. *Appl. Ocean. Res.* **2014**, *48*, 236–243. [CrossRef]

12. Falcão, A.F.; Henriques, J.C. Effect of non-ideal power take-off efficiency on performance of single-and two-body reactively controlled wave energy converters. *J. Ocean. Eng. Mar. Energy* **2015**, *1*, 273–286.

13. Bacelli, G.; Genest, R.; Ringwood, J.V. Nonlinear control of flap-type wave energy converter with a non-ideal power take-off system. *Annu. Rev. Control* **2015**, *40*, 116–126. [CrossRef]

14. Sánchez, E.V.; Hansen, R.H.; Kramer, M.M. Control performance assessment and design of optimal control to harvest ocean energy. *IEEE J. Ocean. Eng.* **2014**, *40*, 15–26. [CrossRef]

15. Tom, N.M.; Madhi, F.; Yeung, R.W. Power-to-load balancing for heaving asymmetric wave-energy converters with nonideal power take-off. *Renew. Energy* **2019**, *131*, 1208–1225. [CrossRef]

16. Tona, P.; Nguyen, H.N.; Sabiron, G.; Creff, Y. An Efficiency-Aware Model Predictive Control Strategy for a Heaving Buoy Wave Energy Converter. In Proceedings of the European Wave and Tidal Energy Conference (EWTEC) 2015, Nantes, France, 6–11 September 2015.

17. Saupe, F.; Creff, Y.; Gilloteaux, J.C.; Bozonnet, P.; Tona, P. *Latching Control Strategies for a Heaving Buoy Wave Energy Generator in a Random Sea*; IFAC World Congress: Prague, Czech Republic, 2014; Volume 19, pp. 7710–7716.

18. Ringwood, J.V.; Ferri, F.; Ruehl, K.; Yu, Y.H.; Coe, R.G.; Bacelli, G.; Weber, J.; Kramer, M.M. A competition for WEC control systems. In Proceedings of the 12th European Wave and Tidal Energy ConferencEuropean Wave and Tidal Energy Conference, Cork, Ireland, 27 August–1 September 2017.

19. Neary, V.S.; Lawson, M.; Previsic, M.; Copping, A.; Hallett, K.C.; LaBonte, A.; Rieks, J.; Murray, D. *Methodology for Design and Economic Analysis of Marine Energy Conversion (MEC) Technologies*; Technical Report; Sandia National Lab. (SNL-NM): Albuquerque, NM, USA, 2014.

20. Mérigaud, A. A Harmonic Balance Framework for the Numerical Simulation of Non-Linear Wave Energy Converter Models in Random Seas. Ph.D. Thesis, Maynooth University, Maynooth, Ireland, 2018.

21. Cummins, W. The impulse response function and ship motions. *Schiffstechnik* **1962**, *9*, 101–109.

22. Ogilvie, T. Recent progress toward the understanding and prediction of ship motions. In Proceedings of the Sixth Symposium on Naval Hydrodynamics, Bergen, Norway, 10–12 September 1964.

23. Hasselmann, K.; Barnett, T.; Bouws, E.; Carlson, H.; Cartwright, D.; Enke, K.; Ewing, J.; Gienapp, H.; Hasselmann, D.; Kruseman, P. *Measurements of Wind-Wave Growth and Swell Decay During the Joint North Sea Wave Project (JONSWAP)*; Technical Report; Deutches Hydrographisches Institut: Hamburg, Germany, 1973.

24. Olaya, S. Contribution à la Modélisation Multi-Physique et au Contrôle Optimal d'un Générateur Houlomoteur: Application à un Système 'deux corps'. Ph.D. Thesis, Université de Bretagne occidentale, Brest, France, 2016. (In French)

Journal of
*Marine Science and Engineering*



# The Influence of Sizing of Wave Energy Converters on the Techno-Economic Performance

Jian Tan *, Henk Polinder , Antonio Jarquin Laguna , Peter Wellens and Sape A. Miedema

Department of Maritime & Transport Technology, Delft University of Technology,
2628 CD Delft, The Netherlands; H.Polinder@tudelft.nl (H.P.); A.JarquinLaguna@tudelft.nl (A.J.L.);
P.R.Wellens@tudelft.nl (P.W.); S.A.Miedema@tudelft.nl (S.A.M.)
* Correspondence: J.Tan-2@tudelft.nl

**Abstract:** Currently, the techno-economic performance of Wave Energy Converters (WECs) is not competitive with other renewable technologies. Size optimization could make a difference. However, the impact of sizing on the techno-economic performance of WECs still remains unclear, especially when sizing of the buoy and Power Take-Off (PTO) are considered collectively. In this paper, an optimization method for the buoy and PTO sizing is proposed for a generic heaving point absorber to reduce the Levelized Cost Of Energy (LCOE). Frequency domain modeling is used to calculate the power absorption of WECs with different buoy and PTO sizes. Force constraints are used to represent the effects of PTO sizing on the absorbed power, in which the passive and reactive control strategy are considered, respectively. A preliminary economic model is established to calculate the cost of WECs. The proposed method is implemented for three realistic sea sites, and the dependence of the optimal size of WECs on wave resources and control strategies is analyzed. The results show that PTO sizing has a limited effect on the buoy size determination, while it can reduce the LCOE by 24% to 31%. Besides, the higher mean wave power density of wave resources does not necessarily correspond to the larger optimal buoy or PTO sizes, but it contributes to the lower LCOE. In addition, the optimal PTO force limit converges at around 0.4 to 0.5 times the maximum required PTO force for the corresponding sea sites. Compared with other methods, this proposed method shows a better potential in sizing and reducing LCOE.

**Keywords:** WECs; size optimization; techno-economic performance; PTO sizing

**Citation:** Tan, J.; Polinder, H.; Laguna, A.J.; Wellens, P.; Miedema, S.A. The Influence of Sizing of Wave Energy Converters on the Techno-Economic Performance. *J. Mar. Sci. Eng.* **2021**, *9*, 52. https://doi.org/10.3390/jmse9010052

Received: 25 November 2020
Accepted: 29 December 2020
Published: 5 January 2021

**Publisher's Note:** MDPI stays neutral with regard to jurisdictional claims in published maps and institutional affiliations.

## 1. Introduction

Wave energy has been highlighted as a renewable energy resource for decades. However, large scale utilization of wave energy is still far from commercialization [1,2]. An important obstacle in the development of Wave Energy Converters (WECs) is that their techno-economic performance is not competitive with offshore wind and other renewable technologies [3]. Another challenge is the wide variety of WEC types [4], which makes it hard to converge the attention and investment. During the last 40 years, over 200 WEC concepts have been proposed [1]. To select the promising WECs, it is necessary to compare and evaluate their viability at different potential sea sites.

The feasibility of WECs has already been evaluated in existing literature. A set of technical and economic indicators of various WEC concepts have been presented [5]. During the operation of WECs, power production relies not only on the wave resources available but also on the operating principles of the particular device [4]. In Reference [5], a benchmarking of 8 types of WECs was established considering 5 different representative sea sites, and a series of cost-related metrics of different WECs was analyzed and compared. In Reference [6], based on a cost estimation model, a techno-economic evaluation for 6 types of WECs was conducted for different devices. The sensitivity of the Levelized Cost Of Energy (LCOE) to wave resources, array configurations, Capital Expenditure (CAPEX), and Operational Expenditure (OPEX) was also discussed.

175

Apart from the variation of wave resources and types of WECs, the size determination also has a significant impact on the techno-economic evaluation of WECs. Firstly, the power performance of WECs depends strongly on their size [4]. Secondly, given the variation of operating principles, the sensitivity of power performance to the size of WECs also varies considerably [7]. Thirdly, the cost is highly related to the size of WECs [8]. Thus, the size optimization of WECs is expected to make a difference in the evaluation results. However, the optimization is not widely considered in evaluation studies even though the original sizes of WECs are usually designed to suit only some specific sea sites.

Independently or integrated with evaluation studies, size optimization of WECs has been discussed in existing literature. In Reference [9,10], the theoretical ratio of absorbed power to the buoy volume was derived based on linear wave theory. It was concluded that the smaller the volume of the floating WEC is, the higher this ratio is. In Reference [11], a theoretical method for determining the suitable buoy size of WECs for different sea sites was proposed based on Budal diagram. By means of that method, the volume of the buoy could be selected to guarantee the required working time at full capacity of devices for a certain sea site. In Reference [4,7,12], the optimal characteristic lengths of WECs were demonstrated from the perspective of maximizing Capture Width Ratio (CWR), concluding that the optimal characteristic length depends on sea sites. Furthermore, the effect of sizing of WECs on their techno-economic performance has been investigated [8,13], and the results showed that size optimization is able to reduce the LCOE dramatically. However, above-mentioned literature regarding size optimization of WECs concentrated mainly on buoy sizing without considering Power Take-Off (PTO) sizing. Mostly, the PTO size is simply scaled with the same scaling factor used for the buoy.

As a core component, the PTO system is significantly influential to the techno-economic performance of WECs [14]. On the one hand, its cost normally accounts for over 20% of the total CAPEX [15]. On the other hand, the PTO size is highly related to the rated power and the force constraint, which can directly affect the absorbed power. The affecting factors on PTO sizing have been investigated, as well. In Reference [16,17], a series of studies on the influence of the generator rating and control strategies on the power performance was conducted through simulations and experiments. In Reference [14,18], the impact of the control strategies on the PTO sizing of point absorbers was investigated. Although these studies made a profound contribution to presenting the impact of PTO sizing on the absorbed power, buoy sizing was not taken into account collectively. Recently, the influence of PTO sizes has started to be considered in a few of studies regarding buoy geometry optimization. In Reference [19], the geometry optimization of a cylindrical WEC with only a few of different force constraints was performed. The paper indicates that the PTO force constraint has a notable influence on the optimization results, and this finding is highly valuable for the design optimization of WECs. However, there are some limitations in Reference [19]. Firstly, the force constraint is not treated as an optimization variable and only a limited number of force constraints were included. Secondly, the optimization is conducted to maximize the absorbed power for the specific sea states, which is not necessarily beneficial for improving the techno-economic performance at realistic sea sites.

To the authors' knowledge, there is a lack of studies considering both PTO and buoy sizing of WECs. Thus, the aim of this paper was to investigate the collective influence of PTO sizing and buoy sizing on the LCOE. To achieve this aim, four research questions were addressed. Firstly, what are the optimal buoy sizes and PTO sizes for minimizing the LCOE in different wave resources and control strategies? Secondly, how much reduction in the LCOE can be gained by including PTO sizing? Thirdly, how does PTO sizing interact with buoy sizing from the techno-economic perspective? Fourthly, how much difference on the size determination and the optimized LCOE will there be when using different size optimization methods?

It should be acknowledged that one of the important factors affecting the LCOE is related to the system survivability and safety of WECs. Therefore, the techno-economic performance in practice is strongly related to the recurrence of extreme conditions, such as the roughest sea state for 50 or 100 years in the deployment site. In this way, the total LCOE could be significantly higher than the optimized value based on the current study, which could mitigate the effects of sizing on the techno-economic performance. However, a better understanding on the collective influence of buoy and PTO sizing of WECs can clearly contribute to a more suitable WEC design. In addition, sizing of a single WEC could make a difference in the total number of units selected for a wave farm or array, where the operation and maintenance costs are directly influenced by the number of WECs. For instance, a wave farm with a large number of small individual WECs is expected to result in the increased number of operation and maintenance activities, leading to higher operation and maintenance costs. On the other hand, larger WECs can reduce the number of units in the wave farm, but their requirements for service operation vessels could be higher [8]. Thus, as stated in Reference [20], estimating an accurate value for OPEX is a difficult task since there is not enough available information in practical projects. For the sake of simplicity this study will be limited to the analysis of a single device using general values of OPEX costs derived from existing literature.

In this paper, a size optimization method considering both buoy and PTO sizing is established, and it is applied to a generic heaving point absorber. The optimization is performed based on an exhaustive search algorithm. Firstly, frequency domain modeling is used to calculate the power performance of WECs with different buoy and PTO sizes. The implementation of the PTO sizing using passive and reactive control is demonstrated, respectively. In addition, a preliminary economic model is described to build a cost function with the aim to minimize the LCOE. Next, based on the proposed method, size optimization of WECs is carried out for three realistic sea sites. The interaction between buoy and PTO sizing is analyzed, and the dependence of size determination on wave resources and control strategies is presented. Furthermore, a comparison between this proposed size optimization method and other methods is performed. Finally, conclusions are drawn based on the obtained results.

## 2. Methodology

This section starts with the description of the heaving spherical WEC concept and chosen sea sites. Next, the equations of motion and frequency domain modeling of WECs are presented. Finally, the size optimization method is established, in which the approaches to perform PTO sizing, buoy sizing, and the cost estimation of WECs are explained in detail.

### 2.1. WEC Concept Description

A generic heaving point absorber [21,22] is used in this study as WEC reference. The diameter of the buoy in the original size is 5 m. The average density of the buoy in all sizes is assumed to be identical and with a value of half of the water density (1025 kg/m$^3$). The schematics of the concept are shown in Figure 1. In practice, the amplitude of the buoy motion has to be limited to protect the mechanical structure, and the displacement limit of this WEC is set as 0.8 times the radius of the buoy. In addition, WECs have to be stopped from operating at severe sea states. Thus, there must be a maximum operational wave height for protection. A similar prototype to the WEC in this work is WaveStar, which is a semi-spherical heaving point absorber [23]. The maximum operational wave height of WaveStar is 6 m and the diameter of WaveStar is 5 m. Therefore, the maximum operational wave height for the original WEC in this case is estimated as $H_s = 5$ m, in a conservative way.

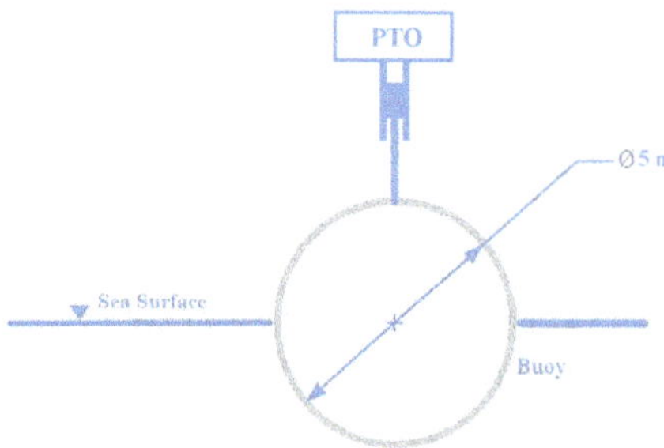


**Figure 1.** Schematic of the heaving point absorber concept.

Three realistic sea sites are selected to investigate the techno-economic performance of WECs. They are Yeu Island, in the oceanic territory of France, Biscay Marine Energy Park (BIMEP) in Spain, and DK North Sea Point 2 (DK2) in Denmark. The scatter diagrams of these sea sites are taken from Reference [8], which are shown in Tables 1–3. In these scatter diagrams, $H_s$ and $T_z$ represent the significant wave height and the mean zero-crossing wave period, respectively, and the time of the corresponding occurrence are depicted in each cell as the number of occurrence hours in a year for that particular sea state. It should be pointed out that Yeu, BIMEP, and DK2 are geographically far from each other and their most frequent wave heights and periods differ significantly. Hence, these three sea sites are chosen to be the representatives of European wave characteristics.

**Table 1.** Scatter diagram showing yearly hours of occurrence of sea states in the Yeu Island (Yeu), France.

| Hours $T_z$ (s) / $H_s$ (m) | 2.14 | 2.86 | 3.57 | 4.29 | 5.00 | 5.71 | 6.43 | 7.14 | 7.89 | 8.57 | 9.29 | 10.00 | 10.71 | 11.43 | 12.14 |
|---|---|---|---|---|---|---|---|---|---|---|---|---|---|---|---|
| 0.50 | 1.27 | 55.70 | 79.75 | 67.09 | 64.56 | 62.03 | 37.98 | 12.66 | 1.27 | | | | | | |
| 1.00 | | 56.97 | 219.00 | 268.37 | 337.99 | 369.64 | 244.32 | 135.45 | 62.03 | 3.80 | 1.27 | 1.27 | | | |
| 1.50 | | 1.27 | 36.71 | 163.30 | 278.50 | 281.03 | 215.20 | 145.58 | 124.06 | 86.08 | 17.72 | 3.80 | 2.53 | | |
| 2.00 | | | 69.62 | 225.33 | 235.46 | 175.98 | 146.84 | 64.56 | 62.03 | 22.79 | 11.39 | 6.33 | | | |
| 2.50 | | | | 81.02 | 345.59 | 324.07 | 272.17 | 106.34 | 72.16 | 43.04 | 11.39 | 2.53 | | | |
| 3.00 | | | | 15.19 | 281.03 | 340.53 | 250.66 | 112.66 | 87.35 | 58.23 | 31.66 | 10.13 | | | |
| 3.50 | | | | | 69.62 | 293.69 | 192.42 | 118.99 | 65.83 | 51.90 | 36.71 | 22.79 | 7.60 | 1.27 | |
| 4.00 | | | | | 3.80 | 110.13 | 127.86 | 65.83 | 49.37 | 27.85 | 27.85 | 22.79 | 15.19 | 3.80 | |
| 4.50 | | | | | | 26.58 | 43.04 | 60.76 | 21.52 | 12.66 | 6.33 | 5.06 | | 1.27 | |
| 5.00 | | | | | | 2.53 | 20.25 | 34.18 | 15.19 | 7.60 | 1.27 | | | | |
| 5.50 | | | | | | | 7.60 | 24.05 | 12.66 | 6.33 | | | | | |
| 6.00 | | | | | | | 3.80 | 13.92 | 8.86 | | 1.27 | | | | |
| 6.50 | | | | | | | | 2.53 | 7.60 | 1.27 | | | | | |

**Table 2.** Scatter diagram showing yearly hours of occurrence of sea states in Biscay Marine Energy Park (BIMEP), Spain.

| Hours $T_z$ (s) / $H_s$ (m) | 5.00 | 7.00 | 9.00 | 11.00 | 13.00 | 15.00 | 17.00 |
|---|---|---|---|---|---|---|---|
| 0.75 | 148.92 | 219.00 | 78.84 | 17.52 | | | |
| 1.50 | 858.48 | 2664.52 | 1445.40 | 508.08 | 78.84 | 8.76 | |
| 2.50 | | 744.60 | 560.64 | 324.12 | 157.68 | 35.04 | |
| 3.50 | | 87.60 | 262.80 | 61.32 | 52.56 | 35.04 | 8.76 |
| 4.50 | | | 105.12 | 35.04 | 8.76 | 8.76 | 8.76 |
| 5.50 | | | 8.76 | 26.28 | | | |

**Table 3.** Scatter diagram showing yearly hours of occurrence of sea states in DK North Sea Point 2 (DK2), Denmark.

| Hours $T_z$ (s)<br>$H_s$ (m) | 2.50 | 3.50 | 4.50 | 5.50 | 6.50 | 7.50 | 8.50 | 9.50 |
|---|---|---|---|---|---|---|---|---|
| 0.25 | 584.00 | 634.00 | 113.00 | 29.00 | 7.00 | 1.00 | | |
| 0.75 | 20.00 | 1552.00 | 610.00 | 153.00 | 56.00 | 16.00 | 6.00 | 2.00 |
| 1.25 | | 188.00 | 1397.00 | 123.00 | 25.00 | 9.00 | 7.00 | 3.00 |
| 1.75 | | 1.00 | 621.00 | 501.00 | 8.00 | 2.00 | 1.00 | 1.00 |
| 2.25 | | | 14.00 | 709.00 | 18.00 | 1.00 | | |
| 2.75 | | | | 286.00 | 224.00 | 1.00 | | |
| 3.25 | | | | 10.00 | 314.00 | 2.00 | | |
| 3.75 | | | | | 190.00 | 34.00 | | |
| 4.25 | | | | | 17.00 | 121.00 | | |
| 4.75 | | | | | | 77.00 | 1.00 | |
| 5.25 | | | | | | 26.00 | 11.00 | |
| 5.75 | | | | | | 2.00 | 16.00 | |
| 6.25 | | | | | | | 10.00 | |
| 6.75 | | | | | | | 4.00 | 1.00 |
| 7.25 | | | | | | | | 1.00 |

*2.2. Frequency Domain Modeling*

In this subsection, the frequency domain model of WECs is presented based on linear wave theory. As the device in this paper is assumed to oscillate only in heave motion, the frequency domain modeling of the WEC is only discussed for the heaving degree of freedom. According to Newton's law, the motion of the WEC as a rigid body can be described through Equation (1).

$$ma(t) = F_{hs}(t) + F_e(t) + F_{pto}(t) + F_r(t), \tag{1}$$

where textitm is the mass of the oscillating body, $F_{hs}$ is the hydrostatic force, $F_e$ is the wave excitation force, $F_r$ is the wave radiation force, $F_{pto}$ is the PTO force, and $a(t)$ is the acceleration. If the body is assumed to perform harmonic motion and a linear PTO model is used to simulate the behavior of the PTO system, (1) could be rewritten in the form of complex amplitudes [10], as

$$\hat{F}_e(\omega) = [R_i(\omega) + R_{pto}]\hat{u} + i\omega\hat{u}[m + M_r(\omega)] + i\hat{u}[-\frac{K_{pto}}{\omega} - \frac{S_{wl}}{\omega}], \tag{2}$$

where $R_i(\omega)$ is the hydrodynamic damping coefficient, $R_{pto}$ is the PTO damping coefficient, $\omega$ is the wave frequency, $m$ and $M_r(\omega)$ are the mass and the added mass of the WEC, $\hat{u}$ is complex amplitude of the vertical velocity, $K_{pto}$ is the PTO stiffness coefficient, and $S_{wl}$ is the hydrostatic stiffness. The intrinsic impedance of the heaving buoy and PTO impedance can be introduced as

$$Z_i(\omega) = R_i(\omega) + iX_i(\omega), \tag{3}$$

$$X_i(\omega) = \omega[m + M_r(\omega)] - \frac{S_{wl}}{\omega}, \tag{4}$$

where $Z_i(\omega)$ is the intrinsic impedance of the heaving buoy, and $X_i(\omega)$ is the intrinsic reactance. Similarly, the impedance of PTO can be given as:

$$Z_{pto}(\omega) = R_{pto}(\omega) + iX_{pto}(\omega), \tag{5}$$

$$X_{pto}(\omega) = -\frac{K_{pto}}{\omega}, \tag{6}$$

where $Z_{pto}(\omega)$ is the PTO impendence, and $X_{pto}(\omega)$ is the PTO reactance. So, (2) is rewritten as

$$\hat{F}_e(\omega) = [Z_i(\omega) + Z_{pto}(\omega)]\hat{u}. \tag{7}$$

The hydrodynamic characteristics of WECs, including $M_r(\omega)$, $R_i(\omega)$, and $F_e(\omega)$, are calculated using the Boundary Element Method through the open source software Nemoh [24]. Then, by solving (7), the complex amplitude of velocity $\hat{u}$ could be obtained as

$$\hat{u} = \frac{\hat{F}_e}{Z_i + Z_{pto}}. \tag{8}$$

Then, the complex amplitude of the motion displacement is expressed as

$$\hat{z} = \frac{\hat{F}_e}{i\omega(Z_i + Z_{pto})}. \tag{9}$$

For regular wave conditions, the time averaged absorbed power can be obtained and expressed as

$$P_{ave\_re} = \frac{1}{2}R_{pto}|\hat{u}|^2. \tag{10}$$

The above analysis is based on the assumption of harmonic motion, but incoming waves in real sea states are always irregular. In this work, the calculation of power absorption in irregular waves is conducted based on the superposition of regular waves [25]. So, the power absorption in each sea state is calculated by integrating over frequency the product of the spectrum density with the power absorption in regular waves which can be expressed as

$$P_{ave\_irr} = \int_0^{\infty} R_{pto}(\frac{|\hat{z}|}{\zeta_a}, \omega)^2 S_{\zeta_a}(\omega)d\omega, \tag{11}$$

where $\zeta_a$ is the wave amplitude of the regular wave components, and $S_{\zeta_a}(\omega)$ is the spectral density of the defined unidirectional wave spectrum. In this work, JONSWAP spectrum, together with the peakedness factor of 3.3, is used to represent the irregular waves for the North Sea [26].

It must be acknowledged that frequency domain modeling has limited applicability. Firstly, it is restricted to the linear theory. The accuracy of this approach around the resonance of WECs is limited where the motion amplitude is too high and the linear assumption is violated [27]. However, the displacement limit is considered in this paper, which could ease this problem [11]. Secondly, frequency domain modeling does not allow the implementation of real-time control strategies by which PTO parameters can be adjusted instantaneously with the PTO force saturation and buoy displacement constraints [19,28]. Although there are limitations in frequency domain modeling, it is considered reasonable given the purpose of this paper to get insight into the influence of sizing on the techno-economic performance. Frequency domain models are more computationally efficient compared to time domain approaches, which makes it highly suitable in the optimization studies that requires a large number of iterations. In addition, the energy production of WECs in different buoy and PTO sizes is calculated based on the same frequency domain model, which is fair for the size determination and techno-economic analysis.

### 2.3. Size Optimization Method

A proposed size optimization method for improving the techno-economic performance of WECs is presented in the following part, and the methods to conduct PTO sizing and buoy sizing are explained, respectively. An economic model is established to calculate the corresponding costs. The flowchart of this size optimization method is shown in the Figure 2. The cost function for the size optimization adopted in this paper is LCOE, which is introduced in more detail in Section 2.3.3. An exhaustive search algorithm is used in the optimization. The buoy scale factor $\lambda$ and a normalized factor for PTO sizing, namely

PTO sizing ratio, are treated as the optimization variables. Before defining the PTO sizing ratio, it is necessary to introduce the unconstrained PTO force and the maximum required PTO force. The unconstrained PTO force is defined based on the sea state, and it represents the PTO forces required to maximize the power absorption without any force constraint for the particular sea state. The maximum required PTO force is defined based on the sea site and is the largest value of unconstrained PTO forces for all operational sea states in a particular sea site. This largest value occurs at the operational sea state with largest wave power density. Then, the PTO sizing ratio is equal to the PTO force limit divided by the maximum required PTO force at the corresponding sea site, and the PTO force limit is the force constraint for the PTO system with the particular size. The maximum required PTO force varies with the wave resource, the buoy size, and the control strategy. So, in each case, the maximum required PTO force is recalculated for each sea site and buoy scale factor $\lambda$. The initial bounds of the buoy scale factor $\lambda$ and PTO sizing ratio are set as 0.3,2.0 and 0.1,1.0, respectively. If the the optimal solution is not found within these ranges during iterations, the bounds would be automatically extended until a solution is obtained. A discrete iteration step of 0.1 is used in the exhaustive searching algorithm for both buoy scale factors $\lambda$ and PTO sizing ratios. The proposed size optimization method is also compatible with other optimization algorithms, which may provide more precise solutions or save computational costs. However, it is beyond the scope of this paper to discuss the relative impacts of optimization algorithms in detail.

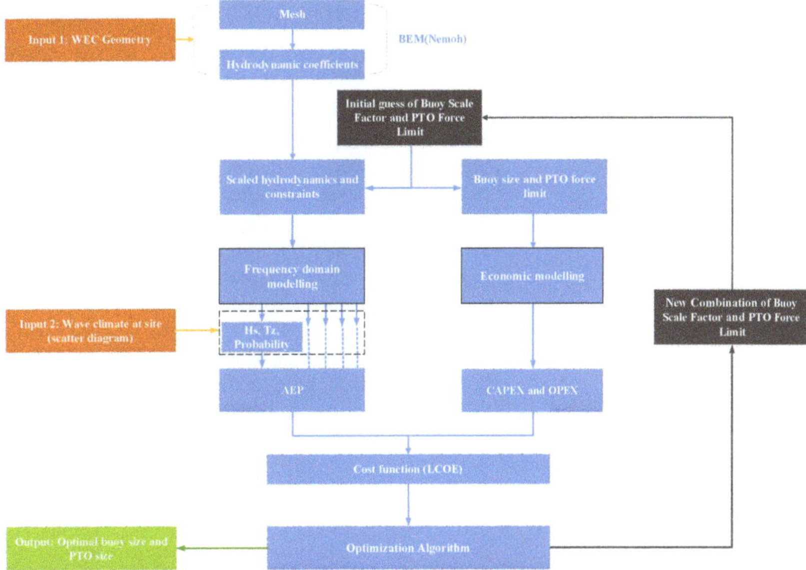


**Figure 2.** Flowchart of the size optimization method of Wave Energy Converters (WECs).

### 2.3.1. PTO Sizing

Here, the PTO sizing implies the implementation of determining the optimal PTO size for different sea sites. The PTO size is directly related to the rated power, PTO force limits, displacement limits, and PTO peak power constraints of WECs [29]. In this work, the PTO force limit is used to represent the PTO size. According to (5) and (8), the PTO force amplitude can be expressed as

$$\left|\hat{f}_{pto}\right| = \left|\hat{u}Z_{pto}\right| = \frac{\left|Z_{pto}\right|}{\left|Z_i + Z_{pto}\right|}\left|\hat{F}_e\right| = \left|\hat{F}_e\right|\frac{\sqrt{R_{pto}^2 + X_{pto}^2}}{\sqrt{(R_{pto} + R_i)^2 + (X_i + X_{pto})^2}}. \tag{12}$$

As is shown in (12), the PTO force amplitude is a function of $R_{pto}$ and $X_{pto}$. Therefore, one approach to restrict the PTO force amplitude is to adjust PTO parameters. Reactive control and passive control are typical control strategies in WECs. In the reactive control, both the PTO reactance and the PTO damping coefficient could be varied to tune the device. However, in the passive control, only the PTO resistance load (damping force) is provided. Next, the methods to determine the PTO parameters for limiting the PTO force amplitude in the passive control strategy and the reactive control strategy are explained, respectively.

- The determination of PTO parameters in passive control:
  Here, the PTO force amplitude is constrained by means of adjusting the PTO parameters. Besides, the PTO parameters are also expected to be determined to limit the stroke and maximize the absorbed power.
  First, let us discuss the PTO force constraints. In the passive control strategy, only PTO damping can be varied and the PTO reactance equals zero. Therefore, the PTO force amplitude expressed in (12) can be simplified as

$$|\hat{F}_{pto}| = |\hat{F}_e| \frac{R_{pto}}{\sqrt{(R_{pto} + R_i)^2 + X_i^2}}. \tag{13}$$

To reveal the relationship between $|\hat{F}_{pto}|$ and $R_{pto}$, the derivation of (13) with respect to $R_{pto}$ is calculated and gives

$$\frac{d(|\hat{F}_{pto}|)}{dR_{pto}} = |\hat{F}_e| \frac{R_{pto}R_i + X_i^2 + R_{pto}^2}{\left[(R_{pto} + R_i)^2 + X_i^2\right]^{\frac{3}{2}}}. \tag{14}$$

It can be deduced that (14) is always positive as $R_{pto}$ and $R_i$ are greater than 0, which also implies that $|\hat{F}_{pto}|$ is a monotonic function of $R_{pto}$. In other words, limiting the PTO damping can directly constrain the PTO force amplitude. During size optimization, the PTO sizing ratio and buoy scale factor $\lambda$ are used as optimization variables. However, to determine PTO parameters with the force limit, the PTO force limit should be derived to be explicit. According to the definition of the PTO sizing ratio, the PTO force limit can be calculated by multiplying the given PTO sizing ratio with the maximum required PTO force for the particular buoy scale factor and considered sea site. Therefore, the PTO force limit is directly related to each set of optimization variables. Here, the PTO force limit is represented by $F_{pto\_limit}$, and the maximum allowed PTO damping $R_{pto\_force}$ can be obtained by solving (15), in which only the positive solution should be retained.

$$|\hat{F}_e| \frac{R_{pto\_force}}{\sqrt{(R_{pto\_force} + R_i)^2 + X_i^2}} = F_{pto\_limit}. \tag{15}$$

Therefore, for constraining the PTO force amplitude, $R_{pto}$ should satisfy

$$R_{pto} \leq R_{pto\_force}. \tag{16}$$

Secondly, except PTO force constraints, the displacement limit should be considered during the selection of PTO parameters. It can be deduced from (8) that the $\hat{u}$ decreases with $R_{pto}$ increasing. If the stroke constraint of the buoy, $S_m$, comes to play, the PTO resistance should be increased to limit the velocity amplitude, which is shown as

$$|\hat{u}| = \frac{|\hat{F}_e|}{|R_{pto} + Z_i|} \leq |u_m|. \tag{17}$$

In this way, for constraining the stroke amplitude, $R_{pto}$ should satisfy

$$R_{pto} \geq \sqrt{(\frac{\hat{F}_e}{u_m})^2 - X_i^2} - R_i = R_{pto\_stroke}, \tag{18}$$

where $u_m$ is the velocity limit, which is equal to $\omega S_m$. Therefore, it can be seen from (16) and (18) that the upper bound and the lower bound of the available PTO damping are decided by the PTO force limit and the stoke limit, respectively, which could also be expressed as

$$R_{pto\_stroke} \leq R_{pto} \leq R_{pto\_force}. \tag{19}$$

Thus, the PTO damping should be selected from the range expressed in (19) to satisfy the constraints.

According to Reference [28], the optimal PTO damping for maximizing the absorbed power without any constraint is expressed as

$$R_{pto\_opt} = |Z_i| = \sqrt{R_i^2 + X_i^2}. \tag{20}$$

To maximize the absorbed power of WECs, $R_{pto}$ should be as close to $R_{pto\_opt}$ as possible. So, the principle of PTO damping selection in PTO sizing can be presented as:

- If $R_{pto\_stroke} \leq R_{pto\_opt} \leq R_{pto\_force}$, the optimal $R_{pto}$ should be selected as $R_{pto\_opt}$.
- If $R_{pto\_opt} < R_{pto\_stroke}$ or $R_{pto\_opt} > R_{pto\_force}$, the optimal $R_{pto}$ should be selected as the one of $R_{pto\_stroke}$ and $R_{pto\_force}$ which is closer to $R_{pto\_opt}$.
- In case $R_{pto\_force} < R_{pto\_stroke}$, there is no feasible PTO damping coefficient satisfying both of the constraints. This case would happen when the PTO force limit is very low or the wave power is very high, which realistically means the device has to be stopped from operation for protecting itself from frequently violating the physical constraints.

- The determination of PTO parameters in reactive control:
Unlike PTO sizing in passive control, both $R_{pto}$ and $X_{pto}$ can be varied to meet the requirement of motion and PTO force constraints. Given the complexity of the multivariable optimization with nonlinear constraints, a numerical optimization tool is used to select the optimal combination of $R_{pto}$ and $X_{pto}$, and it can be expressed in the form as

$$maximize f = P_{absorbed}(R_{pto}, X_{pto})$$

$$subject\ to \begin{cases} |\hat{F}_{pto}(R_{pto}, X_{pto})| \leq F_{pto\_limit} \\ |\hat{u}(R_{pto}, X_{pto})| \leq u_m \\ R_{pto} \geq 0 \\ X_{pto} \in \mathbb{R}. \end{cases} \tag{21}$$

The optimization is performed based on the "interior point" algorithm in MATLAB environment, and the tolerance of the function is set as 1e-4. To avoid the local optimal solution, the 'MultiStart' solver is adopted. In this solver, iterations start with multiple random points, in which the global optimal solution is expected to be found [30]. In this work, the number of multiple starting points is set as 20, and the bounds of the PTO damping and PTO reactance are set as $[0,10\,R_i(\omega)]$ and $[-10X_i(\omega), 10X_i(\omega)]$ for each sea state. In case that no feasible solution is found in the optimization, the PTO absorbed power would be treated as 0.

Based on the above method, PTO parameters for different PTO force constraints can be obtained for each wave condition. Then, the corresponding power performance of WECs can be calculated. Hence, this method takes into account the effects of PTO sizing on the power performance of WECs. However, it has to be clarified that the above PTO sizing method is established based on regular wave conditions. To constrain the buoy displacement and PTO force in irregular wave conditions, it is necessary to calculate their instantaneous solutions, in which time domain modeling is required. However, the inefficiency of time domain simulation would make the iteration process much more time consuming, which is not preferable for the optimization problem. To simplify this problem, in this paper PTO parameters are selected only to suit the typical characteristics of irregular wave conditions, referring to Reference [18]. According to Reference [10], the time-averaged power transport per unit length of wave front of incoming waves at regular wave conditions and irregular wave conditions can be calculated as

$$P_{wave\_re} = \frac{1}{32\pi}\rho g^2 H^2 T, \tag{22}$$

$$P_{wave\_irr} = \frac{1}{64\pi}\rho g^2 H_s^2 T_e. \tag{23}$$

By equaling (22) to (23) at the case of the same energy period, namely $T$ equaling $T_e$, the corresponding wave height in regular wave condition is solved as $H_s/\sqrt{2}$. To transfer the $T_z$ in scatter diagrams to $T_e$, the wave period ratio between $T_e$ and $T_z$ is selected as 1.18 given the JONSWAP spectrum and the peakedness factor of 3.3 [31]. Then, PTO parameters for irregular wave conditions can be selected to suit regular wave conditions whose period and height correspond to $T_e$ and $H_s/\sqrt{2}$, respectively [18]. The purpose of the transfer between irregular wave conditions and regular wave conditions is to simplify the determination of PTO parameters for PTO sizing. However, the selected PTO parameters based on regular wave condition cannot strictly guarantee that the PTO force and stroke constraints would not be violated at the corresponding irregular wave conditions. Considering the purpose of this paper to investigate the impacts of sizing on the techno-economic performance, it is considered to be acceptable. In this paper, all the power absorption of WECs are calculated based on irregular wave conditions, and the PTO parameters are optimized for each sea state. As an example, the optimized PTO parameters of WEC in the original buoy size for different sea states are shown in Figures 3 and 4.

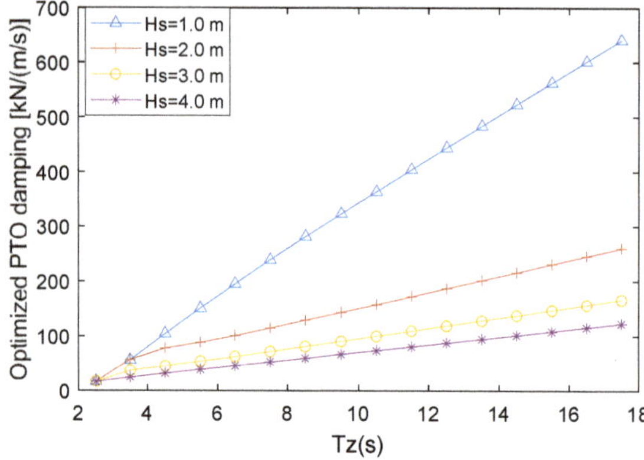


**Figure 3.** Optimized Power Take-Off (PTO) damping of the WEC with passive control for various sea states ($\lambda = 1$ and PTO force limit = 50 kN).

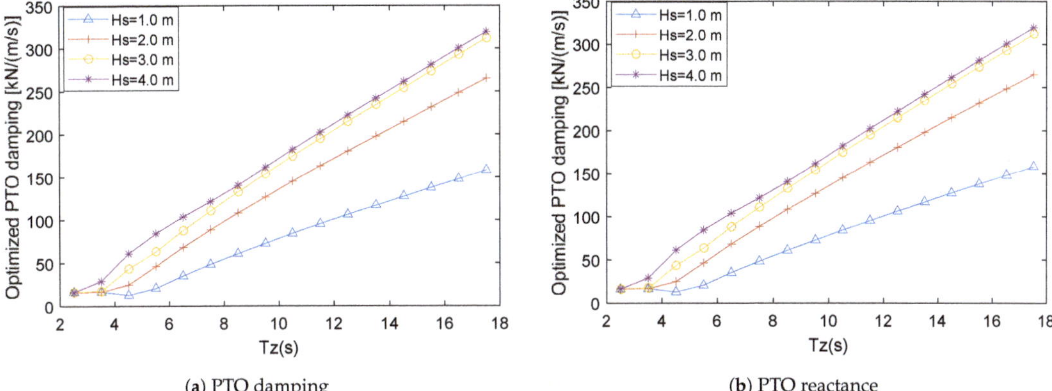

(a) PTO damping

(b) PTO reactance

**Figure 4.** Optimized PTO parameters of the WEC with reactive control for various sea states ($\lambda = 1$ and PTO force limit = 250 kN).

Therefore, as the PTO parameters are determined, the absorbed power of WECs at each sea state can be obtained and the AEP (Annual Energy Production) at the specific sea site is calculated as

$$AEP = A \cdot \eta \cdot \sum_{x=1}^{x=n} P_{absorbed}(x) \cdot T(x), \tag{24}$$

where $\eta$ is the overall conversion efficiency from the annual absorbed energy to the AEP and is assumed 70% [32]; $A$ is the availability of WECs to work, and it is set as 90% due to the necessary operation and maintenance [23]; $T$ represents the total hours of the appearance of a certain sea state, which is presented in the scatter diagram in Tables 1–3; $x$ represents the sea state in scatter diagrams.

### 2.3.2. Buoy Sizing

During the iterations of the size optimization, the buoy size of WECs are scaled following geometrical similarity. Therefore, the hydrodynamic coefficients of buoys in different sizes can be obtained by means of Froude scaling law [33],

$$\lambda = \frac{L_s}{L_o} = \frac{H_s}{H_o}$$
$$\omega_s = \omega_o \lambda^{-0.5}$$
$$F_{e\_s} = F_{e\_o} \lambda^3, \tag{25}$$
$$B_{r\_s} = B_{r\_o} \lambda^{2.5}$$
$$M_{r\_s} = M_{r\_o} \lambda^3$$

where $\lambda$ is the buoy scale factor; $L$ is the geometrical length of the buoy; $\omega$ is the wave frequency; $F_e$, $B_r$, and $M_r$ are the excitation force, the radiation damping, and the added mass coefficients, respectively; and the subscript $s$ and $o$ represent the "scaled device" and the "original device", respectively. Hence, this allows the hydrodynamic coefficients of the original buoy to be transferred to the scaled buoy instead of using BEM method to calculate the hydrodynamic coefficients in each iteration. In this way, the computing efficiency can be significantly improved. The density of the buoy structure in different sizes are assumed to be same. The maximum operation wave height and the displacement limit of WECs are scaled with the buoy scale factor $\lambda$.

### 2.3.3. Economic Modeling for Cost Estimation

LCOE is an important techno-economic metric of WECs. For evaluating the LCOE, it is essential to establish an economic model to estimate the CAPEX and OPEX of WECs.

Following Reference [34], the steel price is selected as 1.6 GBP (British Pounds)/kg and the structure cost is calculated by assuming that all the structure cost comes from the steel cost. Based on the inflation calculator tool [35], the cumulative inflation rate of GBP from 2017 to 2020 is 5.89% and the exchange rate of Euros to GBP is set as 0.87. Referring to Reference [34], the statistical percentages of CAPEX-related components in total LCOE can be found. The percentage values are recalculated as the average percentage in total CAPEX, shown in Table 4. According to Table 4, the cost of "Foundation and Mooring" and "Installation" accounts for 19.1% and 10.2% averagely of CAPEX, respectively. Comparatively, the cost of the structure accounts for 38.2% of CAPEX in average. Therefore, mass-related capital cost can be calculated as

$$C_{Mass} = C_S + C_F + C_I = (\frac{P_{F\&M}}{P_S} + \frac{P_I}{P_S} + 1)C_S, \tag{26}$$

where $C_{Mass}$ represents the Mass-related-capital-cost; $C_S$, $C_F$, and $C_I$ are the cost of the structure, foundation, and the installation, respectively, and $P_s$, $P_{F\&M}$, and $P_I$ are their corresponding percentages in the total LCOE . It can be seen from Table 4 that the cost of "Connection" and "PTO" averagely accounts for 8.3% and 24.2% of CAPEX, respectively. Similarly, power-related capital cost can be calculated as

$$C_{Power} = C_P + C_C = (\frac{P_C}{P_P} + 1)C_{PTO}, \tag{27}$$

where $C_{Power}$ represents the power-related capital cost; $C_P$ and $C_C$ are the cost of the PTO and the connection, respectively; and $P_P$ and $P_C$ are their corresponding percentages in the total LCOE. Therefore, the CAPEX is calculated as

$$CAPEX = C_{Mass} + C_{Power}. \tag{28}$$

**Table 4.** Percentages of Capital Expenditure (CAPEX)-related components of WECs in total CAPEX.

| CAPEX | Categories | Average Percentage |
|---|---|---|
| Mass-related capital cost | Structure<br>Foundation and mooring<br>Installation | $P_S = 38.2\%$<br>$P_{F\&M} = 19.1\%$<br>$P_I = 10.2\%$ |
| Power-related capital cost | PTO component<br>Connection | $P_P = 24.2\%$<br>$P_C = 8.3\%$ |

In this paper, PTO is assumed to be a direct drive generator and all PTO costs come from the generator. The generator cost is divided into the cost of active material and the cost of manufacturing. The amount of active material required is approximately related to the PTO force limit and the force density of generators. Referring to Reference [36], the maximum force density in this work is assumed as 44 kN/m$^2$, which generally ranges from 30 to 60 kN/m$^2$ depending on the design. The cost of active material of this generator in series production is estimated as 12,000 Euros/m$^2$ based on the currency value in 2006. The cumulative inflation rate from 2006 to 2020 is 22.1% [35]. Taking the inflation into account, the active material of this generator in this paper is estimated as 14,655.31 Euros/m$^2$. Regarding the manufacturing cost, it is approximately assumed as half of the total cost of the generator [37]. So, the cost of PTO can be expressed as

$$C_{PTO} = Cost_{Material} + Cost_{Manufacturing}. \tag{29}$$

In this paper, the annual OPEX is assumed as 8% of the CAPEX, and the discount rate $r$ is assumed as 8% with the lifespan of 20 years, referring to Reference [8]. Then, the LCOE of WECs is calculated as

$$LCOE = \frac{CAPEX + \sum_{t=1}^{n} \frac{OPEX_t}{(1+r)^t}}{\sum_{t=1}^{n} \frac{AEP_t}{(1+r)^t}}, \tag{30}$$

where $n$ represents the total years of the lifespan, and $t$ represents the evaluated year.

It has to be clarified that it is a preliminary economic model, and the parameters in the model differ from one to another project in practice. For instance, reactive control is associated with negative power flow, which could lead to larger losses and related wear. Therefore, control strategies in practice is able to affect the OPEX and conversion efficiency. However, the specific effects are related to the PTO design and maintenance strategy, which is outside the scope of this paper. Given the purpose of this paper to identify the influence of sizing on the techno-economic performance, the assumption on the constant OPEX percentage and conversion efficiency for both control strategies is considered reasonable. Furthermore, survivability of WECs in practice is complex and related to many affecting factors. For instance, the increase of the buoy size results in the larger exerted force and input power flow, which could make the WEC more vulnerable. However, it is also dependent on the mooring design, material and even control strategies of WECs. For simplicity, the lifespan for WECs in all sizes is assumed to be constant. Nevertheless, our aim based on the economic analysis is not to give a final judgement of the optimal size of the WEC but to use the LCOE as an indication for providing an insight about the effects of sizing on the techno-economic performance. Overall, the proposed size optimization method has pure theoretical characteristics, and a more complex size optimization study is required in practical applications.

## 3. Results and Discussion

This section starts with the discussion about the effects of buoy sizing and PTO sizing on the performance of the WEC. Next, the size optimization results for the sea sites are presented. The interaction between PTO sizing and buoy sizing and the benefits of downsizing PTO size for decreasing LCOE are analyzed. Finally, a comparison between this proposed method and other existing methods for size optimization is performed.

### 3.1. The Effects of Sizing on the Performance of the Wec

Taking one single sea state ($H_s = 1.5$ m, $T_z = 5$ s) as an example, the effects of PTO sizing and buoy sizing on the performance of the WEC are investigated. In Figures 5 and 6, the effects on the absorbed power, economic performance and PTO parameters of the WEC with passive control and reactive control are presented, respectively. The horizontal axis in Figures 5 and 6 is expressed as "PTO force limit/unconstrained PTO force". Here, the unconstrained PTO force corresponds to the PTO force required to maximize the power absorption for the considered sea state ($H_s = 1.5$ m, $T_z = 5$ s). The PTO parameters corresponding to each PTO size and buoy size were calculated following the method described in Section 2.3.1. From Figures 5 and 6, it is noted that both the power performance and the economic performance are highly related to the sizing of the WEC. In Figures 5a and 6a, it can be seen that the absorbed power of the WEC increases with the PTO size and the buoy size. In Figures 5b and 6b, the proportion of the power-related capital cost to the total CAPEX increases with the PTO size, but it decreases with the rise of the buoy scale factor $\lambda$. Therefore, at a certain PTO sizing ratio, the CAPEX would be more dominated by the mass-related capital cost than power-related capital cost with the increase of the buoy scale factor $\lambda$. Comparing Figures 5 and 6, it can be found that the effects of sizing on the WEC are also related to the control strategy. Firstly, the absorbed power of the WEC with reactive control is significantly higher than that in passive control. Secondly, the proportion of power-related capital cost to the total CAPEX in the WEC with reactive control is much higher than that with passive control. This phenomenon can be explained by that the reactive control strategy is associated with higher PTO force limits than the passive control strategy. The higher PTO force then leads to the increase of the proportion Thirdly, the trends of PTO parameters changing with the force limit depend on the control strategy. In Figure 5c, the PTO damping coefficient increases with the ratio "PTO force limit/unconstrained PTO force" when the passive control strategy is used. This is logical as the PTO force monotonically increases with the PTO damping, which has been explained in (14). Comparatively, it can be seen from Figure 6c,d, with the increase of "PTO force limit/unconstrained PTO force", the PTO damping coefficient tends to decrease, while the PTO reactance increases. The reason is that the PTO in reactive control would act more like a pure damper to reduce its required force, when the force constraint becomes tighter.

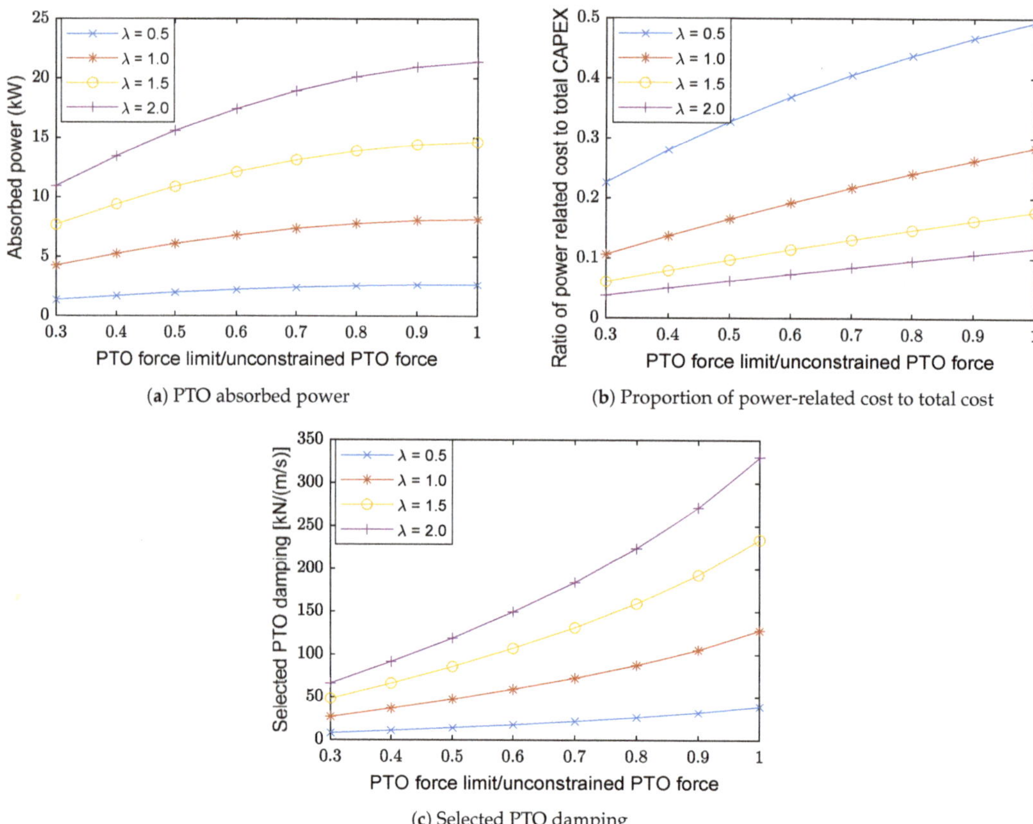

(a) PTO absorbed power

(b) Proportion of power-related cost to total cost

(c) Selected PTO damping

**Figure 5.** Performance of the WEC with passive control under different PTO sizes and buoy scale factor $\lambda$, at $H_s = 1.5$ m and $T_z = 5$ s.

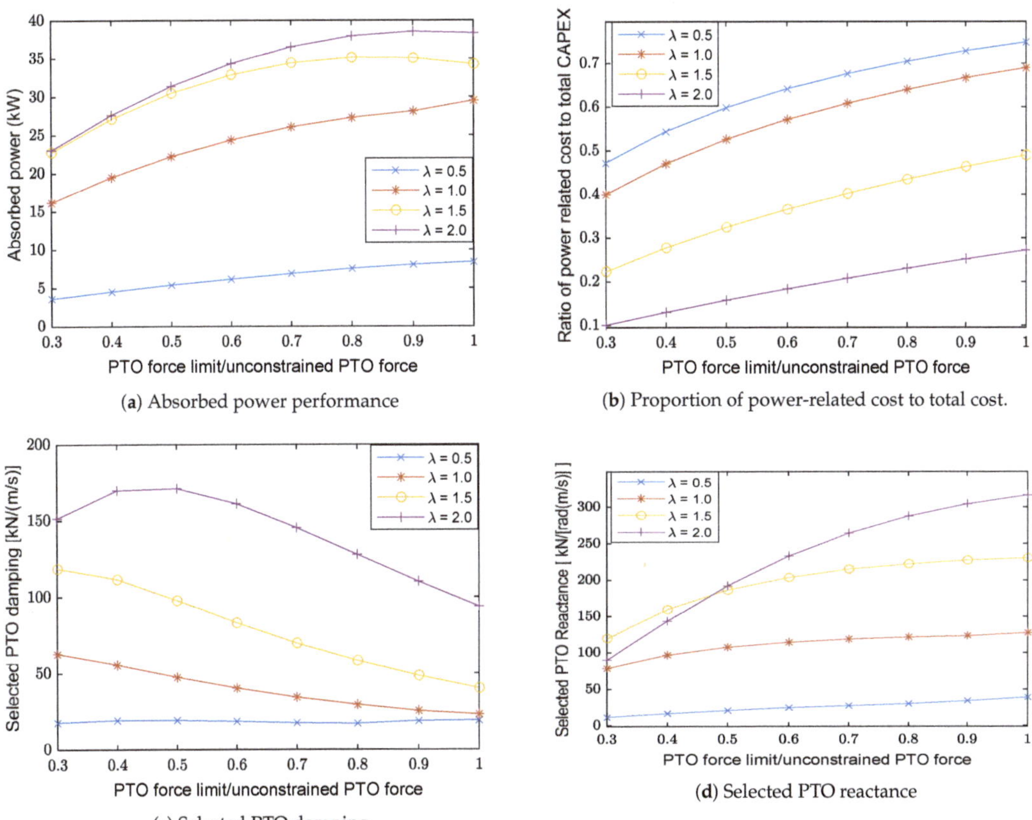

(a) Absorbed power performance

(b) Proportion of power-related cost to total cost.

(c) Selected PTO damping

(d) Selected PTO reactance

**Figure 6.** Performance of the WEC with reactive control under different PTO sizes and buoy scale factor $\lambda$, at $H_s = 1.5$ m and $T_z = 5$ s.

### 3.2. Size Optimization for Typical Realistic Sea Sites

#### 3.2.1. Results of Size Optimization

Based on the proposed method, the size optimization of the WEC for three sea sites is performed. First, to define the PTO sizing ratio, it is necessary to obtain the maximum required PTO forces corresponding to each buoy size. They are calculated by (12), and the results are shown in Figure 7. Figure 7a shows the relation of unconstrained PTO forces of the WEC in original buoy size to sea states, and the maximum required PTO forces for different sea sites are picked. It can be found from Figure 7b that the maximum required PTO force increases dramatically with the increase of the buoy size. In addition, the maximum required PTO forces in the WEC with reactive control are much higher than those with passive control. This is to be expected as PTO forces in reactive control consist of PTO damping-induced forces and PTO reactance-induced forces, while there are only PTO damping-induced forces in the case of passive control.

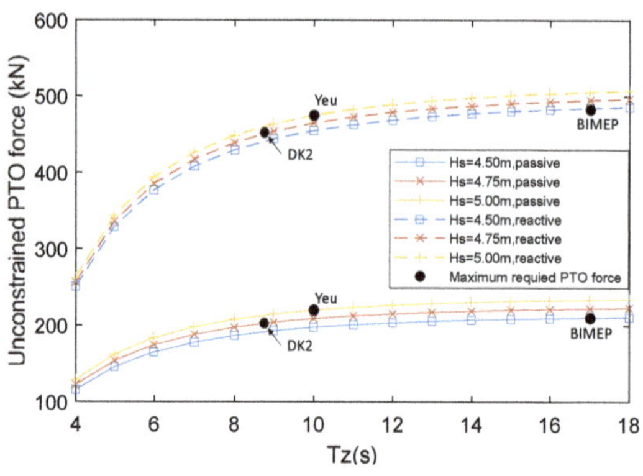

(**a**) Unconstrained PTO force vs $T_z$ ($\lambda = 1$)

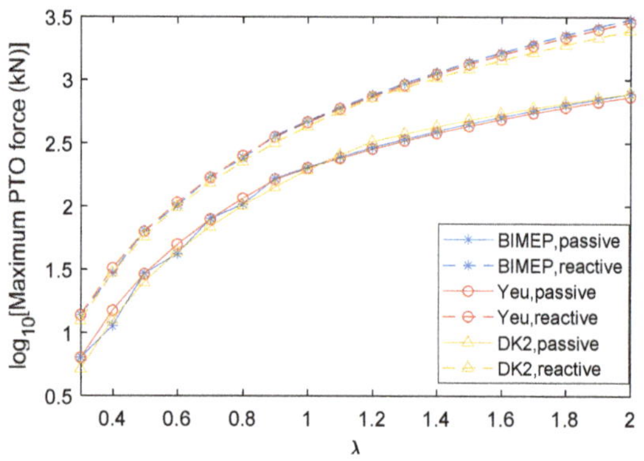

(**b**) Maximum required PTO force vs buoy scale factor $\lambda$

**Figure 7.** Maximum required PTO forces for various sea sites and buoy scale scale factor $\lambda$.

The size optimization results of the WEC using passive and reactive control for the three sea sites are depicted in Figures 8 and 9, respectively. It can be clearly seen from these figures that both the LCOE and the AEP can be significantly influenced by sizing of the WEC, no matter in which sea site or with what kind of control strategies. Therefore, for improving the viability of the WEC, it is highly suggested to conduct size optimization of the WEC for the considered wave resources. Next, it can be noted that upscaling buoy size is able to improve the AEP, while it cannot necessarily reduce the LCOE of the WEC. Similarly, the AEP is highly sensitive to the PTO sizing ratio, and the increase of the PTO size can make a significant contribution to the improvement of the AEP. However, from the techno-economic point of view, enlarging PTO size does not necessarily result in a lower LCOE. In this case, it can be noted that downsizing the PTO size to a suitable level is beneficial for reducing the LCOE. Hence, to improve the techno-economic performance, it is significant to conduct PTO sizing for compromising the AEP and the cost.

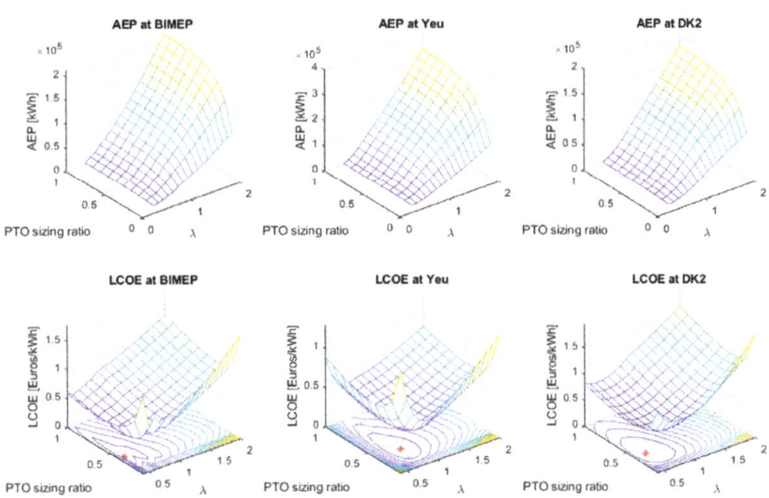

**Figure 8.** Size optimization of the WEC with passive control.

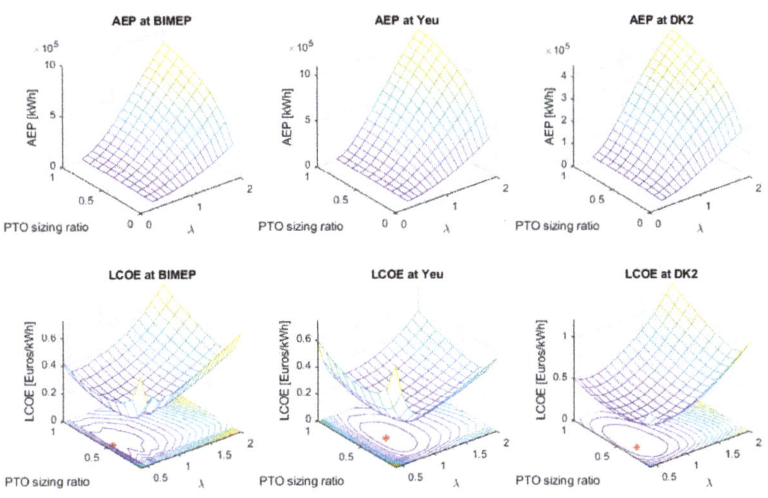

**Figure 9.** Size optimization of the WEC with reactive control.

In Figure 10, the dependence of size optimization of the WEC on wave resources and control strategies is shown. From Figure 10a,d, it can be found that there is not a direct relationship between the buoy size determination and the mean wave power density of wave resources. In other words, the optimal buoy size cannot be indicated by the mean wave power density. For instance, the mean wave power density in BIMEP is almost twice as much as that in DK2, but DK2 corresponds to a higher optimal buoy scale factor $\lambda$. As is seen in Figure 10a, control strategies do not have a notable influence on the buoy size determination for a given sea site. The reason is that the trends of the AEP changing with the buoy scale factor $\lambda$ are comparable in both cases of reactive and passive control. Though control strategies lead to an notable difference in the absolute values of the

AEP. Regarding PTO sizing, it can be found from Figure 10b that the optimal PTO sizing ratios in the WEC with the reactive control are slightly higher than those with passive control in BIMEP and Yeu. The only exception occurs in DK2 where the reactive and the passive control are associated with the same optimal PTO sizing ratio. In addition, it is noteworthy that the optimal PTO sizing ratio is relatively independent of wave resources, and it converges at around 0.4 to 0.5.

Different from the size determination, the optimized LCOE is highly related to wave resources and control strategies. It can be found from Figure 10c that the LCOE of the WEC with reactive control is much lower than the WEC with passive control, and the reduction can reach at 35 % in average for these sea sites. The optimized LCOE of the WEC with reactive control ranges around 0.2 to 0.35 Euros/kWh, while this value ranges around 0.35 to 0.55 Euros/kWh in the case of passive control. This is to be expected since the WEC with reactive control produces much more power than the WEC with passive control at the same sea state. From Figure 10c,d, it can be found that the higher the mean wave power density, the lower the optimal LCOE.

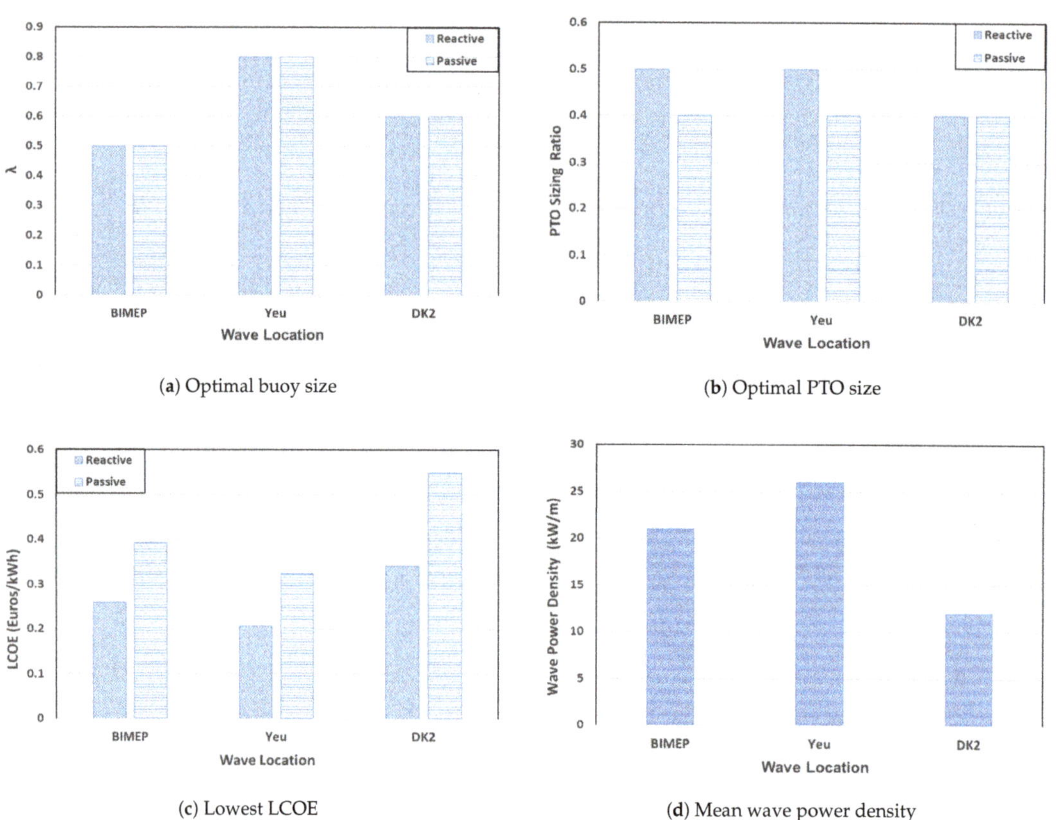

**Figure 10.** The dependence of size optimization on wave resources and control strategies.

After the optimal buoy and PTO size have been determined for each sea site, the optimal cost proportion of the WEC can be obtained. The proportion of the PTO cost to the total CAPEX at the optimal sizing condition is shown in Figure 11. It can be seen that this cost proportion tends to be relatively independent of wave resources, while it is highly related to the control strategy of the WEC. In this case, the PTO cost of the WEC with reactive control accounts for 45% to 50% of the total CAPEX. Comparatively, for the WEC with passive control, this proportion decreases dramatically to around 30%. This can be explained by that the required PTO forces in the WEC with reactive control are much higher than those with passive control. It also implies that the PTO size in the WEC with reactive control should be designed larger than that with passive control. Besides, it is noticed that the optimized cost proportion of PTO is much higher than the statistic value of 24.2% depicted in Table 4. The reason is that the WECs investigated in the literature [34] are generally in large scales, and the costs of the structure are dominating. However, the optimized buoy size of the WEC in this case is relatively small, with the diameter ranging around 2.5 m to 4 m ($\lambda = 0.5 - 0.8$). As a consequence, the PTO cost is more weighted compared with the structure cost. This phenomenon has also been explained in Section 3.1 as the proportion of power-related capital cost decreases with the buoy scale factor $\lambda$.

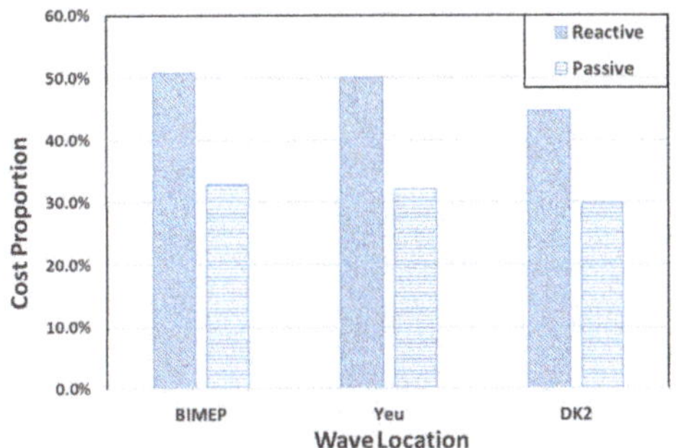

**Figure 11.** The proportion of cost on PTO to total CAPEX at the optimal buoy size and PTO size.

3.2.2. The Benefits of PTO Downsizing for the Techno-Economic Performance

The benefits of PTO sizing for reducing the LCOE of the WEC at the optimal buoy sizes are presented in Figure 12. It can be seen that the LCOE can be significantly reduced by downsizing the PTO size even though the buoy sizes have been optimized. In this case, downsizing the PTO sizing ratio to around 0.4 to 0.5 is preferable to minimize the LCOE. For the WEC with the passive control, downsizing the PTO size is able to reduce the LCOE by 24% to 31%, and it could reduce the LCOE by 24% to 25% in the case of reactive control. Hence, it is essential to take PTO size optimization into account when conducting sizing of the WEC. It also indicates that the techno-economic performance of WECs are generally underestimated due to the absence of PTO sizing in evaluation studies.

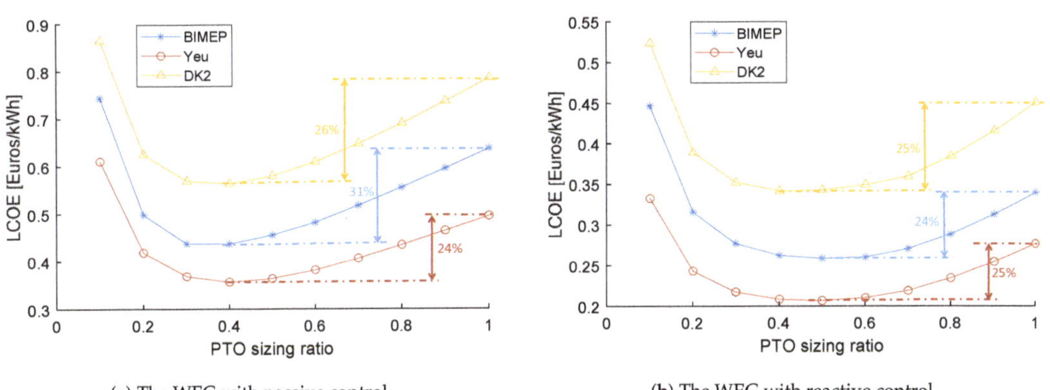

(a) The WEC with passive control          (b) The WEC with reactive control

**Figure 12.** The effects of PTO downsizing on Levelized Cost Of Energy (LCOE) of the WEC at the optimal buoy size.

### 3.2.3. The Interaction between PTO Sizing and Buoy Sizing

The interaction between PTO sizing and buoy sizing is shown in Figure 13. It can be seen that the optimal buoy size generally declines with the corresponding PTO sizing ratio. However, this effect is limited and the optimal buoy size tends to be constant as the PTO sizing ratio is higher than 0.3 or 0.4. Hence, in this case, it can be noted that the buoy size optimization can be influenced by PTO sizing, but only to a limited extent. However, it should be pointed out that the interaction between PTO sizing and buoy sizing is related to wave resources, WEC principles, and economic parameters. Therefore, for avoiding the misestimate of the optimal buoy size, it is suggested to conduct PTO sizing simultaneously with buoy sizing. Besides, comparing Figure 13a,b, it is observed that there is no notable difference between the WEC with passive control and reactive control regarding the impact of PTO sizing on the buoy size determination.

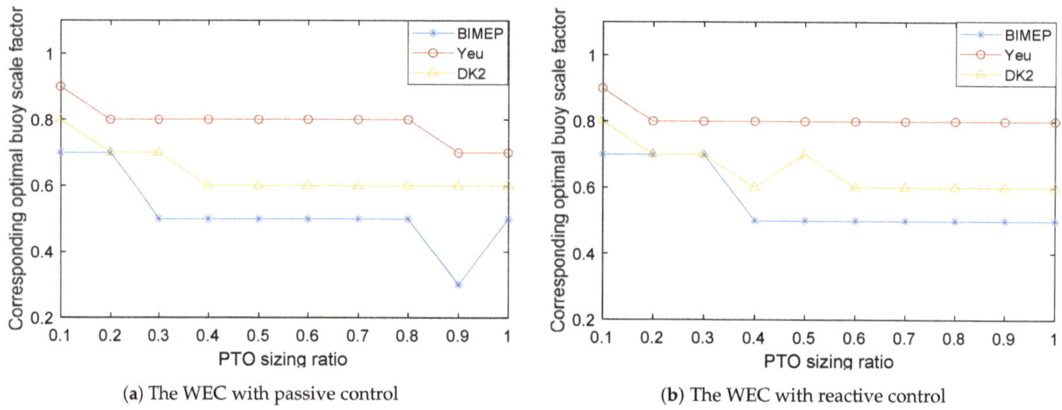


(a) The WEC with passive control

(b) The WEC with reactive control

**Figure 13.** The interaction between PTO and buoy sizing.

### 3.3. The Proposed Method Versus Other Size Optimization Methods

#### 3.3.1. Budal Diagram

Budal diagram is a useful tool to estimate the theoretical absorbed power of the WEC. According to Budal diagram, there are two upper absorbed power bounds in regular wave conditions, namely $P_A$ and $P_B$. $P_A$ bound is related to the maximum amount of power that could be extracted from incoming waves, while $P_B$ reflects the bound of power that could be absorbed by the realistically sized WEC [9,11]. $P_A$ corresponds to the maximum absorbed power at the high wave frequency limit and is expressed as

$$P_A = J/k = \frac{\rho g^3 H^2 T^3}{128 \pi^3};$$ (31)

here, $\rho$ is the water density; $g$ is the gravity acceleration; $H$ is the wave height; $\omega$ is the wave frequency of incoming waves; $k$ is the wave number; and $J$ is the wave-energy transport per unit frontage of the incident wave and deep water condition. Another power bound $P_B$ corresponds to the maximum absorbed power at the low wave frequency limit and is expressed as

$$P_B = \frac{\pi \rho g H V}{4T},$$ (32)

where $V$ is the volume of the buoy. The size of the WEC should match the wave resource to enable the WEC viable. Therefore, combined with the information of wave resources, Budal diagram could be used to select the suitable size of the WEC [9,11]. However, for calculating $V$ in (32), the designed working condition ($H$ and $T$) should be explicit. In Reference [11], the WEC is assumed to be commercially viable if the amount of working time at full capacity exceeds one third of the annual time. Thus, the size of the WEC should match "one third wave power threshold" of the wave resource. Based on the power threshold, the wave height $H_D$ and wave period $T_D$ in the designed wave condition can be calculated. The selection of suitable size of the WEC is conducted following these steps [9]:

1. calculate the wave power threshold $J_T(W/m)$ which is being exceeded one third of the annual time in the concerned wave climate;
2. choose the most frequent wave period in scatter diagrams as the designed wave period $T_D$ which corresponds to $T$ in (31) and (32);
3. since $T_D$ and $J_T$ are already known from step 1 and 2, in harmonic waves, the wave height $H_D$ can be calculated;
4. the suitable volume $V$ can be calculated by solving $P_A = P_B$; and
5. finally, as the buoy volume is determined, the optimal PTO force limit is selected as the value which is required to maximize the absorbed power of the WEC at the designed wave condition ($H_D$ and $T_D$).

Therefore, the suitable size of the WEC could be estimated by Budal diagram without calculating the power performance of the WEC in different sizes. However, in this method, the volume $V$ depends on the assumption of viable conditions, such as "working at full capacity over one third of the annual time" [11]. In addition, Budal diagram is derived as the theoretical power bounds without considering the influence of PTO control strategies. So, the dependence of buoy size determination on the control strategies of the WEC cannot be reflected in this method. The size selection of the WEC is shown in Table 5.

**Table 5.** Size selection based on Budal diagram.

| Sea Site | Wave Power Density Threshold | $T_D$ | $H_D$ | Buoy Volume | PTO Force Limit | |
|---|---|---|---|---|---|---|
| | | | | | Reactive | Passive |
| Yeu Island | 30.4 kW/m | 7.8 s | 2.00 m | 224 m³ | 985 kN | 216 kN |
| BIMEP | 14.6 kW/m | 8.4 s | 1.33 m | 206 m³ | 946 kN | 145 kN |
| DK2 | 8.2 kW/m | 5.4 s | 1.24 m | 33 m³ | 142 kN | 36 kN |

3.3.2. The Size Optimization without PTO Downsizing

As is introduced in Section 1, existing literature regarding size optimization of WECs focused only on buoy sizing without considering PTO sizing. In those studies, the buoy size was optimized for the chosen sea sites, but the PTO size was simply scaled with the buoy scale factor $\lambda$ [8,13]. This kind of buoy size optimization is conducted based on Froude scaling. With the aim to compare different size optimization methods, the size optimization without PTO sizing is conducted as a reference in this paper. The original buoy diameter is still defined as 5 m, the original PTO size is selected to sustain the maximum required force for the considered sea site, namely without PTO downsizing. Then, PTO force limits of the WEC in other buoy scale factors $\lambda$ are scaled following Froude law, as (33).

$$F_{PTO\_limit\_scaled} = F_{PTO\_limit\_original}\lambda^3. \tag{33}$$

The power performance and the cost of the WEC at each buoy size are calculated by frequency domain modeling and the economic modeling, respectively. Therefore, the LCOE of the WEC at each buoy scale factor $\lambda$ can be obtained. The optimization results are shown in Figure 14.

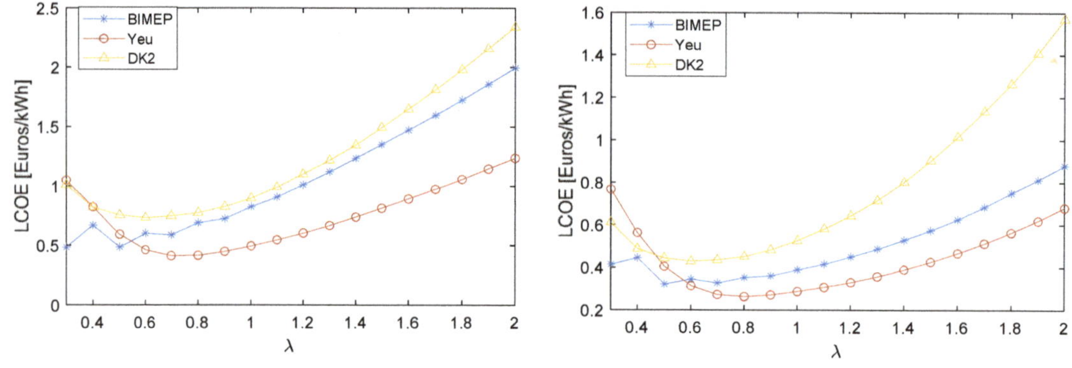

(**a**) LCOE vs buoy scale factor $\lambda$, passive control     (**b**) LCOE vs buoy scale factor $\lambda$, reactive control

**Figure 14.** The size optimization of the WEC without PTO sizing.

### 3.3.3. Comparison of Size Optimization Methods

As is explained above, there are several methods available to conduct the size optimization for improving techno-economic performance of the WEC. A comparison among these methods is performed and the results are shown in Figure 15. Firstly, from Figure 15a,b, it can be observed that Budal diagram is not capable of determining the suitable sizes of the WEC. The deviation of the selected size between Budal diagram and the proposed method differs with wave resources, and this deviation tends to be random. For instance, the selected buoy scale factor $\lambda$ for BIMEP is three times as much as that estimated by this proposed method, while the difference of selected sizes for DK2 is relatively small. However, as a theoretical and efficient approach, Budal diagram can be used to narrow the scope of size selection for potential sea sites. Secondly, compared with Budal diagram, size optimization without PTO downsizing shows a better ability to estimate the suitable buoy size of the WEC. Generally, without PTO sizing, the buoy size optimization can still acquire the suitable buoy size. This phenomenon also verifies the finding in Section 3.2.3. Thirdly, it can be seen from Figure 15c,d that the LCOE of the WEC optimized by this proposed method is clearly lower than those by the other two methods, no matter which control strategy is adopted. Hence, it can be concluded that this proposed method is able to result in a further improvement on the techno-economic performance of the WEC.

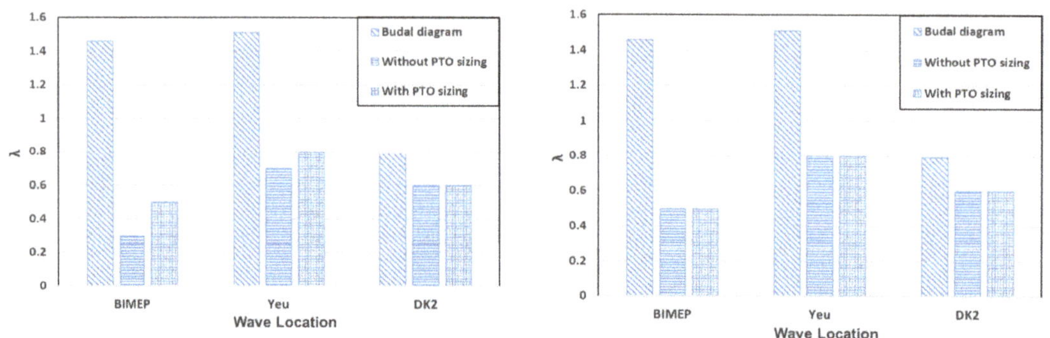

(**a**) Buoy size determination, passive control     (**b**) Buoy size determination, reactive control

**Figure 15.** *Cont.*

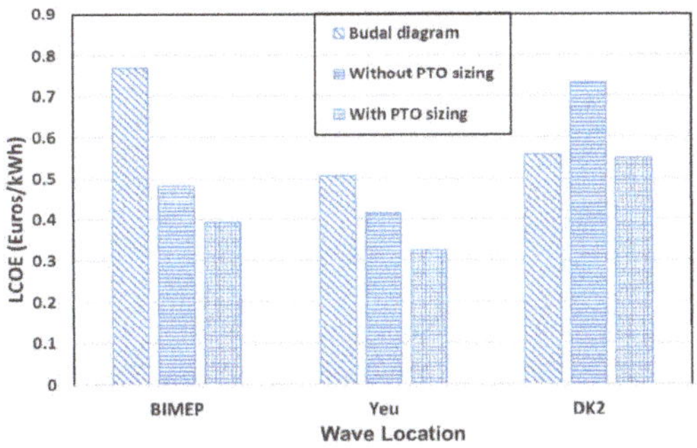

(**c**) Lowest LCOE, passive control

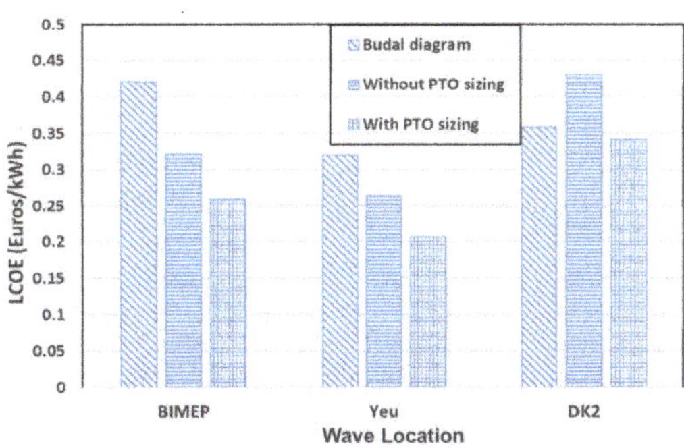

(**d**) Lowest LCOE, reactive control

**Figure 15.** Comparison among different size optimization methods.

## 4. Conclusions

In this paper, a size optimization method of the WEC is proposed for improving the techno-economic performance and applied to a spherical heaving point absorber. Both buoy sizing and PTO sizing are taken into account. A frequency domain model and a preliminary economic model are established to calculate the power performance and economic performance of the WEC in various sizes, respectively. Besides, PTO force limits are used to represent different PTO sizes. For the determination of PTO parameters, a theoretical method is derived to maximize the power absorption under certain force constraints. The reactive control strategy and the passive control strategy are considered, respectively. Based on the proposed method, the size optimization is carried out for three typical sea sites. Furthermore, a comparison between the proposed method and other size optimization methods is performed. The following conclusions are drawn:

Firstly, as expected, both buoy sizing and PTO sizing are able to affect the techno-economic performance of the WEC. To improve the techno-economic performance, it is highly suggested to perform the buoy size optimization and PTO size optimization collectively.

Secondly, the optimal buoy size differs with wave resources, but it is not necessarily proportional to the mean wave power density. In this case, the optimal buoy scale factor $\lambda$ ranges from 0.5 to 0.8. Besides, in most sea sites, the optimal PTO sizing ratios in the WEC with reactive control are slightly higher than those with passive control. The optimal PTO sizing ratios converge at around 0.4 to 0.5 for different sea sites. Furthermore, the higher mean wave power density and reactive control can clearly contribute to the reduction of the LCOE.

Thirdly, downsizing the PTO size would penalize the AEP, but it is beneficial for reducing the LCOE. In this case, the LCOE can be reduced by 24% to 31% through downsizing the PTO sizing ratio from 1 to around 0.4 to 0.5.

Fourthly, the corresponding optimal buoy size tends to slightly decrease with the PTO sizing ratio, but the influence of PTO sizing on the buoy size determination is limited.

Finally, Budal diagram is not able to estimate the suitable buoy size, but size optimization without PTO sizing can make the approximate estimate. Compared with other methods, a further reduction in the LCOE can be achieved by this proposed method.

**Author Contributions:** Conceptualization, J.T. and H.P.; methodology, J.T., H.P., and A.J.L.; software, J.T.; validation, J.T.; formal analysis, J.T., H.P., and A.J.L.; investigation, J.T.; resources, S.A.M.; data curation, J.T.; writing—original draft preparation, J.T.; writing—review and editing, J.T., H.P., and A.J.L.; visualization, J.T.; supervision, H.P., P.W., and S.A.M.; project administration, S.A.M.; funding acquisition, S.A.M. and H.P. All authors have read and agreed to the published version of the manuscript.

**Funding:** This research has received funding from China Scholarship Council under Grant: 20180695 0003.

**Institutional Review Board Statement:** Not applicable.

**Informed Consent Statement:** Not applicable.

**Data Availability Statement:** The data presented in this study are available on request from the corresponding author.

**Acknowledgments:** The authors wish to thank the ODE (Offshore Dredging Engineering) group in Delft University of Technology for supporting this research. The authors also would like to thank Giuseppe Giorgi for informing us about this special issue.

**Conflicts of Interest:** The authors declare no conflict of interest. The funders had no role in the design of the study; in the collection, analyses, or interpretation of data; in the writing of the manuscript; or in the decision to publish the results.

# References


1. Aderinto, T.; Li, H. Ocean Wave energy converters: Status and challenges. *Energies* **2018**, *11*, 1–26. [CrossRef]
2. Lehmann, M.; Karimpour, F.; Goudey, C.A.; Jacobson, P.T.; Alam, M.R. Ocean wave energy in the United States: Current status and future perspectives. *Renew. Sustain. Energy Rev.* **2017**, *74*, 1300–1313. [CrossRef]
3. De Andrés, A.; Macgillivray, A.; Guanche, R.; Jeffrey, H. Factors affecting LCOE of Ocean energy technologies: A study of technology and deployment attractiveness. In Proceedings of the 5th International Conference on Ocean Energy, Halifax, Nova Scotia, 4–6 November 2014; pp. 1–11.
4. Pecher, A. *Handbook of Ocean Wave Energy*; Springer: Berlin/Heidelberg, Germany, 2017; Volume 7. [CrossRef]
5. Babarit, A.; Hals, J.; Muliawan, M.J.; Kurniawan, A.; Moan, T.; Krokstad, J. Numerical benchmarking study of a selection of wave energy converters. *Renew. Energy* **2012**, *41*, 44–63. [CrossRef]
6. Chang, G.; Jones, C.A.; Roberts, J.D.; Neary, V.S. A comprehensive evaluation of factors affecting the levelized cost of wave energy conversion projects. *Renew. Energy* **2018**, *127*, 344–354. [CrossRef]
7. Babarit, A. A database of capture width ratio of wave energy converters. *Renew. Energy* **2015**, *80*, 610–628. doi:10.1016/j.renene.2015.02.049. [CrossRef]
8. De Andres, A.; Maillet, J.; Todalshaug, J.H.; Möller, P.; Bould, D.; Jeffrey, H. Techno-economic related metrics for a wave energy converters feasibility assessment. *Sustainability* **2016**, *8*, 1109. [CrossRef]
9. Sergiienko, N.Y.; Cazzolato, B.S.; Ding, B.; Hardy, P.; Arjomandi, M. Performance comparison of the floating and fully submerged quasi-point absorber wave energy converters. *Renew. Energy* **2017**, *108*, 425–437. [CrossRef]
10. Falnes, J. *Ocean Waves and Oscillating Systems*; Cambridge University Press: Cambridge, UK, 2003; Volume 30, p. 953. [CrossRef]
11. Falnes, J.; Hals, J. Heaving buoys, point absorbers and arrays. *Philos. Trans. R. Soc. Math. Phys. Eng. Sci.* **2012**, *370*, 246–277. [CrossRef]
12. Pecher, A. *Performance Evaluation of Wave Energy Converters*; River Publishers: Hakodate, Denmark, 2014; pp. 1–107. [CrossRef]
13. O'Connor, M.; Lewis, T.; Dalton, G. Techno-economic performance of the Pelamis P1 and Wavestar at different ratings and various locations in Europe. *Renew. Energy* **2013**, *50*, 889–900. [CrossRef]
14. Tai, V.C.; See, P.C.; Merle, S.; Molinas, M. Sizing and control of the electric power take off for a buoy type point absorber wave energy converter. *Renew. Energy Power Qual. J.* **2012**, *1*, 1614–1619. [CrossRef]
15. Tokat, P. *Performance Evaluation and Life Cycle Cost Analysis of the Electrical Generation Unit of a Wave Energy Converter*; Chalmers Tekniska Hogskola: Gothenburg, Sweden, 2018.
16. Shek, J.; Macpherson, D.; Mueller, M. Phase and amplitude control of a linear generator for wave energy conversion. In Proceedings of the 4th IET International Conference on Power Electronics, Machines and Drives (PEMD 2008), York, UK, 2–4 April 2008; pp. 66–70. [CrossRef]
17. Shek, J.; Macpherson, D.; Mueller, M. Experimental verification of linear generator control for direct drive wave energy conversion. *IET Renew. Power Gener.* **2010**, *4*, 395–403. [CrossRef]
18. Tedeschi, E.; Carraro, M.; Molinas, M.; Mattavelli, P. Effect of control strategies and power take-off efficiency on the power capture from sea waves. *IEEE Trans. Energy Convers.* **2011**, *26*, 1088–1098. [CrossRef]
19. Garcia-Rosa, P.; Bacelli, G.; Ringwood, J. Control-informed geometric optimization of wave energy converters: The impact of device motion and force constraints. *Energies* **2015**, *8*, 13672–13687. [CrossRef]
20. Astariz, S.; Iglesias, G. The economics of wave energy: A review. *Renew. Sustain. Energy Rev.* **2015**, *45*, 397–408. [CrossRef]
21. Hals, J.; Bjarte-Larsson, T.; Falnes, J. Optimum Reactive Control and Control by Latching of a Wave-Absorbing Semisubmerged Heaving Sphere. In Proceedings of the 21st International Conference on Offshore Mechanics and Arctic Engineering, Oslo, Norway, 23–28 June 2002; pp. 415–423. [CrossRef]
22. Tedeschi, E.; Molinas, M. Tunable control strategy for wave energy converters with limited power takeoff rating. *IEEE Trans. Ind. Electron.* **2012**, *59*, 3838–3846. [CrossRef]

23. Kramer, M.M.; Marquis, L.; Frigaard, P. Performance Evaluation of the Wavestar Prototype. In Proceedings of the 9th European Wave and Tidal Conference, Southampton, UK, 5–9 September 2011; pp. 5–9.

24. Penalba, M.; Kelly, T.; Ringwood, J. Using NEMOH for Modelling Wave Energy Converters : A Comparative Study with WAMIT. In Proceedings of the 12th European Wave and Tidal Energy Conference, Cork, Ireland, 27 Auguest–1 September 2017; p. 10.

25. Pastor, J.; Liu, Y. Frequency and time domain modeling and power output for a heaving point absorber wave energy converter. *Int. J. Energy Environ. Eng.* **2014**, *5*, 1–13. [CrossRef]

26. Journée, J.M.J.; Massie, W.W.; Huijsmans, R.H.M. *Offshore Hydrodynamics*; Delft University of Technology: Delft, The Netherlands, 2015.

27. Penalba, M.; Giorgi, G.; Ringwood, J.V. Mathematical modelling of wave energy converters: A review of nonlinear approaches. *Renew. Sustain. Energy Rev.* **2017**, *78*, 1188–1207. [CrossRef]

28. Hals, J.; Falnes, J.; Moan, T. Constrained Optimal Control of a Heaving Buoy Wave-Energy Converter. *J. Offshore Mech. Arct. Eng.* **2010**, *133*, 011401. [CrossRef]

29. Prado, M.; Polinder, H. *Direct Drive Wave Energy Conversion Systems: An Introduction*; Woodhead Publishing Limited: Sawston, Cambridge, 2013; pp. 175–194.10.1533/9780857097491.2.175. [CrossRef]

30. Ugray, Z.; Lasdon, L.; Plummer, J.; Glover, F.; Kelly, J.; Martí, R. Scatter Search and Local NLP Solvers : A Multistart Framework for Global Optimization. *Informs J. Comput.* **2007**.10.2139/ssrn.886559. [CrossRef]

31. Cahill, B.; Lewis, A.W. Wave period ratios and the calculation of wave power. In Proceedings of the 2nd Marine Energy Technology Symposium, Seattle, WA, USA, 15–18 April 2014; pp. 1–10.

32. Chozas, J.; Kofoed, J.; Helstrup, N. *The COE Calculation Tool for Wave Energy Converters*; Version 1.6; DCE Technical Reports; No. 161; Aalborg University: Aalborg, Denmark, 2014.

33. Payne, G. Guidance for the experimental tank testing of wave energy converters. *SuperGen Mar.* **2008**, *254*.

34. De Andres, A.; Medina-Lopez, E.; Crooks, D.; Roberts, O.; Jeffrey, H. On the reversed LCOE calculation: Design constraints for wave energy commercialization. *Int. J. Mar. Energy* **2017**, *18*, 88–108. [CrossRef]

35. Bank of England. Inflation Calculator-Bank of England [EB/OL]. Available online: https://www.bankofengland.co.uk/monetary-policy/inflation/inflation-calculator (accessed on 4 October 2020).

36. Polinder, H. *Principles of Electrical Design of Permanent Magnet Generators for Direct Drive Renewable Energy Systems*; Woodhead Publishing Limited: Cambridge, UK, 2013; pp. 30–50. [CrossRef]

37. Tokat, P.; Thiringer, T. Sizing of IPM Generator for a Single Point Absorber Type Wave Energy Converter. *IEEE Trans. Energy Convers.* **2018**, *33*, 10–19. [CrossRef]

Journal of
*Marine Science
and Engineering*

Article

# A Comparative Study of Model Predictive Control and Optimal Causal Control for Heaving Point Absorbers

Mirko Previsic [1,*], Anantha Karthikeyan [1] and Jeff Scruggs [2]

1   RE Vision Consulting, LLC., 1104 Corporate Way, Sacramento, CA 95831, USA; anantha@re-vision.net
2   Civil & Environmental Engineering, University of Michigan, Ann Arbor, MI 48109, USA; jscruggs@umich.edu
*   Correspondence: mirko@re-vision.net

**Abstract:** Efforts by various researchers in recent years to design simple causal control laws that can be applied to WEC devices suggest that these controllers can yield similar levels of energy output as those of more complex non-causal controllers. However, most studies were established without adequately considering device and power conversion system constraints which are relevant design drivers from a cost and economic point of view. It is therefore imperative to understand the benefits of MPC compared to causal control from a performance and constraint handling perspective. In this paper, we compare linear MPC to a casual controller that incorporates constraint handling to benchmark its performance on a one DoF heaving point absorber in a range of wave conditions. Our analysis demonstrates that MPC provides significant performance advantages compared to an optimized causal controller, particularly if significant constraints on device motion and/or forces are imposed. We further demonstrate that distinct control performance regions can be established that correlate well with classical point absorber and volumetric limits of the wave energy conversion device.

**Keywords:** causal control; Model Predictive Control; point absorber; wave-energy converter; absorbed power



**Citation:** Previsic, M.; Karthikeyan, A.; Scruggs, J. A Comparative Study of Model Predictive Control and Optimal Causal Control for Heaving Point Absorbers. *J. Mar. Sci. Eng.* **2021**, *9*, 805. https://doi.org/10.3390/jmse9080805

Academic Editors: Giuseppe Giorgi and Sergej Antonello Sirigu

Received: 2 July 2021
Accepted: 24 July 2021
Published: 27 July 2021

**Publisher's Note:** MDPI stays neutral with regard to jurisdictional claims in published maps and institutional affiliations.



## 1. Introduction

As the field of wave energy conversion transitions from traditional passive control techniques to advanced optimal control strategies for power maximization, it becomes increasingly important to understand the requirements, capabilities, limitations, and benefits of each method to choose the best optimization strategy for a given application.

The control system affects power capture, structural loads, and power-takeoff (PTO) design. To achieve true economic optimality in a wave energy conversion system design, the control system needs to be able to optimize performance given the constraints imposed by the wave energy converter (WEC) system. This is due to the fact that economic outcomes are optimized if a given component or sub-system is continually utilized at its rated operational condition during a majority of its operational life. These rated operational conditions provide constraints within which the device needs to operate, and the WEC control system is responsible for enforcing these constraints because exceeding them will result in mechanical breakdown. As such, the constraint handling of a controller is incredibly important to maximize the economic competitiveness of a WEC device. Constrained optimal control as offered by a Model Predictive Control (MPC) algorithm framework, which is also referred to as non-causal control, is a key tool in this optimization process.

Control systems for wave energy conversion are broadly classified as either causal or non-causal controllers. A causal controller uses information from sensors that monitor the outputs of a dynamical system, in our case the WEC. The sensor information is used as feedback by the controller to follow a desired command signal. Non-causal controllers such as MPC leverage wave prediction to generate control commands in a feed-forward manner rather than using feedback from sensors that measure system output variables. Combination of non-causal (feed-forward) and causal (feedback) control is also possible in a hybrid manner and can be used to correct for modeling and/or wave prediction errors in non-causal controllers.

Within the literature there are two well-published methods for implementing causal control that incorporate constraint handling. The first approach tries to approximate complex conjugate control (CCC) ([1–4]) with a causal feedback law [5]. This approach does well in the unconstrained case where no motion limits are imposed. To allow for constraints, these methods were subsequently augmented to incorporate an MPC controller that works for the sole purpose of enforcing constraints [5]. Performance of the CCC approximating causal control methods is sub-optimal when constraints are imposed, because the impedance matching value obtained from the CCC approximation provides in-sufficient damping for the constrained case.

The second approach for designing a causal controller comes from recognizing that the causal WEC control problems fall into the Linear Quadratic Gaussian (LQG) paradigm ([6–8]). A non-standard LQG optimization problem is established with the objective of maximizing power. The problem treats the input wave excitation force as a stationary stochastic disturbance with a known spectrum. A wave-shaping filter is designed to model the spectrum and this filter model is incorporated in the optimization along with the system dynamics model of the WEC. The LQG optimal controller derived in this manner does not rely on approximations to CCC and will therefore achieve optimality in a broader set of sea states. Stroke protection techniques can be explicitly designed to protect against end-stop violations [7,9]. The control parameters are optimized offline and implemented using a gain schedule to adapt to changing wave conditions. A block-diagram of a casual controller with non-linear stroke protection is shown in Figure 1.

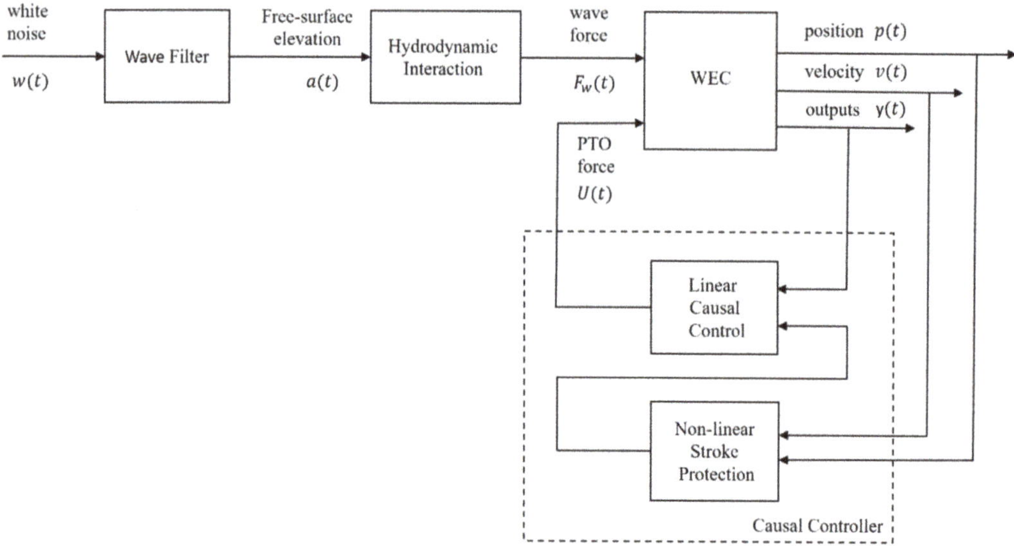


**Figure 1.** Architecture of causal controller.

Non-causal performance optimization of WECs is typically implemented using a receding horizon MPC framework. To guarantee optimality, MPC needs a forecast of oncoming waves 20 to 30 s into the future to plan optimal control commands. Wave forecasting can be conducted by placing probes up-wave, measuring the free-surface elevation in the wave field [10], or using a wave-sensing radar [11,12] to characterize the wave-field propagation. These up-wave measurements are marched forward in space and time to predict the wave elevation at the target WEC location. The predicted wave excitation force input, in addition to the states of a system dynamics model, are used to solve a quadratic optimization problem to find the optimal control trajectory. The states of the device are marched forward in time and the optimization process is repeated at each step with new prediction information. Figure 2 shows the basic MPC control architecture.

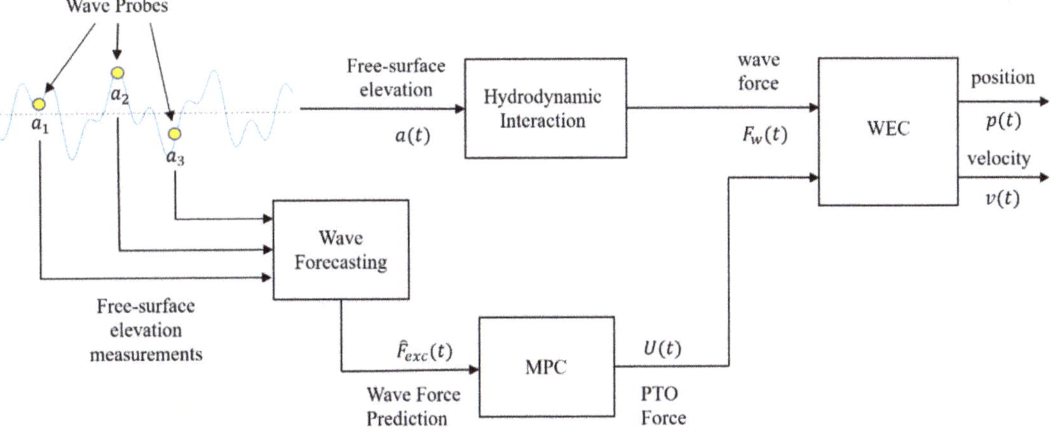


**Figure 2.** Architecture of Model Predictive Control.

Considerable research has been undertaken on expanding the capabilities of MPC to meet the requirements of an entire range of device topologies and PTO configurations ([13–19]). Some of the common constraints that can be handled effectively within the MPC optimization include: (1) PTO force constraints; (2) WEC position, velocity, and acceleration constraints; (3) discrete (ON/OFF) control of PTO forces; and (4) power-flow constraints. MPC can accommodate non-linear device dynamics, such as viscous drag and non-linear models of powertrain losses, which need to be considered in the optimization problem to guarantee optimality [14].

## 2. Glossary

| Variable | Description | Units |
|---|---|---|
| $g$ | Acceleration due to gravity | $m/s^2$ |
| $\rho$ | Density of sea water | $kg/m^3$ |
| $m$ | Mass of the buoy | $kg$ |
| $m_\infty$ | Infinite Added Mass | $kg$ |
| $z$ | Displacement of the buoy in heave | $m$ |
| $\dot{z}$ | Velocity of the buoy in heave | $m/s$ |
| $\ddot{z}$ | Acceleration of the buoy in heave | $m/s^2$ |
| $F_R$ | Radiation damping force | $N$ |
| $F_m$ | PTO Force | $N$ |
| $F_s$ | Hydrostatic spring stiffness | $N$ |
| $F_d$ | Drag force related to viscous effects | $N$ |
| $F_e$ | Wave excitation force | $N$ |
| $k$ | Buoyancy stiffness coefficient | $N/m$ |

| Variable | Description | Units |
|---|---|---|
| $S$ | Water plane area of the buoy | $\text{m}^2$ |
| $B_{vis}$ | Viscous damping coefficient | Ns/m |
| $C_d$ | Drag coefficient | - |
| $A_r$ | A matrix of state space model of the radiation damping sub-system | |
| $B_r$ | B matrix of state space model of the radiation damping sub-system | |
| $C_r$ | C matrix of state space model of the radiation damping sub-system | |
| $D_r$ | D matrix of state space model of the radiation damping sub-system | |
| $X_r$ | State vector comprising of all states used to model the radiation damping sub-system in state space form | |
| $\dot{X}_r$ | Vector comprising of the derivatives of all states used to model the radiation damping sub-system in state space form | |
| $n_r$ | Number of states used to model the radiation damping sub-system. This is equal to the length of the state vector $X_r$ | |
| $0_r$ | Row vector of zeros with length $n_r$ | |
| $A$ | A matrix of WEC state space model | |
| $B$ | B matrix of WEC state space model | |
| $C$ | C matrix of WEC state space model | |
| $D$ | D matrix of WEC state space model | |
| $T_h$ | Prediction horizon | s |
| H | Wave height of a monochromatic wave | m |
| T | Wave period of a monochromatic wave | s |
| $H_s$ | Significant wave height of a polychromatic wave | m |
| $T_p$ | Peak period of a polychromatic wave | s |
| $\lambda$ | Wavelength | m |
| $k_w$ | Wavenumber | |
| $\overline{P}_a$ | Average absorbed power | W |
| $v$ | Buoy heave velocity | m/s |
| $h_r$ | Radiation impulse-response function | Ns/m |
| $h_e$ | Excitation impulse-response function | N/m |
| $t$ | Time instant | s |
| $u(t)$ | Control input at time instant t | N |
| $f_e(t)$ | Wave excitation force at time instant t | N |
| $x(t)$ | State vector at time instant t comprising of all states used to model the device dynamics | |
| $\dot{x}(t)$ | Derivative of the state vector at time instant t that is used to model the device dynamics | |

## 3. The Device Model

We consider a one degree-of-freedom heaving point absorber as an example for our comparison study. A general schematic of the heaving buoy WEC is shown in Figure 3 and key dimensions are provided in Table 1. This device has well-established theoretical limits on power absorption which can be used to identify the design trade-off space for controls optimization.

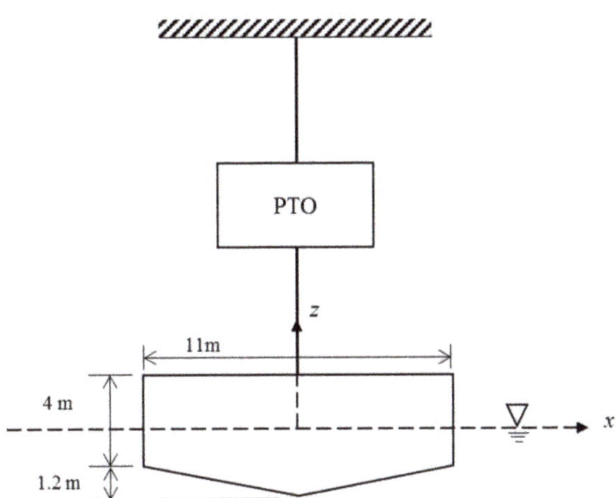

**Figure 3.** Schematic of the heaving wave-energy converter.

**Table 1.** Dimensions and physical parameters of the heaving wave energy converter.

| Quantity | Value | Units |
|---|---|---|
| Buoy diameter | 11 | m |
| Buoy cylinder height | 4 | m |
| Reaction diameter | 11 | m |
| Buoy conical height | 1.2 | m |
| Motion limits | ±2 | m |
| Buoy mass | 228,080 | kg |
| Water Plane Area (S) | 95 | m$^2$ |
| Coefficient of Drag (Cd) | 0.5 | |

The device geometry was modeled using a commercially available Boundary Element code called WAMIT [20]. The added mass, radiation damping, and excitation force kernels were obtained as a function of frequency by post-processing the WAMIT outputs. Figure 4 shows the frequency dependent parameters plotted against the wave period.

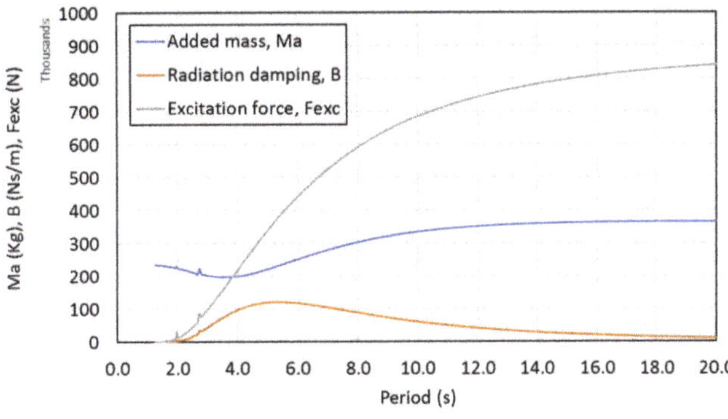


**Figure 4.** Frequency-dependent parameters of the heaving buoy WEC obtained from WAMIT.

Because this floating body is constrained to move in heave, the equation of motion in the vertical direction is given by:

$$m\,\ddot{z}(t) = F_R(t) + F_e\,(t) + F_m(t) + F_s\,(t) + F_d(t) \tag{1}$$

where $z$ is the displacement, $\dot{z}$ is the velocity, and $\ddot{z}$ is the acceleration of the buoy in heave. $m$ is the displaced mass of the buoy, $F_R$ is the radiation force, $F_e$ is the wave-excitation force, $F_m$ is the PTO force, $F_s$ is the hydrostatic restoring force given by $F_s = -k\,z$, where $k$ is the buoyancy stiffness, defined as $k = \rho g S$, where $\rho$ is the water density, $g$ the gravitational constant, and $S$ the water-plane area. $F_d$ is the drag force related to viscous effects in real fluid. The viscous damping coefficient was assumed to be linear according to the relationship $F_d = -B_{vis}\,v$, where $B_{vis} = 0.5\,C_d \rho S$. $C_d$ is the drag coefficient and $\rho$ is the water density. The radiation force $f_R(t)$, can be expressed as:

$$F_R(t) = -m_\infty \ddot{z}(t) - \int_{-\infty}^{t} h_r(t-\tau)\dot{z}(\tau)d\tau \tag{2}$$

where $m_\infty$ is the infinite frequency added mass and $h_r(t)$ is the impulse response function of the radiation force. The excitation force $F_e(t)$ is expressed as:

$$F_e(t) = \int_{-\infty}^{+\infty} h_e(t-\tau)\overline{\eta}(\tau)d\tau \tag{3}$$

where $h_e(t)$ is the excitation force impulse response function and $\overline{\eta}(t)$ is the wave elevation at the device location. The impulse response function relating the wave elevation to the excitation force affecting the device is non-causal. The main reason is that the chosen input, i.e., the wave elevation at the device location, is not the direct cause of the output, i.e., the interaction force between the wavefield and the device.

To recast the system dynamics into state-space form, the radiation term can be modelled as an embedded sub-system through the following state-space realization:

$$\dot{X}_r(t) = A_r X_r(t) + B_r \dot{z}(t) \tag{4}$$

$$f_r(t) = C_r X_r(t) + D_r \dot{z}(t) \tag{5}$$

This leads to the following state-space model:

$$\begin{aligned} \dot{x}(t) &= Ax(t) + Bu(t) + Bf_e(t) \\ y(t) &= Cx(t) + Du(t) \end{aligned} \tag{6}$$

with

$$A = \begin{bmatrix} A_r & 0 & B_r \\ 0 & 0 & 1 \\ -\frac{C_r}{m+m_\infty} & -\frac{k}{m+m_\infty} & -\frac{B_{vis}+D_r}{m+m_\infty} \end{bmatrix}, B = \begin{bmatrix} 0 \\ 0 \\ \frac{1}{m+m_\infty} \end{bmatrix} C = \begin{bmatrix} 0_r & 1 & 0 \\ 0_r & 0 & 1 \end{bmatrix}, D = \begin{bmatrix} 0 \\ 0 \end{bmatrix} \tag{7}$$

## 4. Control Methods

### 4.1. Model Predictive Control (MPC)

A wave energy converter with a linearized device dynamics model can be optimized using a linear MPC framework. In this type of problem, the non-linearities such as viscous drag are approximated using linear relationships. Under the assumption of no loss in the power generation process, optimizing the device average power absorbed $\overline{P}_a$ at a given instant $t_0$ over a defined control horizon $T_h$ can be achieved by determining the optimal control trajectory $u^*(t)$ that maximizes the following cost function:

$$\overline{P}_a = -\frac{1}{T_h} \int_{t_0}^{t_0+T_h} v(t)u(t)dt \tag{8}$$

where $v(t)$ is the instantaneous device velocity. The minus sign is due to the convention of considering absorbed energy with a negative sign. Assuming a fixed time step, we discretize the integral and set up the optimization problem as a minimization by changing the sign in Equation (9). The new optimization objective, therefore, is given by:

$$J = \frac{1}{N} \sum_{k=0}^{N-1} x_{k+1}^T S_v^T u_k \tag{9}$$

where $N$ is the number of time intervals over the control horizon $T_h$ and $S_v$ is a linear operator extracting the WEC velocity from the state-space vector. The control force and state vector, however, are not independent variables, and are constrained by the system dynamics equation of the WEC, which in discrete time is defined as:

$$x_{k+1} = A_d x_k + B_d u_k + B_d f_{e_k}, \; k = 0, \dots, N-1 \tag{10}$$

with initial condition $x_0 = \overline{x}_0$.

To preserve mechanical and structural integrity, motion constraints and force limits are imposed. These constraints limit the maximum actuation force and the WEC device velocity and vertical displacement, i.e.,

$$u_{min} \leq u_k \leq u_{max}, \; k = 0, \dots, N-1 \tag{11}$$

$$p_{min} \leq S_p x_k \leq p_{max}, \; k = 0, \dots, N \tag{12}$$

$$v_{min} \leq S_v x_k \leq v_{max}, \; k = 0, \dots, N \tag{13}$$

where $S_p$ is a linear operator extracting the WEC displacement from the state vector. The cost function, together with the constraints, represents a linear MPC problem in its standard formulation.

Let us define the following consolidated vectors based on the sequence of states and control commands. This will help in simplifying the notation and problem setup:

$$X = \begin{bmatrix} x_1^T & x_2^T & \dots & x_N^T \end{bmatrix}^T \tag{14}$$

$$U = \begin{bmatrix} u_1^T & u_2^T & \dots & u_N^T \end{bmatrix}^T \tag{15}$$

Let us define $\Lambda_v$, $\Lambda_p$ as block diagonal matrices having the velocity extraction matrix $S_v$ and the position extraction matrix $\Lambda_p$, repeated $N$ times respectively on the main block diagonal. Furthermore, let $I$ be the $N \times N$ identity matrix and let $\zeta$ be a $N \times 1$ vector of ones. In this manner, the cost function can then be expressed as follows:

$$J = \frac{1}{N} X^T \Lambda_v^T U \tag{16}$$

The inequality constraints can be reformulated using this vector notation as follows:

$$D_u U \leq d_u \tag{17}$$

$$D_x X \leq d_x \tag{18}$$

where

$$D_u = \begin{bmatrix} I \\ -I \end{bmatrix} d_u = \begin{bmatrix} u_{max} * \zeta \\ -u_{min} * \zeta \end{bmatrix} \tag{19}$$

$$D_x = \begin{bmatrix} \Lambda_p \\ -\Lambda_p \\ \Lambda_v \\ -\Lambda_v \end{bmatrix} \quad d_x = \begin{bmatrix} p_{max} * \zeta \\ -p_{min} * \zeta \\ v_{max} * \zeta \\ -v_{min} * \zeta \end{bmatrix} \tag{20}$$

By recursively applying the discrete-time dynamics equations, it is possible to express $X$ as a function of the control vector $U$, the excitation force vector $F_e$, and the initial condition $\bar{x}_0$:

$$X = Y_d \bar{x}_0 + \Gamma_d U + \Gamma_d F_e \tag{21}$$

where

$$Y_d = \begin{bmatrix} A_d \\ A_d^2 \\ \vdots \\ A_d^N \end{bmatrix} \tag{22}$$

$$\Gamma_d = \begin{bmatrix} B_d & 0 & 0 & 0 \\ A_d B_d & B_d & 0 & 0 \\ \vdots & \vdots & \ddots & 0 \\ A_d^{N-1} B_d & A_d^{N-2} B_d & \cdots & B_d \end{bmatrix} \tag{23}$$

$$F_e = \begin{bmatrix} f_{e_0}^T & f_{e_1}^T & \cdots & f_{e_{N-1}}^T \end{bmatrix}^T \tag{24}$$

This allows us to rewrite the MPC problem as the following cost function and constraint equations:

$$\min_U U^T \Gamma_d^T \Lambda_v^T U + \left( \Lambda_v Y_d \bar{x}_0 + \Lambda_v \Gamma_d F_e \right)^T U \tag{25}$$

$$\begin{bmatrix} D_u \\ D_x \Gamma_d \end{bmatrix} U \leq \begin{bmatrix} d_u \\ d_x - D_x Y_d \bar{x}_0 - D_x \Gamma_d F_e \end{bmatrix} \tag{26}$$

The maximization of absorbed power requires the solution of a constrained convex optimization problem, for which well-consolidated routines, such as interior-point-convex or active-set methods, are available in literature [21,22]. Positive definiteness of the Hessian is, in general, guaranteed for the optimization of a point absorber device, unless the time step chosen for the conversion of the continuous time model into discrete time is too large to represent the actual behavior of the WEC device. At each timestep, an MPC problem needs to be solved, and the first value of the optimal solution vector $U^*$ is applied to the system. The system is marched forward in time by integrating the system dynamics using a standard Runge–Kutta (RK) scheme. The above MPC formulation also assumes that a perfect excitation force prediction is available 20 to 30 s into the future using up-wave measurement probes.

*4.2. Causal Feedback Control*

The first step in optimal causal feedback control design is to determine an analytically tractable stochastic disturbance model to represent the incident wave force $f_e(t)$. This step is nontrivial for two reasons. Firstly, the power spectral density typically used to characterize the free surface elevation $\bar{\eta}(t)$, denoted $S_\eta(\omega)$, does not constitute a rational spectrum (i.e., it is not a ratio of rational polynomials of frequency $\omega$). Secondly, the transfer function from the free surface elevation to the incident wave force, equal to

$$H_e(s) = \int_{-\infty}^{\infty} e^{-st} h_e(t) dt$$

involves the solution to a partial differential equation (due to fluid–structure interaction), and consequently it also does not evaluate to a rational transfer function. Consequently, the resultant power spectral density of the incident wave force, denoted as

$$S_e(\omega) = S_\eta(\omega)|H_e(j\omega)|^2$$

is an irrational function of frequency. To make the causal control design problem tractable, this spectrum must be approximated, as

$$S_e(\omega) \approx |G(j\omega)|^2$$

where the transfer function $G(s)$ is a stable, minimum-phase, strictly-proper, rational transfer function. Equivalently, we seek to find matrices $A_e$, $B_e$, and $C_e$ such that in the above equation,

$$G(s) = C_e[sI - A_e]^{-1}B_e$$

With such a formulation for $G(s)$, we may equivalently represent the incident wave force as filtered white noise; i.e.,

$$\dot{x}_e(t) = A_e x_e(t) + B_e w(t)$$

$$f_e(t) = C_e x_e(t)$$

where $w(t)$ is a white noise process with unit spectral intensity.

There are a number of ways to find appropriate matrices $A_e$, $B_e$, and $C_e$ for the noise filter model described above. Here, we make use of the subspace-based spectral factorization approach. Because this approach is applied exactly as described in [23], we will forgo the details here in the interest of brevity.

With the accomplishment of the finite-dimensional approximation of the incident wave force, we then append the internal states $x_e(t)$ to the state $x(t)$ of the physical system; i.e., we define $x_a(t) = \begin{bmatrix} x^T(t) & x_e^T(t) \end{bmatrix}^T$ to arrive at an augmented state space model:

$$\dot{x}_a(t) = A_a x_a(t) + B_a u(t) + E_a w(t)$$
$$y(t) = C_a x_a(t) + D_a u(t)$$

where matrices $A_a$, $B_a$, $E_a$, $C_a$, and $D_a$ are appropriately defined. Our objective is to design a causal feedback law that maps $y$ into $u$, to maximize the objective

$$\overline{P}_a = -E\left\{vu + Ru^2\right\}$$

where $E\{.\}$ denotes the expectation in stationarity, and $R$ is a nonnegative parameter. For the case in which $R = 0$, the above expression is simply the mean power absorption. In design, $R$ is typically chosen to be greater than zero, for two reasons. Firstly, it can be used to (approximately) quantify the power dissipation in the power train of the WEC, thus changing the performance objective from absorbed power to generated power. Secondly, it can be used as a tuning parameter to balance the power generation objective against the need to keep the mean-square control force magnitude below a desired threshold.

If there were no constraints on the response of the WEC, the above problem could be solved in closed form. More specifically, it distills to a sign-indefinite Linear Quadratic Gaussian (LQG) control problem, which is guaranteed to have a unique, stabilizing solution assuming that the open-loop mapping from $u$ to $\dot{z}$ is passive [6]. However, for the WEC under consideration here, there are constraints on both the displacement $z$ and the control force $u$. In the presence of these constraints, the closed-form LQG control solution cannot be used without modification. To address these constraints, in this paper we implement the methodology outlined in detail in [9]. Because the methodology used here is identical to that one, we only provide an overview here of the basic steps in the nonlinear control design procedure.

The first step is to design a linear feedback controller of the form

$$\dot{\xi} = A_c \xi + B_c y$$
$$u = C_c \xi$$

in which $A_c$, $B_c$, and $C_c$ are optimized to maximize $\overline{P}_a$, subject to two constraints. The first of these constraints is a relaxed version of the WEC displacement constraint, in which we require only that the mean-square displacement be below a bound related to $p_{max}$, i.e.,

$$E\left\{z^2\right\} \le \sigma p_{max}^2$$

where $\sigma < 1$ is a design parameter, chosen here to be 0.25. The second constraint is the restriction that the open-loop controller must be stable; i.e., that $A_c$ be a Hurwitz matrix. Convex techniques for accomplishing this constrained optimization are covered extensively in [9]. The resultant linear controller is guaranteed to result in positive power generation, be robust to saturations in the input $u$, and maintain mean-square displacements below $p_{max}^2$. However, it does not guarantee that the displacement limit is satisfied at all times.

The second step is to augment the above linear control design with a nonlinear stroke-protection feedback loop that ensures that the displacement limit is satisfied at all times (rather than just in the mean-square sense). This stroke protection loop, and its relationship to the linear control design, is illustrated in Figure 1. It accepts as inputs the position and velocity, and outputs a supplemental nonlinear force $q$. This quantity $q$ is sent back to the linear controller through an augmented input matrix, i.e., via the equations:

$$\dot{\xi} = A_c \xi + B_c y + E_c q$$
$$u = C_c \xi + q$$

where $E_c$ is a matrix of design parameters. The design of the feedback function that maps $\{z, \dot{z}\}$ into $q$, and the design of $E_c$, are described in fully in [9]. Here it suffices to say that $q$ can be viewed as being similar to the restoring force of a "virtual hardening spring," providing no force when the displacement is far from its limit, but increasing to infinity when the displacement approaches its limit. The participation of velocity $\dot{z}$ in the function can be thought of as providing damping. Meanwhile, the matrix $E_c$ is designed such that the open-loop transfer function from $q$ to $\dot{z}$ is passive, which guarantees unconditional stability of the closed-loop system.

The controller resulting from the above design framework is guaranteed to protect the displacement from reaching its limit for the case in which the force $u$ is unbounded. When saturation limits are imposed on $u$, of the form

$$u = \begin{cases} u_{min} : & C_c \xi + q < u_{min} \\ C_c \xi + q : & u_{min} \le C_c \xi + q \le u_{max} \\ u_{max} : & C_c \xi + q > u_{max} \end{cases}$$

then the displacement protection is not strictly guaranteed. Indeed, when both the force and displacement are constrained, there may not exist a controller that can honor both. However, proper tuning of the various design parameters in the control design can be undertaken to assure that displacement constraints are satisfied in all but extremely rare cases in stationary response.

## 5. Theoretical Limits on the Average Absorbed Power

### 5.1. Point Absorber Limit

Consider a heaving axisymmetric body oscillating without constraints in resonance with an incoming regular wave of period T (s) and height H (m). Let $J_E$ (W/m) denote the wave power level, $k_w(1/m)$ denote the wavenumber, and $\lambda$ (m) the wavelength. In this case, the point absorber effect in wave-energy extraction [24] imposes a fundamental limit on the average absorbed power $\overline{P}_a$. As described in [24,25], this limit can be derived from the relationship between absorption width $d_a = \overline{P}_a / J_E(m)$ and wavelength $\lambda$, as shown below:

$$d_a \leq \frac{1}{k_w} = \frac{\lambda}{2\pi} \tag{27}$$

Or equivalently:

$$\overline{P}_a \leq \frac{1}{k} J_E = \frac{\rho}{128}\left(\frac{g}{\pi}\right)^3 T^3 H^2 \tag{28}$$

The last equality in Equation (28) is valid for deep water conditions. This is referred to as the Point Absorber Limit. The theoretical limit may be reached if the average absorbed power equals half the average excitation power, which happens when the radiated power is equal to the absorbed power. Furthermore, it is known that the Point Absorber Limit can only be reached up to a certain wave period and height. This depends on the motion amplitude constraints on the absorber. Beyond, only a lower relative power absorption can be realized, as defined by a device's volumetric limit.

### 5.2. Volumetric Limit

The heaving body sweeps a finite volume during its oscillation cycle based on physical limits. This volumetric constraint imposes a limit on the average power absorbed from each oscillation cycle. As shown in [25], the average power $P_a$ in the motion-constrained case is limited by the expression:

$$\overline{P}_a < (\pi \rho g H V / 4T) \tag{29}$$

This is referred to as the volumetric limit. With advanced control, we expect to achieve the active theoretical limit based on the device's operational region.

### 5.3. Controller Design Trade-Off Space

The controller design trade-off space for a WEC is defined by the Point Absorber and Volumetric Limit. For any given wave period, we will call the lesser of the two the "active theoretical limit" or the "theoretical limit" for simplicity. The controller performance cannot exceed the theoretical limit, and this defines the design trade-off space for controller optimization. As shown in Figure 5, the Point Absorber Limit is "active" for shorter wave periods. We will call this set of wave periods Region I. The Volumetric Limit is "active" for longer wave periods and we will call this set of wave periods Region II. The cross-over period, where the Point Absorber and Volumetric Limits are equal, demarcates the two regions.

Equations (28) and (29) show that both upper limits are wave height dependent and describe the WEC device performance in sinusoidal waves. To understand theoretical upper limits for an irregular wave train, we compute the wave height for a sinusoidal wave with the equivalent average power density of the irregular wave train as follows. The power in a monochromatic wave of height H and period T is given by:

$$P_{mono} = H^2 T \tag{30}$$

The equivalent wave height (H) of a monochromatic wave with the same power as the polychromatic wave can be found by equating the two. Assuming the period of the monochromatic wave to be the same as the peak period of the polychromatic sea state, we obtain the wave height of the equivalent sinusoidal wave as:

$$H = 0.64\, H_s \tag{31}$$

This approach allows us to classify regions in the scatter diagram based on point absorber theory, and to draw meaningful conclusions regarding the importance of MPC-based control. For this purpose, we divided the entire set of input wave conditions into three distinct regions based on theoretical limits. Figure 6 illustrates the partitioning of sea states, where Region I corresponds to wave conditions where performance is dominated by the Point Absorber Limit. Region II corresponds to wave conditions where performance is dominated by the Volumetric Limit. Finally, Region III is associated with large waves where controllers are expected to maintain rated power production and minimize structural loads on the system. Region III is not important from a performance optimization perspective, but it is important that a control law active in this region will continue to produce power at the rated capacity, while minimizing structural loads. Although slightly counter-intuitive, structural loads in Region III can be smaller than those in Regions I and II because we are effectively de-tuning the system to the incident wave conditions.

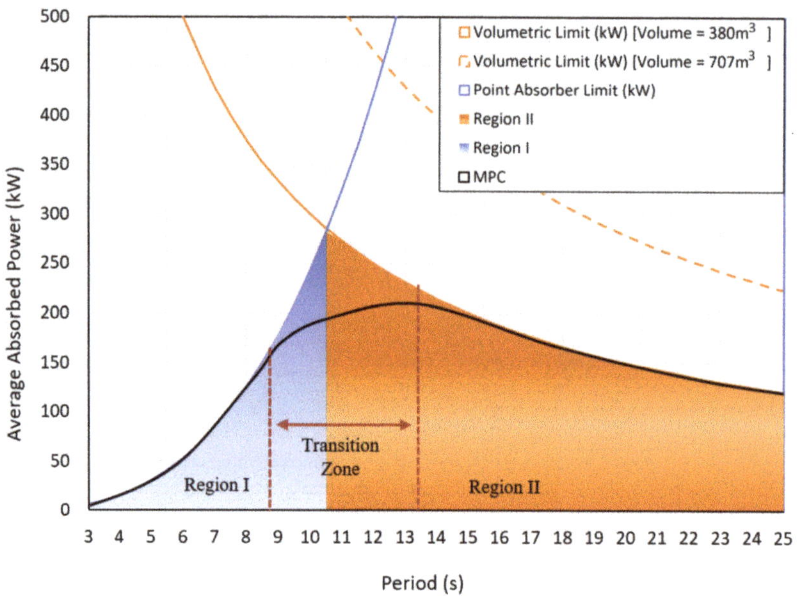


**Figure 5.** Controller design trade-off space based on theoretical limits for sinusoidal wave inputs of height H = 1 m.

Applying the above region-based approach to a small and a large heaving point absorber provides the following regions. To understand their significance in context, we simply superimposed the regions onto the frequency distribution of the DoE's reference resource in Humboldt Bay, California. As shown in Figure 7, the reference geometry used for this paper has only a few sea states, where causal control has the potential to provide similar performance as that of the MPC-based controller. This is in stark contrast to the SANDIA Wavebot, which has a much larger volumetric displacement and as a result performs well in most sea-states using causal control (see Figure 8).

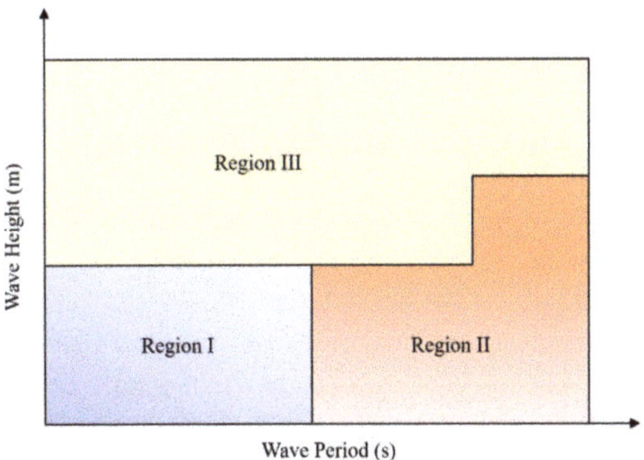

**Figure 6.** Partitioning of sea states based on fundamental limits.

% of Total Occurrence — Peak Period - Tp (s), center of bin

| HmO (m), | 0.6 | 1.7 | 2.9 | 4.1 | 5.2 | 6.4 | 7.5 | 8.7 | 9.9 | 11.0 | 12.2 | 13.3 | 14.5 | 15.7 | 16.8 | 18.0 | 19.1 | 20.3 | 21.5 | 22.6 | 23.8 |
|---|---|---|---|---|---|---|---|---|---|---|---|---|---|---|---|---|---|---|---|---|---|
| 0.25 |  |  |  |  |  |  | 0.0 | 0.0 |  |  |  |  |  |  |  |  |  |  |  |  |  |  |
| 0.75 |  |  |  | 0.0 | 0.5 | 1.5 | 2.7 | 1.9 | 1.1 | 0.5 | 0.2 | 0.0 |  |  |  |  |  |  |  |  |  |  |
| 1.25 |  |  | Region I | 0.0 | 0.6 | 4.1 | 5.6 | 4.5 | 2.7 | 1.3 | 0.7 | 0.3 | 0.1 | 0.0 | 0.0 |  |  | Region II |  |  |  |  |
| 1.75 |  |  |  |  | 0.1 | 3.3 | 5.1 | 4.6 | 3.9 | 2.1 | 1.2 | 0.8 | 0.3 | 0.1 | 0.0 |  |  |  |  |  |  |  |
| 2.25 |  |  |  |  |  | 0.9 | 5.3 | 3.7 | 4.1 | 2.9 | 1.3 | 0.8 | 0.4 | 0.2 | 0.1 | 0.0 |  |  |  |  |  |  |
| 2.75 |  |  |  |  |  | 0.1 | 2.4 | 2.6 | 2.8 | 2.8 | 1.6 | 0.8 | 0.3 | 0.1 | 0.1 | 0.0 |  |  |  |  |  |  |
| 3.25 |  |  |  |  |  |  | 0.4 | 1.5 | 1.5 | 2.0 | 1.4 | 0.8 | 0.3 | 0.1 | 0.0 | 0.0 | 0.0 | 0.0 |  |  |  |  |
| 3.75 |  |  |  |  |  |  | 0.0 | 0.5 | 0.6 | 1.1 | 1.0 | 0.6 | 0.3 | 0.1 | 0.0 | 0.0 |  |  |  |  |  |  |
| 4.25 |  |  |  |  |  |  |  | 0.1 | 0.2 | 0.4 | 0.6 | 0.4 | 0.2 | 0.1 | 0.0 | 0.0 |  |  |  |  |  |  |
| 4.75 |  |  |  |  |  |  |  | 0.0 | 0.1 | 0.1 | 0.3 | 0.3 | 0.2 | 0.1 | 0.0 | 0.0 |  |  |  |  |  |  |
| 5.25 |  |  |  |  |  |  |  |  | 0.0 | 0.0 | 0.1 | 0.2 | 0.1 | 0.1 | 0.0 |  |  |  |  |  |  |  |
| 5.75 |  |  |  |  |  |  |  |  |  |  | 0.0 | 0.1 | 0.1 | 0.0 | 0.0 |  |  |  |  |  |  |  |
| 6.25 |  |  |  |  |  |  |  |  |  |  |  | 0.0 | 0.0 | 0.0 | 0.0 |  |  |  |  |  |  |  |
| 6.75 |  |  |  |  |  |  |  |  |  |  |  | 0.0 | 0.0 |  |  |  |  |  |  |  |  |  |
| 7.25 |  |  |  |  |  |  |  |  |  |  |  |  |  |  |  |  |  |  |  |  |  |  |

**Figure 7.** % of Total Occurrences at DOE reference site overlaid on Operating Region for the Heaving Buoy.

% of Total Occurrence — Peak Period - Tp (s), center of bin

| HmO (m), | 0.6 | 1.7 | 2.9 | 4.1 | 5.2 | 6.4 | 7.5 | 8.7 | 9.9 | 11.0 | 12.2 | 13.3 | 14.5 | 15.7 | 16.8 | 18.0 | 19.1 | 20.3 | 21.5 | 22.6 | 23.8 |
|---|---|---|---|---|---|---|---|---|---|---|---|---|---|---|---|---|---|---|---|---|---|
| 0.25 |  |  |  |  |  |  | 0.0 | 0.0 |  |  |  |  |  |  |  |  |  |  |  |  |  |  |
| 0.75 |  |  |  | 0.0 | 0.5 | 1.5 | 2.7 | 1.9 | 1.1 | 0.5 | 0.2 | 0.0 |  |  |  |  |  |  |  |  |  |  |
| 1.25 |  |  |  | 0.0 | 0.6 | 4.1 | 5.6 | 4.5 | 2.7 | 1.3 | 0.7 | 0.3 | 0.1 | 0.0 | 0.0 |  |  |  |  |  |  |  |
| 1.75 |  |  |  |  | 0.1 | 3.3 | 5.1 | 4.6 | 3.9 | 2.1 | 1.2 | 0.8 | 0.3 | 0.1 | 0.0 |  |  |  |  |  |  |  |
| 2.25 |  |  | Region I |  |  | 0.9 | 5.3 | 3.7 | 4.1 | 2.9 | 1.3 | 0.8 | 0.4 | 0.2 | 0.1 | 0.0 |  |  | Region II |  |  |  |
| 2.75 |  |  |  |  |  | 0.1 | 2.4 | 2.6 | 2.8 | 2.8 | 1.6 | 0.8 | 0.3 | 0.1 | 0.1 | 0.0 |  |  |  |  |  |  |
| 3.25 |  |  |  |  |  |  | 0.4 | 1.5 | 1.5 | 2.0 | 1.4 | 0.8 | 0.3 | 0.1 | 0.0 | 0.0 | 0.0 | 0.0 |  |  |  |  |
| 3.75 |  |  |  |  |  |  | 0.0 | 0.5 | 0.6 | 1.1 | 1.0 | 0.6 | 0.3 | 0.1 | 0.0 | 0.0 |  |  |  |  |  |  |
| 4.25 |  |  |  |  |  |  |  | 0.1 | 0.2 | 0.4 | 0.6 | 0.4 | 0.2 | 0.1 | 0.0 | 0.0 |  |  |  |  |  |  |
| 4.75 |  |  |  |  |  |  |  | 0.0 | 0.1 | 0.1 | 0.3 | 0.3 | 0.2 | 0.1 | 0.0 | 0.0 |  |  |  |  |  |  |
| 5.25 |  |  |  |  |  |  |  |  | 0.0 | 0.0 | 0.1 | 0.2 | 0.1 | 0.1 | 0.0 |  |  |  |  |  |  |  |
| 5.75 |  |  |  |  |  |  |  |  |  |  | 0.0 | 0.1 | 0.1 | 0.0 | 0.0 |  |  |  |  |  |  |  |
| 6.25 |  |  |  |  |  |  |  |  |  |  |  | 0.0 | 0.0 | 0.0 | 0.0 |  |  |  |  |  |  |  |
| 6.75 |  |  |  |  |  |  |  |  |  |  |  |  |  |  |  |  |  |  |  |  |  |  |
| 7.25 |  |  |  |  |  |  |  |  |  |  |  |  |  |  |  |  |  |  |  |  |  |  |

**Figure 8.** Percentage of total occurrences overlaid on operating region-SANDIA WaveBot.

The comparison in Table 2 shows a buoy with a much larger volumetric displacement, similar to that tested by SANDIA [26]. It shows that fewer regions are present where volumetric limits dominate and, therefore, MPC becomes less important.

**Table 2.** Percentage of total sea state occurrence in Regions I and II for the heaving buoy and SANDIA WaveBot.

| % of Total Occurrence | Heaving Buoy (Swept Volume = 380 m$^3$) | SANDIA WaveBot (Swept Volume = 4303 m$^3$) |
|---|---|---|
| Region I | 32% | 94% |
| Region II | 68% | 6% |

## 6. Results

### 6.1. Benchmarking Performance of Causal Control Design Methods

To provide a fundamental understanding of the unconstrained controls' performance, we compared Optimal Causal Control, which is based on the LQG paradigm [7], with SANDIA's causal controller designed based on the CCC approximation principle–Proportional Integral (PI) and Feedback Resonator (MPC-FBR). A set of MATLAB and Simulink files containing methods for implementing PI and FBR controllers is available online through the Marine Hydro-Kinetic Data Repository [26]. Both controllers were simulated to compare performance in different wave conditions using no wave prediction information. MPC was also simulated with the same inputs to serve as an upper limit for benchmarking performance of the two causal controller design methods.

The results of this initial trade-off study demonstrated that the algorithm provided by SANDIA provided significantly lower power capture across all sea states, but especially in longer-period waves, where the impedance-matching condition of the underlying controls law is physically impossible due to the device's volumetric constraints. Because we found the LQG method outperformed the CCC controller, the remainder of comparisons in this paper were carried out between the LQG causal control method and a linear MPC controller. Time-domain simulations were carried out using Optimal Causal Control using the LQG method outlined in [9] and linear Model Predictive Control for a complete set of wave conditions. A discrete time-step of 0.01 s and a simulation length of 3600 s was utilized. A Pierson–Moskowitz (PM) wave spectrum with the following formulation was chosen to generate the wave train using the WAFO Toolbox [27]. A plot of the input PM spectrum is shown in Figure 9

$$S(\omega) = \frac{5H_{m0}^2}{\omega_p \omega_n^5}\, e^{-\frac{5}{4}\,\omega_n^{-4}} \tag{32}$$

where $\omega_p = \frac{2\pi}{T_p}$ and $\omega_n = \frac{\omega}{\omega_p}$.

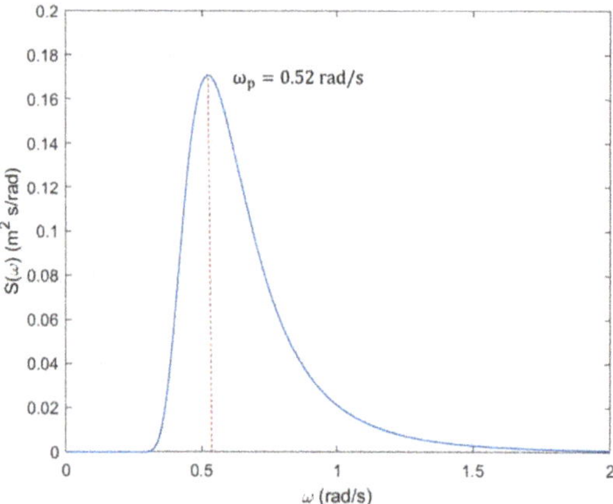


**Figure 9.** Spectral density computed using a PM spectrum for Hm0 = 1 m, Tp = 12 s.

*6.2. Controller Performance with Unconstrained PTO Force*

First, let us look at the unconstrained case where no PTO force limits are imposed on the controller. Motion constraints are still enforced because the displacement of the device relative to the equilibrium is limited to ±2 m. Figure 10 shows a plot of the ratio of average absorbed power using causal control normalized to the average absorbed power using MPC. For short period waves, the ratio of causal controller response closely follows that of MPC and the performance ratio is within 90%. This corresponds to Region I, where power absorption is dominated by the Point Absorber Limit. In long-period waves, the performance ratio starts to deteriorate for wave periods >10 s. This corresponds to Region II, where power absorption is dominated by the volumetric constraint. The results show that, unlike in the case of MPC, the stroke-limited power absorption capability of causal controllers is sub-optimal.

| Causal Control / MPC | | Peak Period - Tp (s), center of bin | | | | | | | | | | | | | | | | | | |
|---|---|---|---|---|---|---|---|---|---|---|---|---|---|---|---|---|---|---|---|---|
| | | 0.6 | 1.7 | 2.9 | 4.1 | 5.2 | 6.4 | 7.5 | 8.7 | 9.9 | 11.0 | 12.2 | 13.3 | 14.5 | 15.7 | 16.8 | 18.0 | 19.1 | 20.3 | 21.5 |
| Hm0 (m), center of bin | 0.25 | | | | | | | | | | | | | | | | | | | |
| | 0.75 | | | | | | | | | | | | | | | | | | | |
| | 1.25 | | | | | 100% | 100% | 96% | 60% | 59% | 59% | 57% | 55% | 58% | 64% | 66% | 67% | | | |
| | 1.75 | | | | | 100% | 100% | 82% | 71% | 61% | 72% | 77% | 74% | 76% | 76% | 75% | 76% | | | |
| | 2.25 | | | | | 100% | 99% | 82% | 80% | 80% | 77% | 82% | 79% | 79% | 78% | 77% | 77% | | | |
| | 2.75 | | | | | 100% | 93% | 82% | 83% | 81% | 82% | 82% | 81% | 80% | 78% | 77% | 76% | | | |
| | 3.25 | | | | | 100% | 92% | 83% | 83% | 82% | 83% | 81% | 81% | 79% | 77% | 77% | 73% | | | |
| | 3.75 | | | | | | | | | | | | | | | | | | | |
| | 4.25 | Region I | | | | | | | Region II | | | | | | | | | | | |
| | 4.75 | | | | | | | | | | | | | | | | | | | |
| | 5.25 | | | | | | | | | | | | | | | | | | | |

**Figure 10.** Avg. power (causal control)/avg. power (MPC) for a set of power-producing sea states.

Figures 11–14 show the time domain comparison of the device response with MPC and causal control for a significant wave height of 1.0 m and Tp = 15 s. The comparison is shown for a wave input in Region II where the volumetric constraint is active. Note that the MPC motion profile often covers the full stroke as opposed to causal control (Figure 13). MPC's "latching" behavior can also be seen in the velocity response profile.

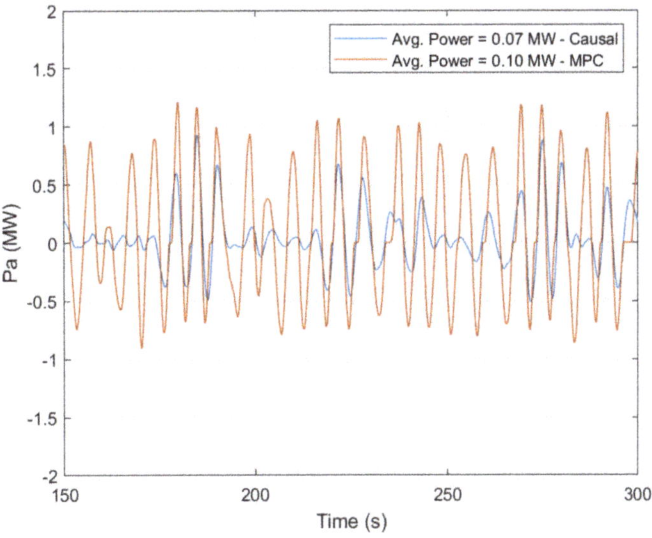


**Figure 11.** Comparison of average power using causal control vs. MPC for an input wave with Hs = 1.0 m, Tp = 15 s.

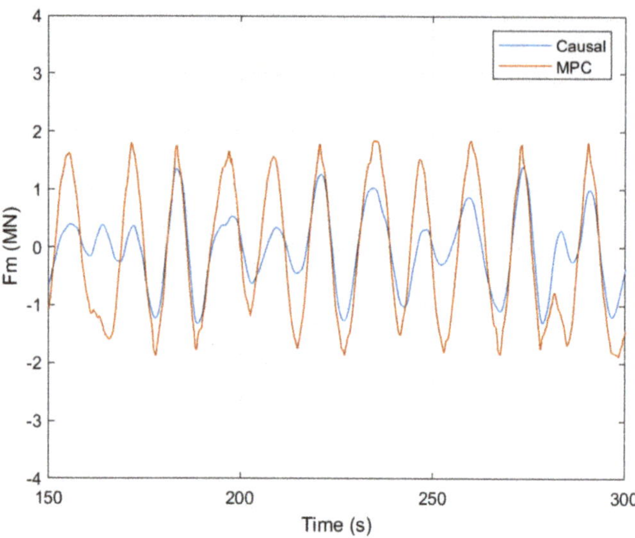

**Figure 12.** Comparison of PTO force using causal control vs. MPC for an input wave with Hs = 1.0 m, Tp = 15 s.

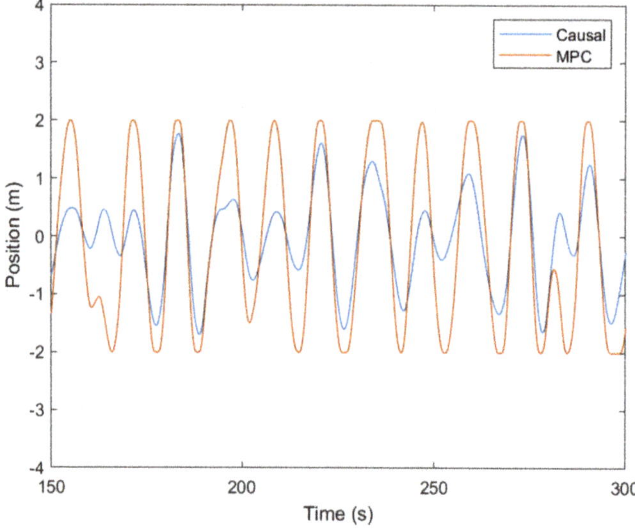

**Figure 13.** Comparison of position response using causal control vs. MPC for an input wave with Hs = 1.0 m, Tp = 15 s.

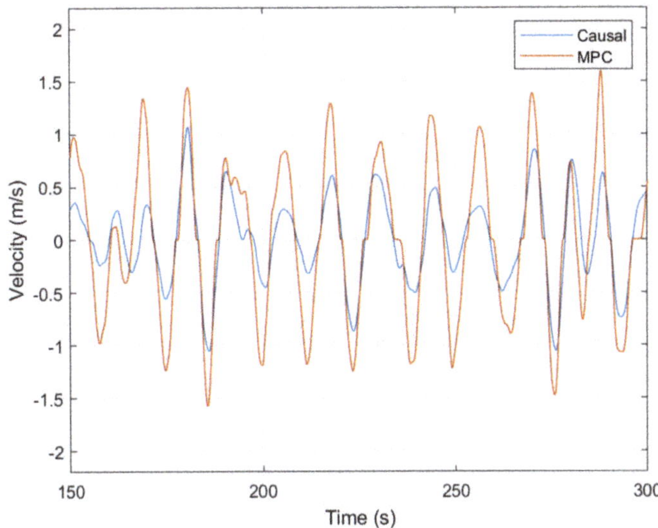

**Figure 14.** Comparison of velocity response using causal control vs. MPC for an input wave with Hs = 1.0 m, Tp = 15 s.

### 6.3. Comparison of Annual Average Energy Capture for Reference Site

A simulation study was carried out to benchmark performance of the two control approaches by calculating the overall annual average energy capture for a deep-water reference site. The heaving buoy WEC was simulated for all of the wave conditions in the scatter diagram shown in Figure 7 for the reference site at Humboldt Bay, CA. Both controllers were simulated with a stroke limit of ±2 m, and no limits on the peak PTO force were imposed.

Assuming a plant capacity factor of 30%, we obtained rated power of 473 kW using MPC and 375 kW with causal control (see Table 3). At this site, the causal controller captures only 79% of the MPC-based controller.

**Table 3.** Performance comparison of MPC and causal control for the DoE reference site at Humboldt Bay, CA, USA.

| | | **MPC** | **Causal Control** |
|---|---|---|---|
| Rated Power (kW) | | 473 | 375 |
| Average Power (kW) | | 142 | 113 |
| Capacity Factor | | 30% | 30% |
| % Annual Energy | Region I | 19% | 18% |
| Captured | Region II | 81% | 82% |

Figure 10 shows the ratio of average absorbed power between causal control and MPC at each sea state. Note that in the regions which are point absorber limited (Region I), the two controllers perform equally well. For long-period waves where performance is upper bound by the Volumetric Limit (Region II), as expected, MPC outperforms causal control. Overall, we note that the percentage of sea state occurrences in Region II is 67.8% and is responsible for close to 81% of the respective annual energy captured from all sea states.

### 6.4. Controller Performance with Constrained PTO Force

Controllers must often operate with limits imposed on the maximum PTO force due to hardware limitations. Therefore, it is important to determine how well the two control approaches handle force constraints. In this situation, the controller must attempt to maximize power absorption while simultaneously limiting the maximum forces used on the PTO. Not all controllers can achieve this dual objective effectively, which typically results in sub-optimal performance for causal controllers. Figure 15 shows a plot of average power capture vs. force constraint for both control approaches. Note that for the same choice of PTO force limit, the average power capture with MPC significantly exceeds causal control. Another way of interpreting these results is that MPC requires less PTO force to achieve the same power capture performance as that of causal control. Figures 16–19 show time domain comparison of the device response with causal control and MPC when the PTO force is limited to 0.5 MN.

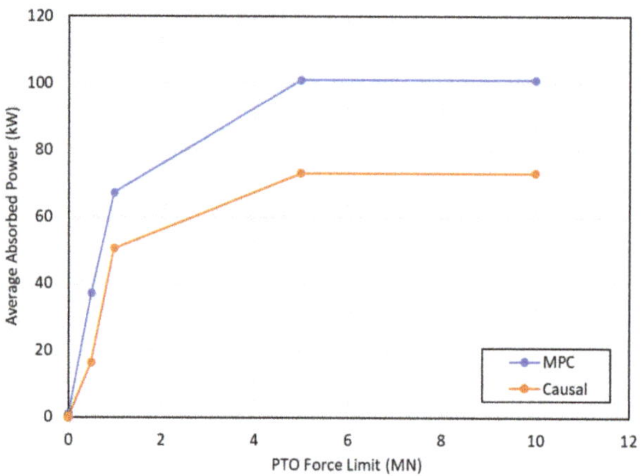

**Figure 15.** Average absorbed power vs. PTO force limit for wave input with Hs = 1 m, Tp = 16 s.

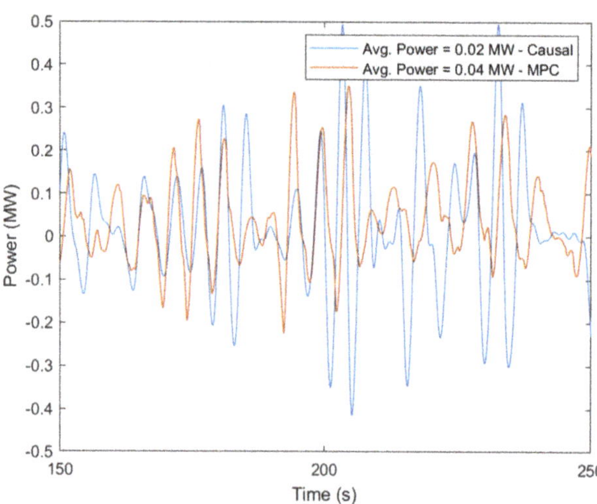

**Figure 16.** Time domain response comparison for absorbed power with PTO force limit = 0.5 MN. Significant wave height Hs = 1.0 m, peak period Tp = 16 s.

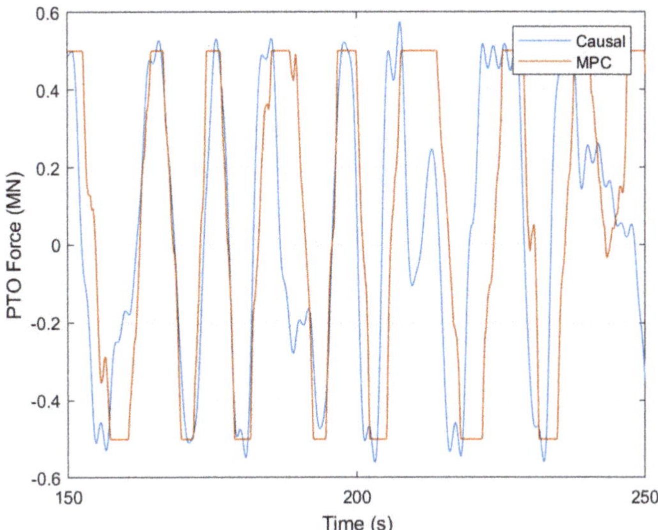

**Figure 17.** Time domain response comparison of PTO force with force limit = 0.5 MN. Significant wave height Hs = 1.0 m, peak period Tp = 16 s.

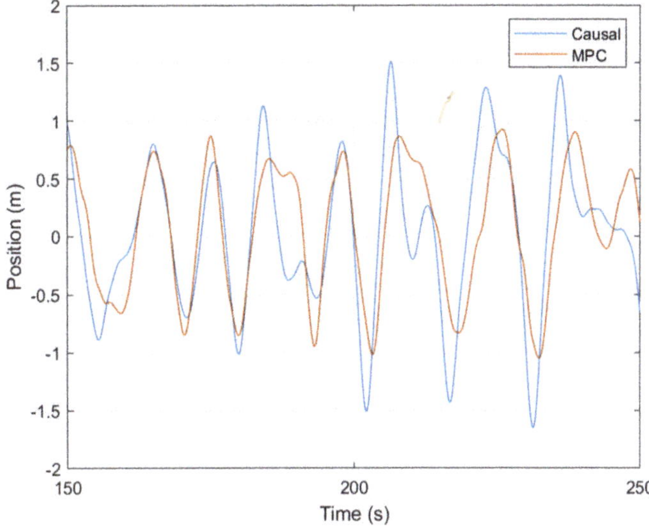

**Figure 18.** Time domain response comparison of WEC position with PTO force limit = 0.5 MN. Significant wave height Hs = 1.0 m, peak period Tp = 16 s.

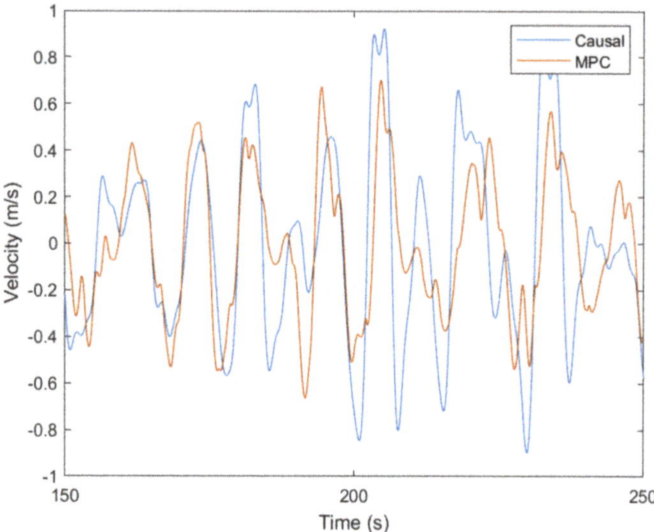

**Figure 19.** Time domain response comparison of WEC velocity with PTO force limit = 0.5 MN. Significant wave height Hs = 1.0 m, peak period Tp = 16 s.

In addition to the control commands issued by the linear causal controller, the causal controller also has a non-linear stroke protection block which adds to the requirements of the PTO machinery force. Causal controllers require careful tuning of the non-linear control parameters to reduce spikes in PTO machinery force. Tuning the stroke protection parameters helps to smooth the PTO force response but it comes at the cost of sacrificing the amount of power that is absorbed. Figure 17 shows a comparison of the PTO force response with causal control and MPC. Note that MPC is able to limit the PTO force effectively, and does not violate the force limit. In contrast, causal control occasionally violates the force limit.

## 7. Conclusions

In this paper, we presented results of a performance comparison study between causal and non-causal methods in a design trade-off space formed by theoretical upper limits on power absorption. While exploring this design trade-off space, we show that identification of operating regions based on device displaced volume and site-specific annual spectrum data are key drivers in identifying suitable control methods for a given WEC application. MPC and causal control performance is similar for Region I, where performance is constrained by Point Absorber Limits and assuming that the PTO is capable of full four-quadrant control. For operating Region II, where performance is limited by Volumetric Limits, MPC shows a significant performance advantage compared to the causal control method.

It should be noted that the cost of most in-ocean systems is directly proportional to their displaced volume. This means that the ratio of average power to volume (P/V) is a key indicator of structural efficiency of a WEC device. Small point absorbers tend to have better P/V ratios, which drives economically viable solutions towards smaller devices. As shown in this study, these smaller point absorbers will produce power predominantly in Region II, where volumetric constraints dominate and MPC provides significant performance advantages. Furthermore, constrained optimal control offered by MPC provides benefits beyond pure performance gains. The ability to directly handle a wide range of constraints as part of the optimization problem ensures constraints can be gently applied, thus reducing transient loads that are harmful to the PTO and device structural components, while maximizing power production.

**Author Contributions:** Conceptualization, M.P. and A.K.; methodology, M.P., A.K., J.S.; software, A.K., J.S., M.P.; validation, M.P., A.K. and J.S.; formal analysis, M.P., A.K. and J.S.; investigation, M.P., A.K. and J.S.; writing—original draft preparation, M.P., A.K. and J.S.; writing—review and editing, M.P.; visualization, M.P., A.K. and J.S.; project administration, M.P.; funding acquisition, M.P. All authors have read and agreed to the published version of the manuscript.

**Funding:** This research was funded by the US Department of Energy, grant number DE-EE0007173.

**Institutional Review Board Statement:** Not applicable.

**Informed Consent Statement:** Not applicable.

**Acknowledgments:** This work was supported by the U.S. Department of Energy under Award No. DE-EE0007173.

**Conflicts of Interest:** The authors declare no conflict of interest.

# References

1. Nebel, P. Maximizing the Efficiency of Wave-Energy Plant Using Complex-Conjugate Control. *J. Syst. Control Eng.* **1992**, *206*, 225–236. [CrossRef]
2. Budal, K.; Falnes, J. Interacting Point Absorbers with Controlled Motion. In *Power from Sea Waves*; Count, B., Ed.; Academic: London, UK, 1980; pp. 381–399.
3. Evans, D.V. A theory for wave-power absorption by oscillating bodies. *J. Fluid Mech.* **1976**, *77*, 1–25. [CrossRef]
4. Falnes, J.; Kurniawan, A. *Ocean. Waves and Oscillating Systems: Linear Interactions Including Wave Energy Extraction*; Cambridge University Press: Cambridge, UK, 2002.
5. Coe, R.; Bacelli, G.; Nevarez, V.; Cho, H.; Wilches-Bernal, F. *A Comparative Study on Wave Prediction for WECs. SANDIA REPORT SAND2018-8602*; Sandia National Lab.(SNL-NM): Albuquerque, NM, USA, 2018.
6. Scruggs, J.; Lattanzio, S.; Taflanidis, A.; Cassidy, I. Optimal causal control of a wave energy converter in a random sea. *Appl. Ocean Res.* **2013**, *42*, 1–15. [CrossRef]
7. Scruggs, J. Causal control design for wave energy converters with finite stroke. *IFAC-PapersOnLine* **2017**, *50*, 15678–15685. [CrossRef]
8. Nie, R.; Scruggs, J.; Chertok, A.; Clabby, D.; Previsic, M.; Karthikeyan, A. Optimal causal control of wave energy converters in stochastic waves—Accommodating nonlinear dynamic and loss models. *Int. J. Mar. Energy* **2016**, *15*, 41–55. [CrossRef]
9. Lao, Y.; Scruggs, J.; Karthikeyan, A.; Previsic, M. Discrete-time causal control of WECs with finite stroke in stochastic waves. In Proceedings of the 13th European Wave and Tidal Energy Conference, Napoli, Italy, 1–6 September 2019.
10. Blondel, E.; Bonnefoy, F.; Ferrant, P. Deterministic non-linear wave prediction using probe data. *Ocean Eng.* **2010**, *37*, 913–926. [CrossRef]
11. Lyzenga, D.; Nwogu, O.; Trizna, D.; Hathaway, K. Ocean wave field measurements using X-band Doppler radars at low grazing angles. In Proceedings of the 2010 IEEE International Geoscience and Remote Sensing Symposium, Honolulu, HI, USA, 25–30 July 2010.
12. Lyzenga, D.R.; Walker, D.T. A Simple Model for Marine Radar Images of the Ocean Surface. *IEEE Geosci. Remote Sens. Lett.* **2015**, *12*, 2389–2392. [CrossRef]
13. Gieske, P. Model Predictive Control of a Wave Energy Converter: Archimedes Wave Swing. Master's Thesis, Delft University of Technology, Delft, The Netherlands, 2007.
14. Karthikeyan, A.; Previsic, M.; Scruggs, J.; Chertok, A. Non-linear Model Predictive Control of Wave Energy Converters with Realistic Power Take-off Configurations and Loss Model. In Proceedings of the 2019 IEEE Conference on Control Technology and Applications (CCTA), Hong Kong, China, 19–21 August 2019.
15. Eriksson, C. Model Predictive Control of CorPower Ocean Wave Energy Converter. Ph.D. Thesis, KTH, Stockholm, Sweden, 2016.
16. Cretel, J.; Lewis, A.; Lightbody, G.; Thomas, G. An application of model predictive control to a wave energy point absorber. In Proceedings of the IFAC Conference on Control Methodologies and Technology for Energy Efficiency, Vilamoura, Portugal, 29–31 March 2010.
17. Nathan, T.; Yeung, R. Nonlinear model predicitve control applied to a generic ocean-wave energy extractor. *J. Offshore Mech. Arct. Eng.* **2014**, *135*, 41901.
18. Richter, M.; Magana, M.E.; Sawodny, O.; Brekken, T.K.A. Nonlinear Model Predictive Control of a Point Absorber Wave Energy Converter. *IEEE Trans. Sustain. Energy* **2013**, *4*, 118–126. [CrossRef]
19. Faedo, N.; Olaya, S.; Ringwood, J.V. Optimal control, MPC and MPC-like algorithms for wave energy systems: An overview. *IFAC J. Syst. Control* **2017**, *1*, 37–56. [CrossRef]
20. WAMIT. Available online: https://www.wamit.com/ (accessed on 20 June 2020).
21. Coleman, T.F.; Li, Y. A Reflective Newton Method for Minimizing a Quadratic Function Subject to Bounds on Some of the Variables. *SIAM J. Optim.* **1996**, *6*, 1040–1058. [CrossRef]
22. Gill, P.E.; Murray, W.; Wright, M.H. *Practical Optimization*; Academic Press: London, UK, 1981.

23. Lao, Y.; Scruggs, J.T. A Modified Technique for Spectral Factorization of Infinite-Dimensional Systems Using Subspace Techniques. In Proceedings of the 2019 IEEE 58th Conference on Decision and Control (CDC), Nice, France, 11–13 December 2019.
24. Budal, K.; Falnes, J. A Resonant Point Absorber of Ocean Waves. *Nature* **1975**, *256*, 478–479.
25. Hals, J.; Falnes, J.; Moan, T. Constrained Optimal Control of a Heaving Buoy Wave-Energy Converter. *J. Offshore Mech. Arct. Eng.* **2010**, *133*, 011401. [CrossRef]
26. Laboratories, S.N. MASK3 for Advanced WEC Dynamics and Controls [Data Set]. Available online: https://dx.doi.org/10.15473/1581762 (accessed on 20 May 2020).
27. WAFO-Group. *WAFO—A Matlab Toolbox for Analysis of Random Waves and Loads—A Tutorial*; Center for Mathematical Sciences, Lund University: Lund, Sweden, 2000.

Journal of
*Marine Science and Engineering*

*Article*

# Real-Time Nonlinear Model Predictive Controller for Multiple Degrees of Freedom Wave Energy Converters with Non-Ideal Power Take-Off

Ali S. Haider [1,*], Ted K. A. Brekken [1] and Alan McCall [2]

[1] Electrical Engineering and Computer Science, Oregon State University, Corvallis, OR 97331, USA; brekken@eecs.oregonstate.edu
[2] Dehlsen Associates, LLC., Santa Barbara, CA 93101, USA; amccall@ecomerittech.com
* Correspondence: haidera@oregonstate.edu

**Abstract:** An increase in wave energy converter (WEC) efficiency requires not only consideration of the nonlinear effects in the WEC dynamics and the power take-off (PTO) mechanisms, but also more integrated treatment of the whole system, i.e., the buoy dynamics, the PTO system, and the control strategy. It results in an optimization formulation that has a nonquadratic and nonstandard cost functional. This article presents the application of real-time nonlinear model predictive controller (NMPC) to two degrees of freedom point absorber type WEC with highly nonlinear PTO characteristics. The nonlinear effects, such as the fluid viscous drag, are also included in the plant dynamics. The controller is implemented on a real-time target machine, and the WEC device is emulated in real-time using the WECSIM toolbox. The results for the successful performance of the design are presented for irregular waves under linear and nonlinear hydrodynamic conditions.

**Keywords:** nonlinear model predictive control; two degrees of freedom wave energy converter; nonlinear hydrodynamics; nonlinear power take-off

check for
**updates**

**Citation:** Haider, A.S.; Brekken, T.K.A.; McCall, A. Real-Time Nonlinear Model Predictive Controller for Multiple Degrees of Freedom Wave Energy Converters with Non-Ideal Power Take-Off. *J. Mar. Sci. Eng.* **2021**, *9*, 890. https://doi.org/10.3390/jmse9080890

Academic Editor: Constantine Michailides

Received: 3 August 2021
Accepted: 16 August 2021
Published: 18 August 2021

**Publisher's Note:** MDPI stays neutral with regard to jurisdictional claims in published maps and institutional affiliations.

## 1. Introduction

Renewable energy technologies present a viable and sustainable contribution to the world's growing energy demands, and the ocean provides potential for an enormous untapped energy resource for the world's energy portfolio [1,2]. The prospect of ocean wave energy has triggered research in optimal power capture techniques for wave energy converters, including non-ideal operating conditions, such as the non-ideal PTO system constraints [3] and nonlinear sea conditions. Achieving optimal power capture by a WEC in practice is a multifaceted objective. It depends on various factors, such as the physical design of the WEC, the design of the PTO system, the ocean conditions, and the control techniques.

Model predictive control (MPC) is a promising control approach for wave energy converters' relatively slow plant dynamics because it maximizes energy capture while respecting the system's mechanical limits. MPC is a look-ahead control strategy that predicts future system behavior to solve a constrained optimization problem and determines the best control action to maximize the output power of WEC. MPC and other optimal control schemes, such as pseudospectral methods and MPC-like algorithms, have been comprehensively studied in the literature for a single WEC device and an array of wave energy converters [4–9]. An MPC algorithm uses an internal model of the plant to predict the system's future states [10]. Nonlinear control algorithms can consider the non-ideal operating conditions and nonlinear effects, including but not limited to non-ideal power take-off mechanism [11], nonlinear viscous drag terms [12,13], and nonlinear mooring dynamics [2]. The non-ideality of PTO systems in most literature is limited to the efficiency of the PTO mechanism [13–16]. One of the motivations for this research is to consider higher-order nonlinear PTO characteristics as an optimization objective for the NMPC problem. The economic MPC techniques consider a general economic cost function directly in real time [17–19]. However, we have deployed a real-time iterative (RTI) algorithm [20,21] to optimize a more general class of non-ideal PTO mechanisms using pseudo-quadratic formulations [3]; this method also supports nonlinearities in the plant dynamics, such as mooring and fluid viscous drag. Another motivation for this work is investigating nonlinear multiple degrees of freedom WEC coupled to non-ideal PTO.

Lots of work has been focused on studying multiple degrees of freedom WEC devices that prove a significant improvement of power capture by the WEC device. Multi-resonant feedback control of a three degree of freedom WEC is presented [22], where a linear hydrodynamic model is considered, and multi-resonant proportional-derivative control law is proposed where the focus is linear plant dynamics under unconstrained control. An analysis of a multi-degree-of-freedom point absorber WEC in the surge, heave, and pitch directions is presented in [23], and frequency and time-domain formulations are presented for the linear plant dynamics. A time-domain model for a point absorber WEC in six degrees of freedom is developed in [24] with an optimal resistive loading. The three degrees of a freedom model of a WEC is presented in [25], where the capture performance of various PTO systems is investigated for a linear plant model. An active control strategy based on the optimal velocity trajectory tracking for a multi-DoF submerged point absorber WEC is presented in [26], where a linearized dynamic system model is considered along with an ideal PTO mechanism. A nonlinear MPC design and implementation based on differential flatness parameterization has been proposed in [27]. Given that most of the work focused on linear plant dynamics for multiple degrees of freedom WEC or ideal PTO mechanisms, and the lack of application of NMPC for such class or problems, we have investigated the application of NMPC to nonlinear multiple DoF WEC plant with a non-ideal PTO mechanism, and focus on the real-time implementation of the control algorithm on a real-time target machine.

This research presents the maximization of power extraction by a 2-DoF WEC device, a WECSIM [28] model of the full-scale version of the Dehlsen Associates, LLC multi-pod CENTIPOD [29]. Although the CENTIPOD device is an array device, the cross-coupling between pods is ignored for this study, which is negligible for the sea conditions of interest in this work and will be investigated in the future. The goal is to optimize the power extracted by the heave and pitch PTOs subject to actuation and velocity constraints. The objective function is a nonstandard and nonquadratic functional of PTO force and velocity, resulting from a practical PTO generator power loss characteristic. The WEC model includes nonlinear viscous drag terms; hence, the resulting plant model is a nonlinear dynamic system. We have implemented an NMPC for the problem. To tackle a free-formed objective function subjected to nonlinear system dynamics, we have used the extended version of the NMPC design from [30], based on pseudo-quadratization using an ACADO toolkit [21]. No prior knowledge of wave excitation is assumed. The WEC model is simulated on a real-time emulator machine, while control is deployed on a Speedgoat real-time performance machine [31], which is interfaced with the WEC emulator machine through an ethernet port. The simulation results for real-time NMPC are presented for the linear and nonlinear hydrodynamics conditions simulated in WECSIM.

## 2. Time Domain Model of a Multiple Degree of Freedom WEC

The WEC device is a full-scale version of the Dehlsen Associates, LLC multi-pod CENTIPOD [29,32]. A 1:35-scale version of the device is shown in Figure 1. This CENTIPOD device has three floating pods and three spars fixed to a backbone structure. The backbone is anchored using mooring lines, as shown in Figure 2. In its 2-DoF version, each pod is attached to a PTO mechanism in the heave and pitch degrees of freedom. All pods in Figure 2 are assumed identical, and since the CENTIPOD device is an array device, the array effect [33] could become prominent as the significant height of the waves increases and the incident angle of the waves is not parallel to the $x$-axis in Figure 2. For this study, incident waves are assumed parallel to the *x*-axis, and for the sea state of interest in this work, the cross-coupling between the pods is very small and is neglected, although it will be investigated in future work.

**Figure 1.** Image of the Dehlsen Associates, LLC, 1:35-scale CENTIPOD WEC.

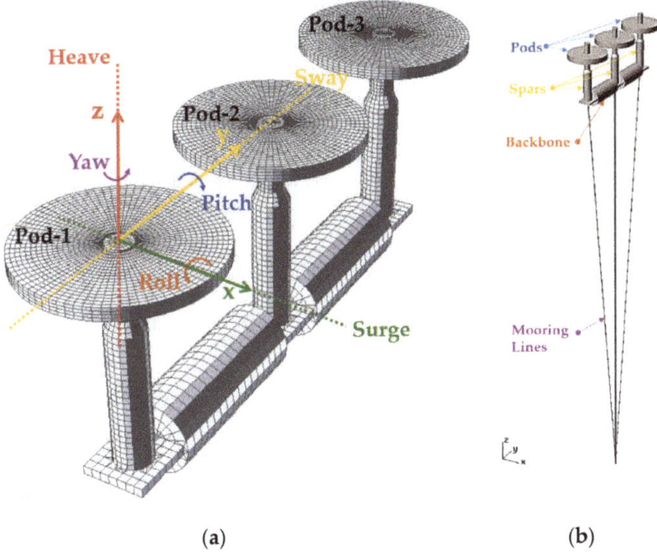



(a)                              (b)

**Figure 2.** Degrees of freedom for dynamic modeling of CENTIPOD WEC: (**a**) baseline configuration; (**b**) model with mooring lines.

We will follow the subscript notation of the WEC-Sim toolbox [28] for the degrees of freedom for WEC, in which the integers from 1, 2, … 6 correspond to surge, sway, heave, roll, pitch, and yaw, respectively. Some other notations and symbols for WEC modeling are given in Table 1.

**Table 1.** Notations and symbols for WEC modeling.

| Variable | Description |
| --- | --- |
| $v_i$ | Velocity (Linear or Angular) in $i^{th}$ DoF |
| $x_i$ | Displacement (Linear or Angular) in $i^{th}$ DoF |
| $\xi_i$ | Intermediate State variables for radiation force State-Space approximation |
| $F_{r,pq}$ | Radiation force in $p^{th}$ DoF due to velocity in $q^{th}$ DoF |
| $F_{hs,i}$ | Hydrostatic force in $i^{th}$ DoF |
| $F_{v,i}$ | Viscous drag force in $i^{th}$ DoF |
| $F_{e,i}$ | Wave excitation force in $i^{th}$ DoF |
| $F_{p,i}$ | PTO force in $i^{th}$ DoF |
| $m$ | Mass of the float |
| $A_{pq}(\infty)$ | Added mass at the infinite frequency in $p^{th}$ DoF due to acceleration in $q^{th}$ DoF |
| $C_i$ | The hydrostatic restoring coefficient in $i^{th}$ DoF |
| $C_{vd,i}$ | Viscous drag coefficient in $i^{th}$ DoF |
| $A_{qp}$ | Frequency-dependent added mass in $p^{th}$ DoF due to acceleration in $q^{th}$ DoF |
| $B_{qp}$ | Frequency-dependent damping in $p^{th}$ DoF due to velocity in $q^{th}$ DoF |
| $K_{pq}$ | Radiation force impulse response without infinite frequency added mass |
| $Z_{qp}$ | WEC Intrinsic impedance response in $p^{th}$ DoF due to velocity in $q^{th}$ DoF |
| $a_i$ | Polynomial coefficients |
| $c_{i,j}$ | Polynomial coefficients for cost functional |
| $I_{p,i}$ | $i^{th}$ PTO current |
| $\eta_{Conv}$ | PTO converter efficiency |
| $K_{Cu}$ | PTO generator copper loss constant |
| $R_\Omega$ | PTO generator winding resistance |

### 2.1. Surge-Pitch-Heave Model of WEC Modeling in State-Space Form

Each pod in Figure 2 is modeling as a wave point absorber device. The Cummins equation for the coupled surge and pitch dynamics for a point absorber pod (assuming a local reference frame) is given by,

$$(m + A_{11}(\infty))\dot{v}_1 + A_{15}(\infty)\dot{v}_5 = -F_{r,11}(t) - F_{r,15}(t) - F_{v,1}(t) + F_{e,1}(t), \tag{1}$$

$$(m + A_{55}(\infty))\dot{v}_5 + A_{51}(\infty)\dot{v}_1 = -F_{r,55}(t) - F_{r,51}(t) - F_{v,5}(t) - F_{hs,5}(t) - F_{p,5}(t) + F_{e,5}(t). \tag{2}$$

The Cummins equation for the heave dynamics of a point absorber pod is given by,

$$(m + A_{33}(\infty))\dot{v}_3(t) = -F_{r,33}(t) - F_{hs,3}(t) - F_{v,3}(t) - F_{p,3}(t) + F_{e,3}(t), \tag{3}$$

The hydrostatic, viscous damping, and radiation force terms in (1) through (3) are given by,

$$F_{r,ij}(t) = \int_{-\infty}^{t} K_{ij}(t - \tau)v_j d\tau, \tag{4}$$

$$F_{hs,i}(t) = C_i x_i, \tag{5}$$

$$F_{v,i}(t) = C_{d,i} v_i |v_i|. \tag{6}$$

A transfer function expression can approximate the convolution integral term in (4),

$$F_{r,pq}(t) = \int_{-\infty}^{t} K_{pq}(t - \tau)v_q d\tau \Leftrightarrow F_{r,pq}(j\omega) = Z_{pq}(j\omega)V_q(j\omega), \tag{7}$$

Using the device data from WAMIT [34], we can approximate the intrinsic impedance $Z_{pq}(j\omega)$ in (7) by a second order transfer function using system identification techniques,

$$Z_{pq}(j\omega) = \left[ j\omega \left( A_{pq}(j\omega) - A_{pq}(\infty) \right) + B_{qp}(j\omega) \right] \approx \left. \frac{\alpha_{pq,1}s + \alpha_{pq,0}}{s^2 + \beta_{pq,1}s + \beta_{pq,0}} \right|_{s=j\omega}, \tag{8}$$

Using (8) in (7) enables us to express the radiation force as a second-order transfer function,

$$F_{r,pq}(s) \approx \frac{\alpha_{pq,1}s + \alpha_{pq,0}}{s^2 + \beta_{pq,1}s + \beta_{pq,0}} V_q(s), \tag{9}$$

The transfer function expression in (9) can be converted to the state-space expressions in the observer canonical forms for each of the radiation force terms,

$$\begin{bmatrix} \dot{\xi}_k(t) \\ \dot{\xi}_{k+1}(t) \end{bmatrix} = \begin{bmatrix} 0 & 1 \\ a_k & a_{k+1} \end{bmatrix} \begin{bmatrix} \xi_k(t) \\ \xi_{k+1}(t) \end{bmatrix} + \begin{bmatrix} b_k \\ b_{k+1} \end{bmatrix} v_q(t), \tag{10}$$

$$y_{pq}(t) = \begin{bmatrix} 1 & 0 \end{bmatrix} \begin{bmatrix} \xi_k(t) \\ \xi_{k+1}(t) \end{bmatrix} \approx F_{r,pq}(t). \tag{11}$$

By the comparison of (9)–(11), we have, $\alpha_{pq,1} = b_k$, $\beta_{pq,1} = -a_{k+1}$, $\beta_{pq,0} = -a_k$, and $\alpha_{pq,0} = b_{k+1} - b_k a_{k+1}$. Making a change of variables in (1),

$$M_{ii} = (m + A_{ii}(\infty)), \tag{12}$$

$$F_{1,net} = -F_{r,11}(t) - F_{r,15}(t) - F_{v,1}(t) + F_{e,1}(t), \tag{13}$$

$$F_{5,net} = -F_{r,55}(t) - F_{r,51}(t) - C_5 x_5 - F_{v,5}(t) - F_{p,5}(t) + F_{e,5}(t). \tag{14}$$

Using (12)–(14) in (1), we get the pitch-surge coupled model of a pod as,

$$\begin{bmatrix} \dot{v}_1 \\ \dot{v}_5 \end{bmatrix} = \begin{bmatrix} M_{11} & A_{15}(\infty) \\ A_{51}(\infty) & M_{55} \end{bmatrix}^{-1} \begin{bmatrix} F_{1,net} \\ F_{5,net} \end{bmatrix} \tag{15}$$

The viscous drag force term $v_i|v_i|$ in (6) is a hard nonlinearity that may lead to convergence issues for the optimization solvers. One solution is to approximate this term with a soft nonlinearity by replacing it with a smooth higher-order polynomial. A third-order polynomial approximation for $v_i|v_i|$ is used in the surge and heave direction, where the range of interest of velocity is $v_i \in (-1.5, 1.5)$ $m/sec$, and a fifth-order polynomial approximation is used for pitch direction, where the range of interest of velocity is $v_i \in (-0.5, 0.5)$ $rad/sec$. With, $p_{i,j}$ being the $j^{th}$ polynomial coefficient for $i^{th}$ degree polynomial curve fit

$$F_{v,i} = C_{d,i} v_i |v_i| \approx C_{d,i} \left( p_{3,3} v_i^3 + p_{3,1} v_i \right), \quad i = 1, 3, \tag{16}$$

$$F_{v,5} = C_{d,5} v_5 |v_5| \approx C_{d,5} \left( p_{5,5} v_5^5 + p_{5,3} v_5^3 + p_{5,1} v_5 \right). \tag{17}$$

The curve fits (16) and (17) are shown in Figure 3a,b, respectively. Using (15) and (3), we get a Surge-Heave-Pitch model of a pod as,

$$\dot{\mathbf{X}} = \mathbf{A}\mathbf{X} + \mathbf{B_p}\mathbf{F_p} + \mathbf{B_v}\mathbf{F_v} + \mathbf{B_e}\mathbf{F_e}, \tag{18}$$

where,

$$\mathbf{F_p} = \begin{bmatrix} F_{p,5} & F_{p,3} \end{bmatrix}^T, \tag{19}$$

$$\mathbf{F_v} = \begin{bmatrix} F_{v,1} & F_{v,5} & F_{v,3} \end{bmatrix}^T, \tag{20}$$

$$\mathbf{F_e} = \begin{bmatrix} F_{e,1} & F_{e,5} & F_{e,3} \end{bmatrix}^T, \tag{21}$$

$$\mathbf{X} = \begin{bmatrix} v_1 & v_5 & x_5 & \zeta_3 & \zeta_4 & \zeta_5 & \zeta_6 & \zeta_7 & \zeta_8 & \zeta_9 & \zeta_{10} & v_3 & x_3 & \zeta_1 & \zeta_2 \end{bmatrix}^T. \tag{22}$$

and the radiation force terms are approximated by following state variables using (10),

$$F_{r,11} = \zeta_3, \; F_{r,15} = \zeta_5, \; F_{r,51} = \zeta_7, \; F_{r,55} = \zeta_9, \; F_{r,33} = \zeta_1. \tag{23}$$

and,

$$\mathbf{A} = \left[ \begin{array}{ccccccccccc|c} 0 & 0 & -m_{15}C_5 & -m_{11} & 0 & -m_{11} & 0 & -m_{15} & 0 & -m_{15} & 0 & \\ 0 & 0 & -m_{55}C_5 & -m_{51} & 0 & -m_{51} & 0 & -m_{55} & 0 & -m_{55} & 0 & \\ 0 & 1 & 0 & 0 & 0 & 0 & 0 & 0 & 0 & 0 & 0 & \\ b_3 & 0 & 0 & 0 & 1 & 0 & 0 & 0 & 0 & 0 & 0 & \\ b_4 & 0 & 0 & a_3 & a_4 & 0 & 0 & 0 & 0 & 0 & 0 & \\ 0 & b_5 & 0 & 0 & 0 & 0 & 1 & 0 & 0 & 0 & 0 & \mathbf{0}_{11 \times 4} \\ 0 & b_6 & 0 & 0 & 0 & a_5 & a_6 & 0 & 0 & 0 & 0 & \\ 0 & b_7 & 0 & 0 & 0 & 0 & 0 & 0 & 1 & 0 & 0 & \\ 0 & b_8 & 0 & 0 & 0 & 0 & 0 & a_7 & a_8 & 0 & 0 & \\ b_9 & 0 & 0 & 0 & 0 & 0 & 0 & 0 & 0 & 0 & 1 & \\ b_{10} & 0 & 0 & 0 & 0 & 0 & 0 & 0 & 0 & a_9 & a_{10} & \\ \hline & & & & & & & & & & & \begin{array}{cccc} 0 & \frac{-C_3}{M_{33}} & \frac{-1}{M_{33}} & 0 \\ 1 & 0 & 0 & 0 \\ b_1 & 0 & 0 & 1 \\ b_2 & 0 & a_1 & a_2 \end{array} \\ & & & \mathbf{0}_{4 \times 11} & & & & & & & & \end{array} \right], \tag{24}$$

$$\mathbf{B_p} = \begin{bmatrix} -m_{15} & 0 \\ -m_{55} & 0 \\ 0 & 0 \\ 0 & 0 \\ 0 & 0 \\ 0 & 0 \\ 0 & 0 \\ 0 & 0 \\ 0 & 0 \\ 0 & 0 \\ 0 & 0 \\ 0 & 0 \\ 0 & \frac{-1}{M_{33}} \\ 0 & 0 \\ 0 & 0 \\ 0 & 0 \end{bmatrix}, \; \mathbf{B_v} = \begin{bmatrix} -m_{11} & -m_{15} & 0 \\ -m_{51} & -m_{55} & 0 \\ 0 & 0 & 0 \\ 0 & 0 & 0 \\ 0 & 0 & 0 \\ 0 & 0 & 0 \\ 0 & 0 & 0 \\ 0 & 0 & 0 \\ 0 & 0 & 0 \\ 0 & 0 & 0 \\ 0 & 0 & 0 \\ 0 & 0 & 0 \\ 0 & 0 & \frac{-1}{M_{33}} \\ 0 & 0 & 0 \\ 0 & 0 & 0 \\ 0 & 0 & 0 \end{bmatrix}, \; \mathbf{B_e} = \begin{bmatrix} m_{11} & m_{15} & 0 \\ m_{51} & m_{55} & 0 \\ 0 & 0 & 0 \\ 0 & 0 & 0 \\ 0 & 0 & 0 \\ 0 & 0 & 0 \\ 0 & 0 & 0 \\ 0 & 0 & 0 \\ 0 & 0 & 0 \\ 0 & 0 & 0 \\ 0 & 0 & 0 \\ 0 & 0 & 0 \\ 0 & 0 & \frac{1}{M_{33}} \\ 0 & 0 & 0 \\ 0 & 0 & 0 \\ 0 & 0 & 0 \end{bmatrix}, \tag{25}$$

Putting (16) and (17) in (18) gives us a 2-DoF (heave and pitch) WEC nonlinear plant model, where the surge is coupled with the pitch and heave is a decoupled DoF. We can use this plant model as a prediction model in NMPC.

### 2.2. Nonquadratic WEC-PTO Model

The electrical power output from the PTO mechanism of the WEC is the difference between the mechanical power input from the waves and the losses in the PTO system. For a given PTO generator, the electrical PTO power cost functional to be maximized, including the electrical losses, is given by,

$$\max_{F_{p,i}} P_{E,i} = \eta_{Conv}\left( P_{Mechanical,i} - P_{Loss,i} \right) = \eta_{Conv}\left( F_{p,i}v_i - K_{Cu}\left[ I_{p,i}\left( F_{p,i} \right) \right]^2 R_{\Omega} \right), \tag{26}$$

The case study scenario is taken from McCleer Power's Linear PTO generator [3] with the PTO force–current characteristics given by Figure 4. We can approximate the experimental data in Figure 4 with a mathematical relation, such as a piecewise linear function or a nonlinear function. We have used polynomial approximation which is a smooth function. This relation is described by a third-order curve fit between the PTO current and the PTO force,

$$I_{p,i}(F_{p,i}) = a_{3,i}F_{p,i}^3 + a_{2,i}F_{p,i}^2 + a_{1,i}F_{p,i} + a_{0,i}, \tag{27}$$

Putting (27) in (16), we get,

$$P_{E,i} = c_{0,i}F_{p,i}v_i - \left( c_{1,i}F_{p,i}^6 + c_{2,i}F_{p,i}^5 + c_{3,i}F_{p,i}^4 + c_{4,i}F_{p,i}^3 + c_{5,i}F_{p,i}^2 + c_{6,i}F_{p,i} + c_{7,i} \right) , \tag{28}$$

The PTO cost functional surface in (28) is plotted in the PTO velocity–force plane, as shown in Figure 5. The surface plot of the mechanical PTO power, $P_{Mechanical,i} = F_{p,i}v_i$ is non-convex, as shown in Figure 5. However, the electrical PTO power surface, $P_{E,i}$ in (26) has a quadratic power loss term, and it gives convexity to the electrical power surface along the PTO force axis in Figure 5.

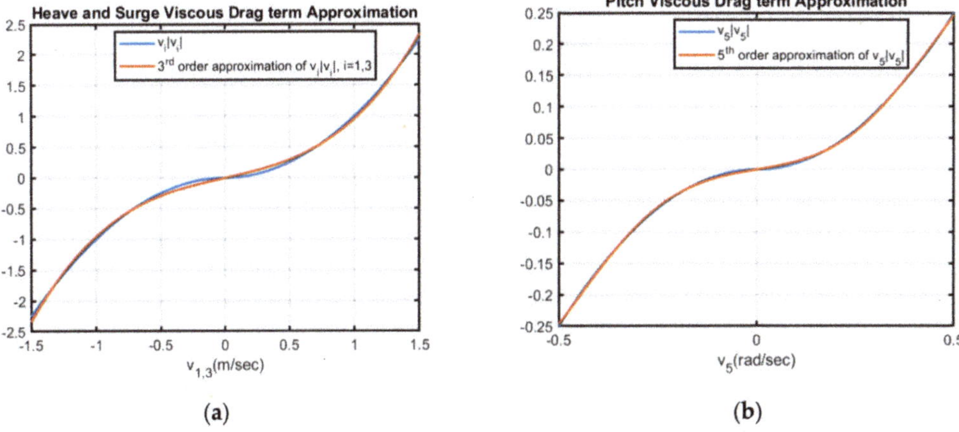


(a)  (b)

**Figure 3.** Polynomial approximations of the quadratic drag term $v_i|v_i|$: (**a**) 3rd order curve fit for heave and surge axes; (**b**) 5th order curve fit for pitch axis.

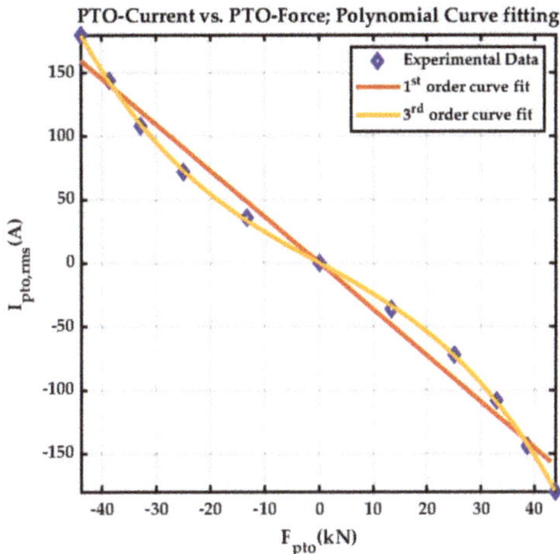


**Figure 4.** Polynomial curve fitting to the PTO force-current experimental data for a PTO generator.

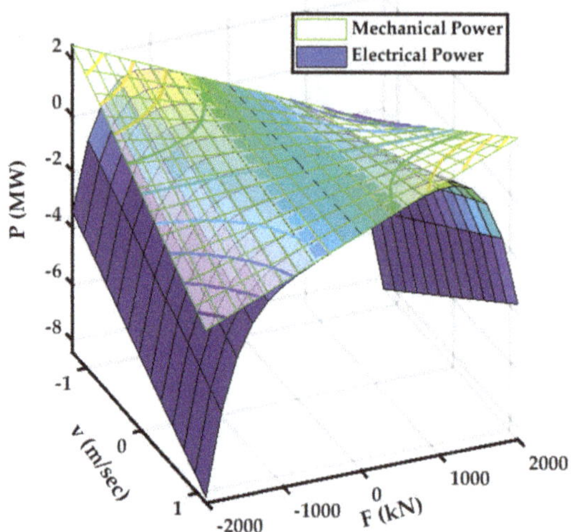



**Figure 5.** Mechanical and electrical PTO power surface plot in PTO velocity-force plane.

## 3. Implementation of NMPC for 2-DoF Heave-Pitch WEC

The optimal control problem of a WEC involves manipulating the PTO force/torque to maximize the power capture while respecting some system constraints. Various optimal control approaches have been developed, and a comprehensive review can be found in [34]. MPC is a model-based online optimal control solution, and a given NMPC problem optimizes a manipulated variable $u(t)$ to maximize some cost functional $P(\cdot)$ while respecting the system constraints. A special class of NMPC problems has been formulated in [30], in which the cost functional takes on a nonlinear piecewise polynomial form. Considering the case of finite horizon optimization control, we can mathematically describe the NMPC problem of such a class as,

$$\underset{\mathbf{u}(t)}{maximize}\ \mathbf{P}\left[t, \dot{\mathbf{X}}(t), \mathbf{X}(t), \mathbf{U}(t), \mathbf{p}(t)\right] \tag{29}$$

$$\text{Where}: \mathbf{P}(\cdot) = \begin{cases} P_1(\cdot) + \rho_{N,1}(\cdot), & q_k(t) < R_1 \\ P_2(\cdot) + \rho_{N,2}(\cdot), & R_1 \leq q_k(t) \leq R_2 \\ \quad \vdots & \quad \vdots \\ P_j(\cdot) + \rho_{N,j}(\cdot), & R_{j-1} \leq q_k(t) \leq R_j \end{cases}, \tag{30}$$

subject to,

$$\text{Dynamic Constraints}: \mathbf{0} = \mathbf{g}\left(t, \dot{\mathbf{X}}(t), \mathbf{X}(t), \mathbf{U}(t), \mathbf{d}(t), \mathbf{p}(t), N\right), \tag{31}$$

$$\text{Boundary Constraint Function}: \mathbf{0} = \mathbf{r}(N, \mathbf{X}(0), \mathbf{U}(0), \mathbf{X}(N), \mathbf{U}(N), \mathbf{p}), \tag{32}$$

$$\text{Path Constraints Function}: \mathbf{0} \geq \mathbf{s}(t, \dot{\mathbf{X}}(t), \mathbf{U}(t), \mathbf{p}(t)). \tag{33}$$

The description of various variables and constants in (298) through (33) is given in Table 2. The wave excitation force $\mathbf{F_e}$ acting on the hull is considered an unmeasured system disturbance, and based on the available measurements, the controller internally estimates $\mathbf{F_e}$.

**Table 2.** Symbols and notations for NMPC formulation.

| Variable | Description |
| --- | --- |
| $N$ | Prediction horizon |
| $\mathbf{X}$ | State vector |
| $\rho_{N,i}$ | Finite horizon terminal cost penalty or Mayer terms |
| $P_i(\cdot)$ | Some Nonlinear functions or Lagrange terms |
| $\mathbf{p}$ | A column vector of time-varying parameters |
| $\mathbf{U}$ | PTO Force manipulated variable vector, $\mathbf{F_p}(N)$ |
| $\mathbf{d}$ | Excitation force disturbance vector, $\mathbf{F_e}\,(N)$ |
| $q_k(t)$ | Cost functional scheduling variable |
| $R_i$ | Some real numbers, such that $R_{k+1} > R_k$ |

For the 2-DoF (heave-pitch) WEC problem, the objective function to be maximized in (28) will be the sum of electrical PTO power output in the heave and pitch DoFs for each pod,

$$P_E = P_{E,3} + P_{E,5} \, , \tag{34}$$

Using the technique developed in [30], we can put (34) into the pseudo-quadratic form by defining a suitable $\mathbf{h_i}$ vector for heave and pitch as,

$$\mathbf{h}_i = \left[ \begin{array}{ccccc} F_{p,i}^3 & F_{p,i}^2 & F_{p,i} & v_i & 1 \end{array} \right]^T, \; i = 3,5 \tag{35}$$

with,

$$\mathbf{h} = \left[ \begin{array}{c} \mathbf{h_3} \\ \mathbf{h_5} \end{array} \right], \tag{36}$$

we can reformulate (34) as,

$$P_E = \frac{1}{2}\mathbf{h}^T \left( 2 \left[ \begin{array}{cc} \mathbf{W_3} & 0 \\ 0 & \mathbf{W_5} \end{array} \right] \right) \mathbf{h} = \frac{1}{2}\mathbf{h}^T (2\mathbf{W})\mathbf{h}, \tag{37}$$

By using (28) in (34), the weighting matrix $\mathbf{W}$ can be obtained by polynomial decomposition of (34) by the vector $\mathbf{h}$ in (36) as the basis vector,

$$\mathbf{W_i} = \frac{1}{2} \left[ \begin{array}{ccccc} -2c_{1,i} & -c_{2,i} & 0 & 0 & 0 \\ -c_{2,i} & -2c_{3,i} & -c_{4,i} & 0 & 0 \\ 0 & -c_{4,i} & -2c_{5,i} & c_{0,i} & -c_{6,i} \\ 0 & 0 & c_{0,i} & 0 & 0 \\ 0 & 0 & -c_{6,i} & 0 & -2c_{7,i} \end{array} \right], \; i = 3,5 \tag{38}$$

The controller is implemented using an ACADO toolkit [21] following the approach developed in [3].

## 4. Results

The schematic diagram of the test setup is shown in Figure 6. The corresponding hardware setup is shown in Figure 7. NMPC is designed in the host machine, which generated code and deployed the controller to the Speedgoat performance real-time target machine [31], model-109100 with Intel Core i3 3.3 GHz, two cores, and 2048 MB DDR3 RAM. The Speedgoat machine is interfaced with a real-time WEC emulator machine through an Ethernet universal data port (UDP) channel. The three WEC pods in Figure 1 are assumed identical, and the same controller is implemented for each pod as shown in Figure 8, while the cross-coupling between pods is ignored for this work. The physical velocity and force constraints of the PTO mechanisms imposed as $|v_3| \leq 2\,m/sec$, $|v_5| \leq 0.5\,rad/sec$ and $|F_{p,i}| \leq 400\,kN$. The emulated WEC-Sim model of CENTIPOD device is shown in Figure 9.

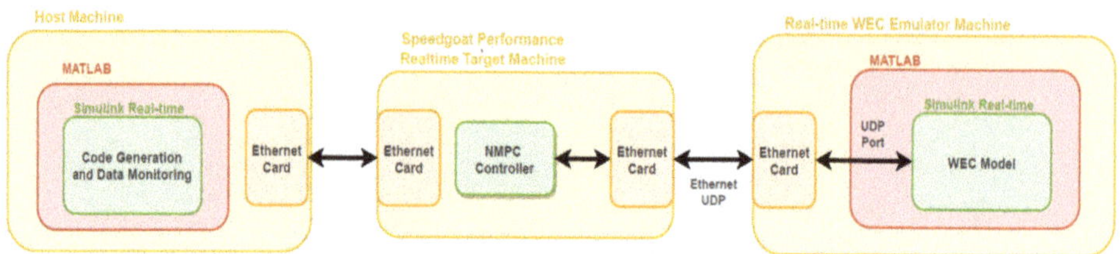

**Figure 6.** Schematic diagram of the test setup.

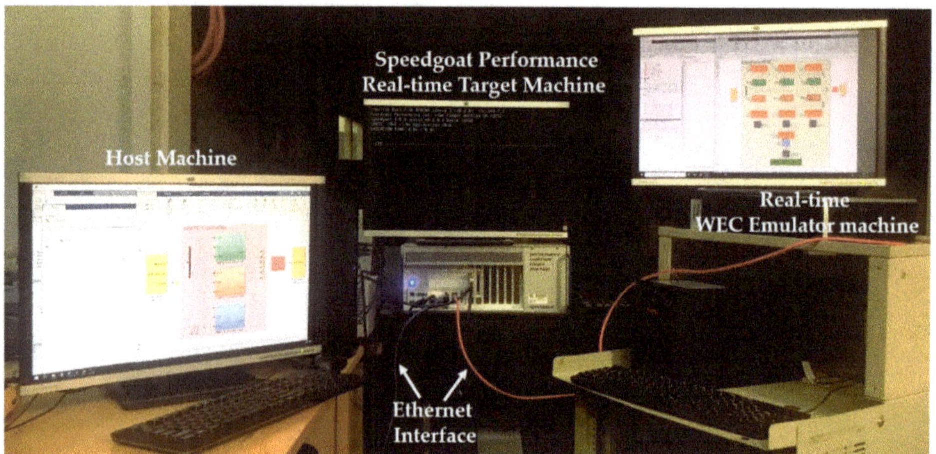


**Figure 7.** Hardware test setup.

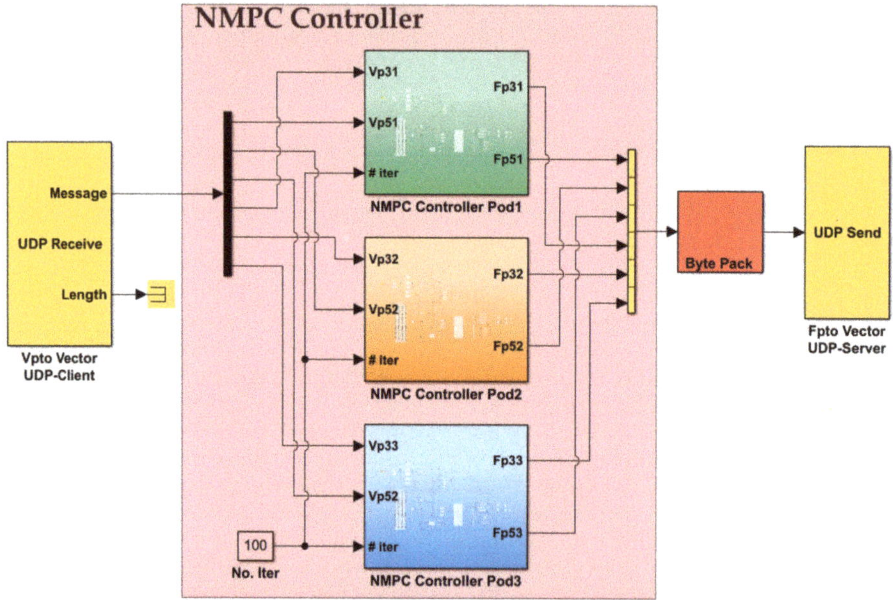


**Figure 8.** NMPC controller for 2-DoF 3-pod CENTIPOD WEC.

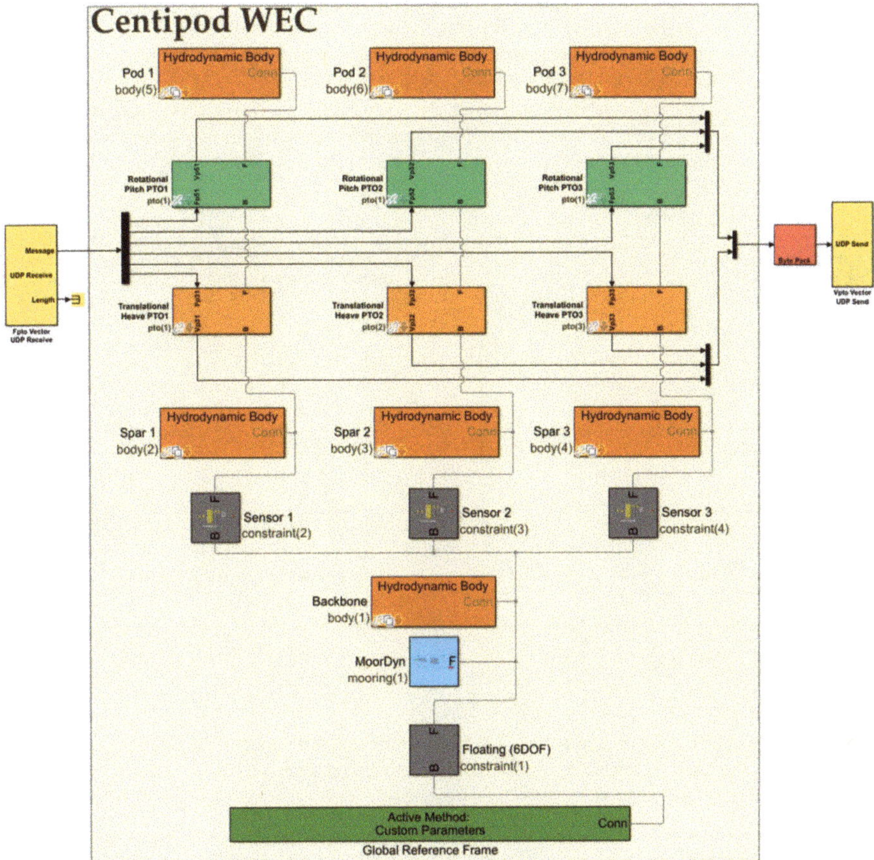


**Figure 9.** WEC-Sim model of Dehlsen's 2-DoF CENTIPOD device with heave and pitch PTOs for each pod.

Since the WEC pods are assumed identical with no cross-coupling, results are presented only for a single pod. The sea state of interest for WEC-Sim is given in Table 3. This particular sea state's selection is based on the future testing site of interest for the WEC device, although the hardware testing and a more elaborated study involving other sea states are planned for the future. A step time of 0.1 *sec* is used for MPC formulation, close to one-tenth of the peak wave period. The performance of NMPC is compared against the linear MPC, and the analysis is performed for the linear and nonlinear hydrodynamics sea conditions.

**Table 3.** Sea states for WEC-Sim simulation.

| WEC-Sim Simulation Parameter | Value |
| --- | --- |
| Significant Wave Height [m] | 2.5 |
| Peak Period [s] | 8 |
| Wave Spectrum Type | Pierson Moskowitz (PM) |
| Wave Class | Irregular |

The average electrical power output results for the heave and pitch PTOs for 2-DoF pod-1 are shown in Figures 10a and 10b, respectively, for linear MPC and NMPC subjected to linear hydrodynamic conditions. Here, we consider the exponentially weighted moving average (EWMA) with the forgetting factor set to unity. The instantaneous electrical power output results corresponding to Figure 10 are shown in Figure 11. The PTO force and wave excitation force profiles for 2-DoF Pod-1 with linear and nonlinear MPC under linear hydrodynamic conditions are shown in Figure 12. The PTO velocity and displacement plots for 2-DoF Pod-1 with linear and nonlinear MPC under linear hydrodynamic conditions are shown in Figure 13.

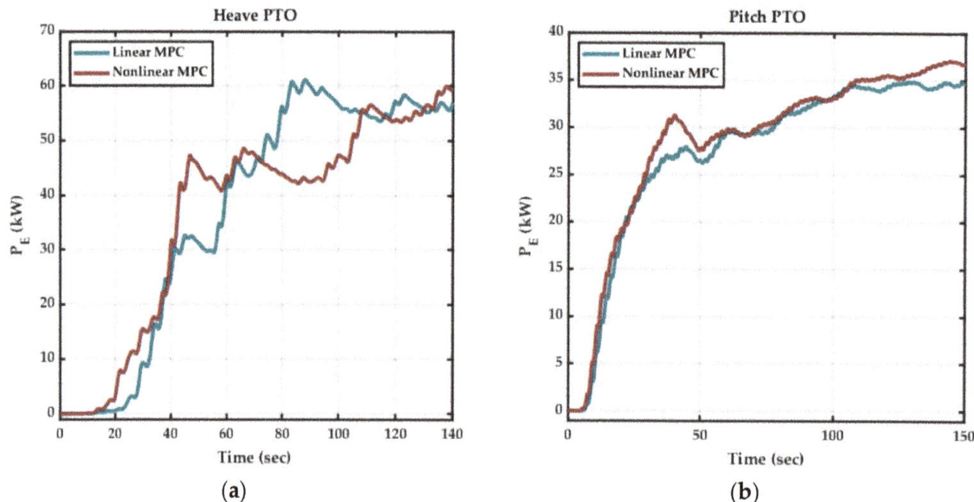

**Figure 10.** Average electrical PTO power output for 2-DoF Pod-1 with linear and nonlinear MPC under linear hydrodynamic conditions in WEC-Sim and $|F_{pto}| \leq 400\ kN$: (**a**) Pod-1 Heave PTO; (**b**) Pod-1 Pitch PTO.

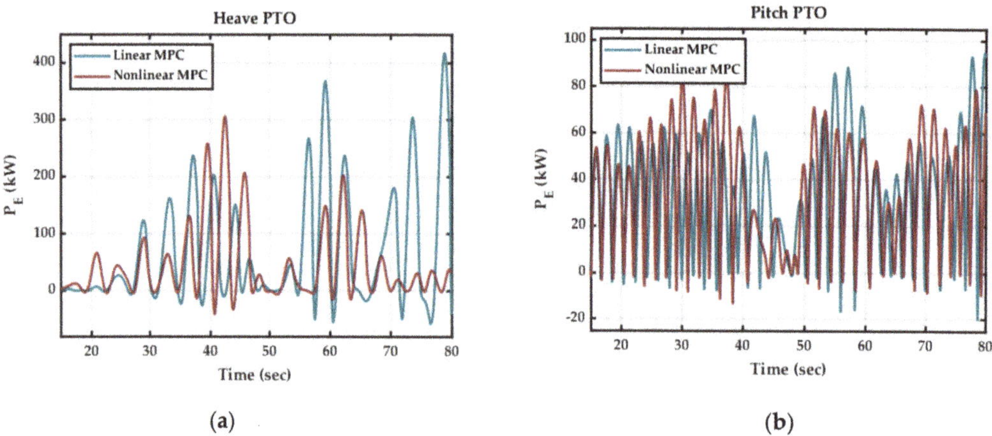

**Figure 11.** Instantaneous electrical PTO power output for 2-DoF Pod-1 with linear and nonlinear MPC under linear hydrodynamic conditions in WEC-Sim and $|F_{pto}| \leq 400\ kN$: (**a**) Pod-1 Heave PTO; (**b**) Pod-1 Pitch PTO.

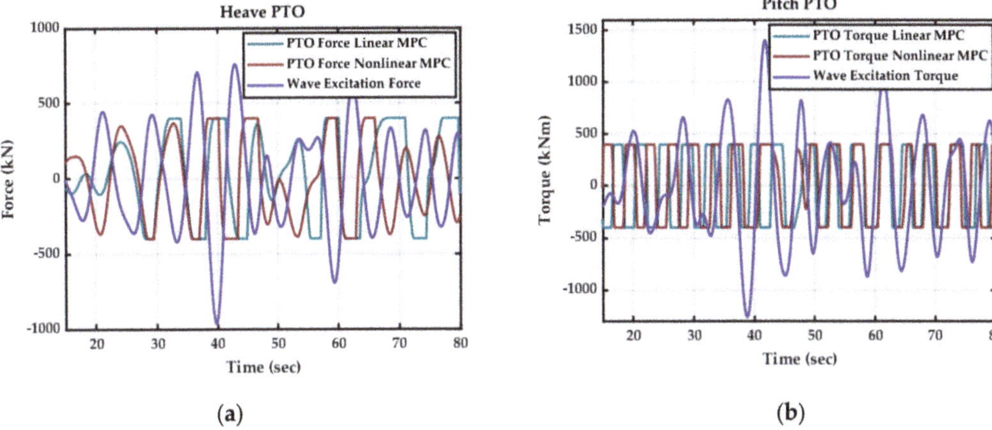

**Figure 12.** The PTO force and wave excitation force profiles for 2-DoF Pod-1 with linear and nonlinear MPC under linear hydrodynamic conditions in WEC-Sim and $\left| F_{pto} \right| \leq 400\,kN$: (**a**) Pod-1 Heave PTO; (**b**) Pod-1 Pitch PTO.

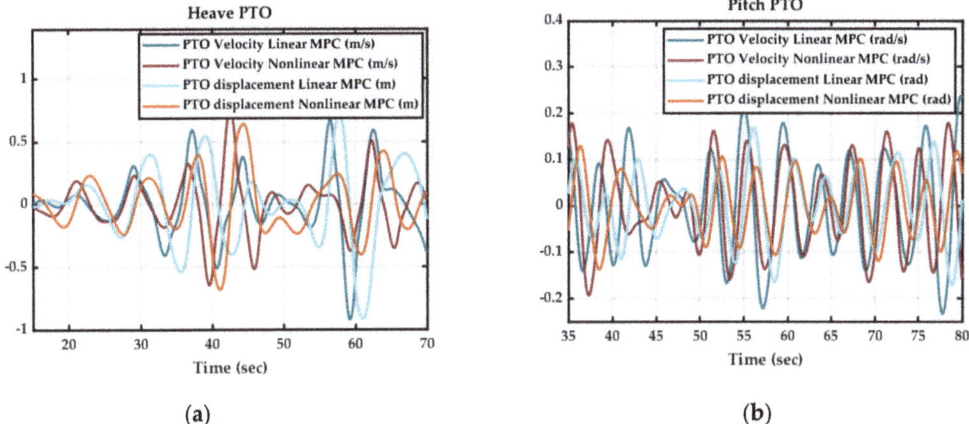

**Figure 13.** The PTO velocity and displacement plots for 2-DoF Pod-1 with linear and nonlinear MPC under linear hydrodynamic conditions in WEC-Sim and $\left| F_{pto} \right| \leq 400\,kN$: (**a**) Pod-1 Heave PTO; (**b**) Pod-1 Pitch PTO.

The average and instantaneous electrical power output results under nonlinear hydrodynamics for 2-DoF Pod-1 with linear and nonlinear MPC are shown in Figures 14 and 15, respectively. The comparison of average electrical PTO power output with NMPC for 1-DoF and 2-DoF Pod-1 is shown in Figure 16 under nonlinear hydrodynamic conditions.

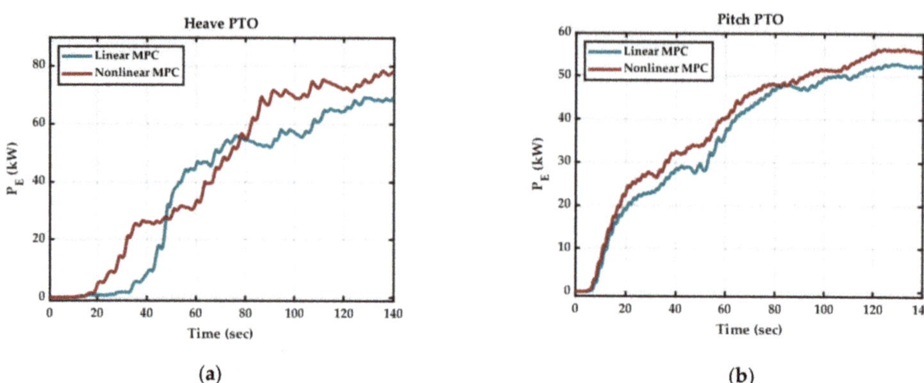

(a)         (b)

**Figure 14.** Average electrical PTO power output for 2-DoF Pod-1 with linear and nonlinear MPC under Nonlinear hydrodynamic conditions in WEC-Sim and $|F_{pto}| \leq 400\,kN$: (**a**) Pod-1 Heave PTO; (**b**) Pod-1 Pitch PTO.

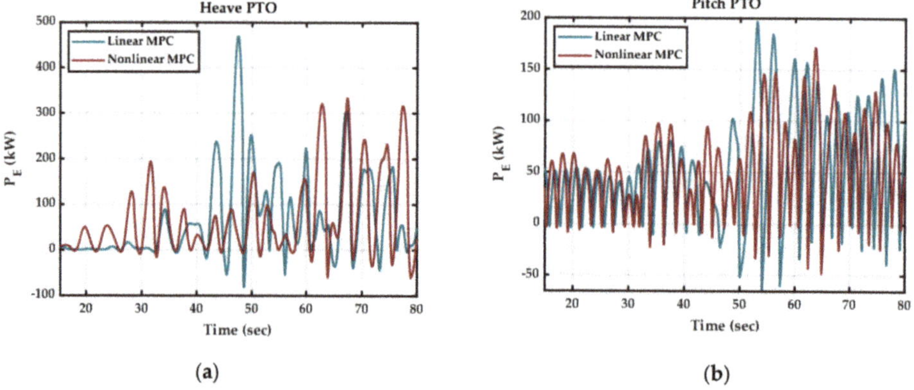

(a)         (b)

**Figure 15.** Instantaneous electrical PTO power output for 2-DoF Pod-1 with linear and nonlinear MPC under nonlinear hydrodynamic conditions in WEC-Sim and $|F_{pto}| \leq 400\,kN$: (**a**) Pod-1 Heave PTO; (**b**) Pod-1 Pitch PTO.

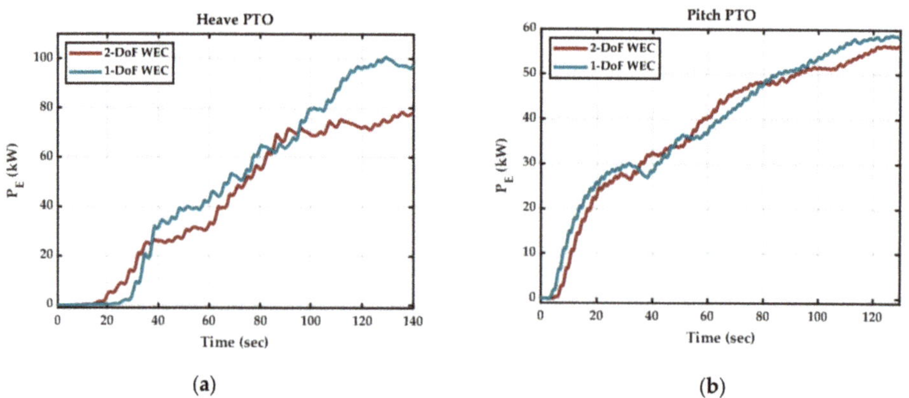

(a)         (b)

**Figure 16.** Average electrical PTO power output for 1-DoF and 2-DoF Pod-1 with nonlinear MPC under nonlinear hydrodynamic conditions in WEC-Sim and $|F_{pto}| \leq 400\,kN$: (**a**) Pod-1 Heave PTO; (**b**) Pod-1 Pitch PTO.

## 5. Discussion

The average electrical power output results in Figures 10 and 14 are summarized in Table 4. NMPC shows a better performance in terms of increased power output when compared to linear MPC. This increase in the output power becomes more prominent under nonlinear hydrodynamic conditions, which are not accounted for by the linear MPC. An overall 5% increase in power by NMPC compared to linear MPC is obtained under linear hydrodynamic conditions. NMPC obtains an overall 5% increase in total power output by pod-1 than linear MPC under linear hydrodynamic conditions and 10.6% under nonlinear hydrodynamic conditions. The corresponding task execution time (TET) stats for the real-time implementations of linear MPC and NMPC in a Speedgoat real-time machine are given in Table 5. Given the controller step time of 0.1 sec, the increase in TET for NMPC compared to linear MPC is not very significant.

**Table 4.** Average electrical power output per PTO for 2-DoF Pod1 with linear MPC and NMPC.

| | Average Electrical Power [kW] | | | | | |
|---|---|---|---|---|---|---|
| | Linear Hydrodynamic Conditions | | | Nonlinear Hydrodynamic Conditions | | |
| Control Algorithm | Heave | Pitch | Total | Heave | Pitch | Total |
| Linear MPC | 57 | 35 | 92 | 70 | 52 | 122 |
| Nonlinear MPC | 60 | 37 | 97 | 79 | 56 | 135 |

**Table 5.** Real-time timings stats for Linear MPC vs. Nonlinear MPC.

| | Task Execution Time (TET) [sec] | | |
|---|---|---|---|
| Control Algorithm | 1-DoF Heave | 1-DoF Pitch | 2-DoF Heave and Pitch |
| Linear MPC | $2.12 \times 10^{-4}$ | $2.67 \times 10^{-4}$ | $5.21 \times 10^{-4}$ |
| Nonlinear MPC | $3.05 \times 10^{-4}$ | $3.21 \times 10^{-4}$ | $6.14 \times 10^{-4}$ |

The average electrical power output results per PTO for 1-DoF and 2-DoF Pod1 with NMPC from Figure 16 are summarized in Table 6. In moving from 1-DoF WEC to 2-DoF WEC, a 35% increase in output power is obtained compared to heave only, and 129% increase compared to pitch only.

**Table 6.** Average electrical power output per PTO for 1-DoF and 2-DoF Pod1 with NMPC.

| | Average Electrical Power [kW] | | |
|---|---|---|---|
| | 1-DoF WEC | | 2-DoF WEC |
| Axis | Heave | Pitch | Heave and Pitch |
| Heave | 98 | 0 | 78 |
| Pitch | 0 | 58 | 55 |
| Net Power | 98 | 58 | 133 |

The locus of electrical PTO power for linear MPC and NMPC under nonlinear hydrodynamic conditions in WEC-Sim, along with the electrical power cost functional surface from Figure 5, are shown in Figure 17.

The locus of electrical PTO power in Figure 17 traverses a trajectory on the cost manifolds and satisfies the cost objective. The cost index formulation in (15) includes a convexifying quadratic term of PTO current, making the resultant electrical PTO surface convex in Figure 5, and with a smooth PTO current profile, the close loop system tends to maintain a stable operation. If the QP problem formulated at a given sample interval is infeasible, the controller will not find a solution. This issue can be handled by monitoring the status of the QP solver during each sampling interval and selecting a suboptimal solution when the QP solver fails. An average of 35% processor load was observed per sampling interval during testing.

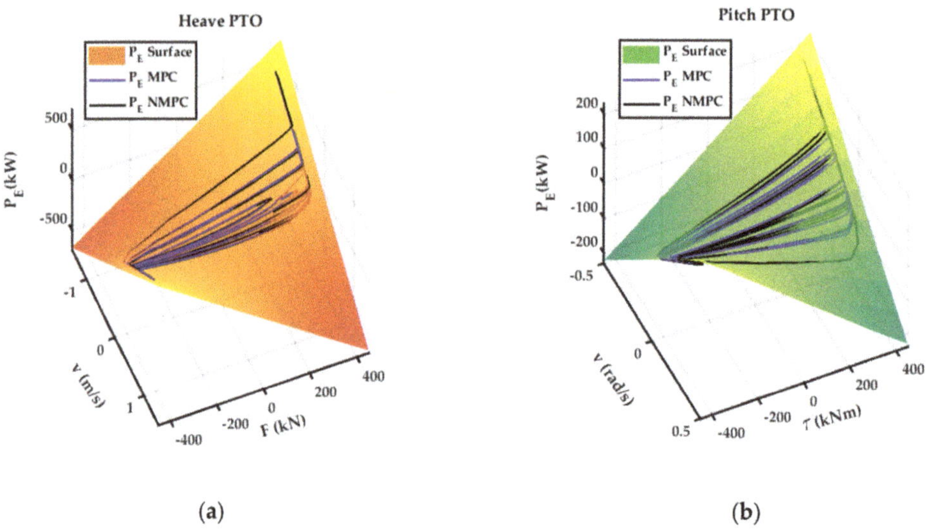



**Figure 17.** The locus of electrical PTO power on the electrical power cost functional surface for linear MPC and NMPC under nonlinear hydrodynamic conditions in WEC-Sim and $|F_{pto}| \leq 400\ kN$: (**a**) Pod-1 Heave PTO; (**b**) Pod-1 Pitch PTO.

### 6. Conclusions

This article presents a real-time implementation of NMPC for a nonlinear 2-DoF WEC based on Dehlsen Associates' CENTIPOD multi-pod WEC device, with non-ideal PTOs in the heave and pitch axes. The three pods of the WEC device are assumed identical, and a nonlinear state-space model of a single pod is developed. An NMPC controller is implemented for a 2-DoF WEC device with the cost functional based on a PTO model case study with a highly nonlinear PTO current–force characteristic. The results of the linear MPC are compared with NMPC for the sea states of interest (irregular waves with Pierson Moskowitz spectrum) under linear and nonlinear hydrodynamic conditions in WEC-Sim. The proposed methodology successfully maintained an overall feasible operation of the real-time NMPC problem in simulation as indicated by the status port of the NMPC QP-solver.

An average of 35% processor load was observed per sampling interval during testing. An overall 5% increase in total power output by a single pod is obtained by NMPC compared to linear MPC under linear hydrodynamic conditions and 10.6% under nonlinear hydrodynamic conditions. Moreover, a 35% increase in net output power is obtained by the 2-DoF WEC device compared to the 1-DoF heave only, and a 129% increase compared to the 1-DoF pitch only. While the result reflects only a single sea state, the improvement is likely to be reflected similarly in annual energy production (AEP). The AEP would have a substantive impact on the levelized cost of energy (LCOE). The present work did not consider the cross-coupling between the three pods of the CENTIPOD device. The cross-coupling would be investigated in future work with anticipation of a further increase in the captured power for the sea conditions where the cross-coupling effect is no longer negligible.

**Author Contributions:** Conceptualization, A.M.; methodology, A.S.H., T.K.A.B. and A.M.; software, A.M. and A.S.H.; validation, A.S.H., T.K.A.B. and A.M.; formal analysis, A.S.H.; investigation, A.S.H., T.K.A.B. and A.M.; resources, T.K.A.B. and A.M.; data curation, A.S.H. and A.M.; writing—A.S.H.; writing—review and editing, T.K.A.B. and A.M.; visualization, A.S.H.; supervision, T.K.A.B.; project administration, A.M. and T.K.A.B.; funding acquisition, A.M. and T.K.A.B. All authors have read and agreed to the published version of the manuscript.

**Funding:** This work was funded under the TEAMER program by U.S. Department of Energy's Energy Efficiency and Renewable Energy (EERE) office within the Water Power Technologies Office (WPTO).

**Institutional Review Board Statement:** Not applicable.

**Informed Consent Statement:** Not applicable.

**Acknowledgments:** The authors would like to thank Dehlsen Associates, LLC and McCleer Power for their technical support and for providing the experimental data.

**Conflicts of Interest:** The authors declare no conflict of interest.

# References


1. Muetze, A.; Vining, J.G. Ocean Wave Energy Conversion—A Survey. In Proceedings of the Conference Record of the 2006 IEEE Industry Applications Conference Forty-First IAS Annual Meeting, Tampa, Fl, USA, 8–12 October 2006; Volume 3, pp. 1410–1417.
2. Richter, M.; Magana, M.E.; Sawodny, O.; Brekken, T.K.A. Nonlinear Model Predictive Control of a Point Absorber Wave Energy Converter. *IEEE Trans. Sustain. Energy* **2013**, *4*, 118–126. [CrossRef]
3. Haider, A.S.; Brekken, T.K.A.; McCall, A. A State-of-the-Art Strategy to Implement Nonlinear Model Predictive Controller with Non-Quadratic Piecewise Discontinuous Cost Index for Ocean Wave Energy Systems. In Proceedings of the 2020 IEEE Energy Conversion Congress and Exposition (ECCE), Detroit, MI, USA, 11–15 October 2020; pp. 1873–1878.
4. Brekken, T.K.A. On Model Predictive Control for a Point Absorber Wave Energy Converter. In Proceedings of the 2011 IEEE Trondheim PowerTech, Trondheim, Norway, 19–23 June 2011; pp. 1–8.
5. Faedo, N.; Olaya, S.; Ringwood, J.V. Optimal Control, MPC and MPC-like Algorithms for Wave Energy Systems: An Overview. *IFAC J. Syst. Control* **2017**, *1*, 37–56. [CrossRef]
6. Zhong, Q.; Yeung, R.W. Model-Predictive Control Strategy for an Array of Wave-Energy Converters. *J. Mar. Sci. Appl.* **2019**, *18*, 26–37. [CrossRef]
7. Genest, R.; Ringwood, J.V. A Critical Comparison of Model-Predictive and Pseudospectral Control for Wave Energy Devices. *J. Ocean Eng. Mar. Energy* **2016**, *2*, 485–499. [CrossRef]
8. Li, G.; Belmont, M.R. Model Predictive Control of Sea Wave Energy Converters—Part I: A Convex Approach for the Case of a Single Device. *Renew. Energy* **2014**, *69*, 453–463. [CrossRef]
9. Li, G.; Belmont, M.R. Model Predictive Control of Sea Wave Energy Converters—Part II: The Case of an Array of Devices. *Renew. Energy* **2014**, *68*, 540–549. [CrossRef]
10. Starrett, M.; So, R.; Brekken, T.K.A.; McCall, A. Increasing Power Capture from Multibody Wave Energy Conversion Systems Using Model Predictive Control. In Proceedings of the 2015 IEEE Conference on Technologies for Sustainability (SusTech), Ogden, UT, USA, 30 July–1 August 2015; pp. 20–26.
11. Falcão, A.F.O.; Henriques, J.C.C. Effect of Non-Ideal Power Take-off Efficiency on Performance of Single- and Two-Body Reactively Controlled Wave Energy Converters. *J. Ocean Eng. Mar. Energy* **2015**, *1*, 273–286. [CrossRef]
12. Davis, A.F.; Thomson, J.; Mundon, T.R.; Fabien, B.C. Modeling and Analysis of a Multi Degree of Freedom Point Absorber Wave Energy Converter. In Proceedings of the OMAE 2014, San Francisco, CA, USA, 8 June 2014; Volume 8A: Ocean Engineering.
13. Al Shami, E.; Wang, X.; Ji, X. A Study of the Effects of Increasing the Degrees of Freedom of a Point-Absorber Wave Energy Converter on Its Harvesting Performance. *Mech. Syst. Signal Process.* **2019**, *133*, 106281. [CrossRef]
14. Strager, T.; Martin dit Neuville, A.; Fernández López, P.; Giorgio, G.; Mureşan, T.; Andersen, P.; Nielsen, K.M.; Pedersen, T.S.; Vidal Sánchez, E. Optimising Reactive Control in Non-Ideal Efficiency Wave Energy Converters. In Proceedings of the OMAE 2014, San Francisco, CA, USA, 8 June 2014; Volume 9A: Ocean Renewable Energy.
15. Perdigão, J.; Sarmento, A. Overall-Efficiency Optimisation in OWC Devices. *Appl. Ocean Res.* **2003**, *25*, 157–166. [CrossRef]
16. Tedeschi, E.; Carraro, M.; Molinas, M.; Mattavelli, P. Effect of Control Strategies and Power Take-Off Efficiency on the Power Capture From Sea Waves. *IEEE Trans. Energy Convers.* **2011**, *26*, 1088–1098. [CrossRef]
17. Jia, Y.; Meng, K.; Dong, L.; Liu, T.; Sun, C.; Dong, Z.Y. Economic Model Predictive Control of a Point Absorber Wave Energy Converter. *IEEE Trans. Sustain. Energy* **2021**, *12*, 578–586. [CrossRef]
18. Ellis, M.; Durand, H.; Christofides, P.D. A Tutorial Review of Economic Model Predictive Control Methods. *J. Process Control* **2014**, *24*, 1156–1178. [CrossRef]
19. Rawlings, J.B.; Angeli, D.; Bates, C.N. Fundamentals of Economic Model Predictive Control. In Proceedings of the 2012 IEEE 51st IEEE Conference on Decision and Control (CDC), Maui, HI, USA, 10–13 December 2012; pp. 3851–3861.

20. Houska, B.; Ferreau, H.J.; Diehl, M. An Auto-Generated Real-Time Iteration Algorithm for Nonlinear MPC in the Microsecond Range. *Automatica* **2011**, *47*, 2279–2285. [CrossRef]

21. Houska, B.; Ferreau, H.J.; Diehl, M. ACADO Toolkit—An Open-Source Framework for Automatic Control and Dynamic Optimization. *Optim. Control. Appl. Methods* **2011**, *32*, 298–312. [CrossRef]

22. Abdelkhalik, O.; Zou, S.; Robinett, R.D.; Bacelli, G.; Wilson, D.G.; Coe, R.; Korde, U. Multiresonant Feedback Control of a Three-Degree-of-Freedom Wave Energy Converter. *IEEE Trans. Sustain. Energy* **2017**, *8*, 1518–1527. [CrossRef]

23. Frequency- and Time-Domain Analysis of a Multi-Degree-of-Freedom Point Absorber Wave Energy Converter–Yizhi Ye, Weidong Chen, 2017. Available online: https://journals.sagepub.com/doi/full/10.1177/1687814017722081 (accessed on 20 July 2021).

24. Coe, R.G.; Bull, D.L. Nonlinear Time-Domain Performance Model for a Wave Energy Converter in Three Dimensions. In Proceedings of the 2014 Oceans, St. John's, NL, Canada, 14–19 September 2014; pp. 1–10.

25. Huang, S.; Shi, H.; Dong, X. Capture Performance of A Multi-Freedom Wave Energy Converter with Different Power Take-off Systems. *China Ocean Eng.* **2019**, *33*, 288–296. [CrossRef]

26. Hillis, A.J.; Whitlam, C.; Brask, A.; Chapman, J.; Plummer, A.R. Active Control for Multi-Degree-of-Freedom Wave Energy Converters with Load Limiting. *Renew. Energy* **2020**, *159*, 1177–1187. [CrossRef]

27. Li, G. Nonlinear Model Predictive Control of a Wave Energy Converter Based on Differential Flatness Parameterisation. *Int. J. Control* **2017**, *90*, 68–77. [CrossRef]

28. WEC-Sim (Wave Energy Converter SIMulator)—WEC-Sim Documentation. Available online: https://wec-sim.github.io/WEC-Sim/ (accessed on 27 March 2021).

29. van Rij, J.; Yu, Y.-H.; McCall, A.; Coe, R.G. Extreme Load Computational Fluid Dynamics Analysis and Verification for a Multibody Wave Energy Converter. In Proceedings of the OMAE 2019, Glasgow, UK, 9–14 June 2019; Volume 10: Ocean Renewable Energy.

30. Haider, A.S.; Brekken, T.K.A.; McCall, A. Application of Real-Time Nonlinear Model Predictive Control for Wave Energy Conversion. *IET Renew. Power Gener.* **2021**. [CrossRef]

31. Speedgoat—The Quickest Path to Real-Time Simulation and Testing. Available online: https://www.speedgoat.com/ (accessed on 23 February 2021).

32. Ecomerit Technologies. Available online: http://www.ecomerittech.com/centipod.php (accessed on 14 August 2021).

33. Wamit, Inc. The State of the Art in Wave Interaction Analysis. Available online: https://www.wamit.com/ (accessed on 28 March 2021).

34. Ringwood, J.V. Wave Energy Control: Status and Perspectives 2020 **This Paper Is Based upon Work Supported by Science Foundation Ireland under Grant No. 13/IA/1886 and Grant No. 12/RC/2302 for the Marine Renewable Ireland (MaREI) Centre. *IFAC-PapersOnLine* **2020**, *53*, 12271–12282. [CrossRef]

Journal of
*Marine Science
and Engineering*

*Article*

# Sliding Mode Control of a Nonlinear Wave Energy Converter Model

Tania Demonte Gonzalez [1,*], Gordon G. Parker [1], Enrico Anderlini [2] and Wayne W. Weaver [1]

1   Department of Mechanical-Engineering-Engineering Mechanics, Michigan Technological University, Houghton, MI 49931, USA; ggparker@mtu.edu (G.G.P.); wwweaver@mtu.edu (W.W.W.)
2   Department of Mechanical Engineering, University College London, London WC1E 6BT, UK; e.anderlini@ucl.ac.uk
*   Correspondence: tsdemont@mtu.edu

**Abstract:** The most accurate wave energy converter models for heaving point absorbers include nonlinearities, which increase as resonance is achieved to maximize the energy capture. Over the power production spectrum and within the physical limits of the devices, the efficiency of wave energy converters can be enhanced by employing a control scheme that accounts for these nonlinearities. This paper proposes a sliding mode control for a heaving point absorber that includes the nonlinear effects of the dynamic and static Froude-Krylov forces. The sliding mode controller tracks a reference velocity that matches the phase of the excitation force to ensure higher energy absorption. This control algorithm is tested in regular linear waves and is compared to a complex-conjugate control and a nonlinear variation of the complex-conjugate control. The results show that the sliding mode control successfully tracks the reference and keeps the device displacement bounded while absorbing more energy than the other control strategies. Furthermore, due to the robustness of the control law, it can also accommodate disturbances and uncertainties in the dynamic model of the wave energy converter.

**Keywords:** wave energy; nonlinear model; Froude-Krylov force; sliding mode control

**Citation:** Demonte Gonzalez, T.;
Parker, G.G.; Anderlini, E.;
Weaver, W.W. Sliding Mode Control
of a Nonlinear Wave Energy
Converter Model. *J. Mar. Sci. Eng.*
**2021**, *9*, 951. https://doi.org/
10.3390/jmse9090951

Academic Editors: Giuseppe Giorgi
and Sergej Antonello Sirigu

Received: 4 August 2021
Accepted: 26 August 2021
Published: 1 September 2021

**Publisher's Note:** MDPI stays neutral
with regard to jurisdictional claims in
published maps and institutional affiliations.

## 1. Introduction

Recently, the generation of electricity from ocean waves has gained special attention. Studies of the potential global market for wave power showed that the world's wave power resource is estimated to be two TW [1]. Additionally, waves have a very high power density, requiring smaller devices to capture the energy carried by the incoming waves. Wave energy can also improve energy security by complementing the output of other renewable energy sources, thus reducing the storage needs. However, although phased development has helped reduce risks, the wave energy sector is still in its infancy, with recent prototype wave energy converters (WECs) having a Levelized Cost of Energy (LCoE) in the range of $120–$470/MWh [2]. Hence, technological advances are required to reduce the LCoE 50–75% to enable the industry to leap from government-funded research to sustainable industrial competition within the energy market. Optimizing the control of WECs has been identified as one of the critical areas with the highest potential to improve the viability of wave energy [3], as the latest control methods can enhance power absorption by up to 20% while reducing structural loads.

Researchers have used well-established control methods in WECs that include Complex-Conjugate Control (CCC), Latching, and Model Predictive Control (MPC) [4]. In complex-conjugate control, optimal energy absorption is sought by tuning the power take-off (PTO) resistance and reactance to cancel the system's inherent hydrodynamic resistance and reactance [5]. This control strategy is conceptually simple and presents a low computational cost; however, it leads to excessive motions and loads of the WEC. This can be avoided by implementing alternative suboptimal control schemes that include

241

physical constraints on the motions and power of the device. Latching control is based on locking the device through dedicated mechanisms during a certain period of time of the wave oscillation cycle to achieve resonance of the WEC [6,7]. Resonance is achieved when the system is being excited at its natural frequency. While latching control solves the PTO limitation problem of CCC, it is prone to failure of the mechanical clamping pieces generating extra costs and reducing its reliability. MPC strategies maximize the energy absorbed by applying the optimal control force to achieve resonance at each time step over a future time horizon [8]. However, a prediction of the wave motion is required, and its complexity of implementation in real-time demands high computational requirements.

Most of these control techniques use the linear model developed by Cummins [9] with hydrodynamic parameters obtained from boundary element methods (BEM) due to their simplicity and high computational efficiency. Although linear models are effective for most ocean engineering applications where the aim is typically to stabilize the structure, linear models are inadequate for WEC control, where the primary objective is to amplify the device's motions to maximize the power absorption [10]. As a result, control strategies based on linear models are often suboptimal and lead to higher LCoE.

Nonlinear models often include the quadratic term of the pressure from the incident flow, the integration of the pressure forces over the instantaneous wetted surface of the device called Froude-Krylov (FK) forces, and the nonlinear incident flow potential [11]. Each of these nonlinearities adds an extra level of complexity and computational load. However, it is understood that nonlinear models are more accurate than linear models, particularly for buoy geometries with varying cross-sectional areas. A previous study [12] that compared different modeling options showed that the dominant sources of nonlinearities are likely to be the nonlinear Froude-Krylov forces for devices with a varying cross-sectional area.

While, in linear modeling, the FK forces are calculated over a constant wetted area of the device, in nonlinear models, the pressure is integrated over the instantaneous wetted surface area. This is generally done by implementing a very fine mesh or a remeshing routine of the surface, which requires more computational effort. As a result, the nonlinear model is more accurate but is very computationally expensive. However, in Reference [13], Giorgi et al. developed a computationally efficient analytical method to compute the nonlinear dynamic and static Froude-Krylov forces based on the instantaneous wetted surface area of the buoy. This analytical method presents only a 2% error compared to the more computationally expensive methods. Therefore, in this study, the nonlinear FK forces are incorporated using a variation of the algebraic solution proposed by Giorgi et al.

This paper proposes a second-order Sliding Mode Control (SMC) strategy that incorporates the nonlinear Froude-Krylov forces to reduce the discrepancies between the mathematical model and the actual system. In addition, this SMC is designed to track the desired reference signals using low computational efforts and rejecting disturbances encountered by the WEC or parametric errors presented in the model. The controller will be investigated in a simulation environment using a standard spherical WEC and regular waves.

Many researchers have studied SMC in oscillating water columns (OWC). A popular SMC for a nonlinear OWC is the super-twisting algorithm presented in [14]. This control strategy has been used in different configurations of OWCs, where it noticeably improved the electric power conversion compared to other strategies [15]. Sliding mode control has also been used to track underwater vehicles' positions [16] due to its robustness characteristics that allow for model uncertainties or environmental disturbances. In this paper, SMC is applied to a heaving point absorber. An SMC strategy for heaving WECs was proposed in Reference [17], where it was shown to provide a viable solution to increase energy extraction. The SMC was designed to track a reference velocity without violating the physical constraints of the device. The results showed that the SMC successfully tracked the reference signal even when system perturbations were present. However, this work was done using a linear approximation of the WEC's dynamic model. This study aims to improve the energy extraction performance of WECs where its SMC can exploit the WEC's nonlinear model terms.

The main contributions of this study consist of first, the derivation of algebraic nonlinear Froude-Krylov forces based on the work of Giorgi et al., second, the application of an SMC to the nonlinear WEC model; and third, the design of a nonlinear variant of complex-conjugate control.

This paper is organized into six sections. First, the following section describes the general dynamics of a heaving WEC and the derivation of the nonlinear algebraic Froude-Krylov forces showing the first contribution mentioned above. Then, the second and third contributions are in Section 3, where the SMC strategy is designed and compared to a complex-conjugate control and a nonlinear variant of complex-conjugate control. In Section 4, the simulation used to evaluate closed-loop performance is described. Section 5 shows the results, and Section 6 provides conclusions and area for further study.

## 2. Model Description

Mathematical models for WECs are essential for the design of model-based control. These models are typically linear because of their low computational requirements, in which they assume small motions. However, there are situations in which the movement is not small while still producing power, and the nonlinearities of the device become significant.

The incoming waves are classified as linear and nonlinear depending on the wave steepness $S$, which is the ratio between the wave height $h$ and the wavelength $\lambda$ [18]. Waves are also classified as regular and irregular, where regular waves are characterized by a specific amplitude $A$, frequency $\omega_0$, and wavelength, whereas irregular waves are composed of multiple frequency waves. In this paper, linear and regular waves are used, representing the wave free surface elevation $\eta(t)$ as a function of time only:

$$\eta(t) = A cos(\omega_0 t) \tag{1}$$

A floating single degree of freedom (SDOF) buoy is shown in Figure 1, where the lower part of the sphere is attached to a linear generator for power take-off (PTO), which is fixed to the seabed. The PTO subsystem converts the wave energy into electrical energy and energizes the system to track the reference profile. The excitation force is shown as $F_{exc}(t)$, and is composed of the hydrodynamic Froude-Krylov force and the diffraction force, the buoy's displacement from its equilibrium position is denoted $z(t)$, and the control force implemented by the PTO system is $F_c(t)$.

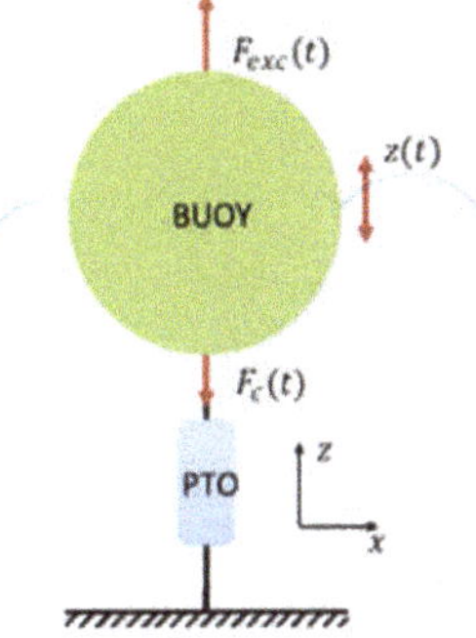


**Figure 1.** Heaving buoy with power take-off (PTO) fixed to the sea bed.

The general heaving WEC hydrodynamic model used in this paper is described as:

$$m\,\ddot{z}(t) = F_{FK} + F_d + F_R - F_c \tag{2}$$

where:

    $m$: Mass of the buoy in the air
    $F_{FK}$: Froude-Krylov Forces
    $F_d$: Diffraction Force
    $F_R$: Radiation Force
    $F_c$: Control Force

### 2.1. Diffraction Force

The diffraction force is the force the incoming waves apply on the buoy as it is held fixed in the water [5]. It can be is represented as the convolution of the product of the wave free surface elevation and the diffraction impulse response function (*IRF*):

$$F_d = \int_{-\infty}^{+\infty} K_{d_{IRF}}(t - \tau)\, \eta(t)\, d\tau \tag{3}$$

where $K_{d_{IRF}}$ is the diffraction *IRF*.

When the buoy diameter is much smaller than the wavelength, the disturbance field generated by the body is small enough to be ignored [19]. Therefore, the diffraction force is neglected in the simulation of this paper. This force could also be accounted as a disturbance to the system.

### 2.2. Radiation Force

The radiation force can be interpreted as the force that the body experiences when it oscillates in calm water. It is common to divide this force into two parts: the added mass force, proportional to the body's acceleration $\ddot{z}(t)$, and the wave damping force, proportional to the body velocity $\dot{z}(t)$ [20]. Thus, the radiation force can be represented as:

$$F_R(t) = -\int_{-\infty}^{+\infty} K_{r_{IRF}}(t - \tau)\, \dot{z}(t)\, d\tau - m_a(\omega_\infty)\ddot{z}(t) \tag{4}$$

where:

    $K_{r_{IRF}}$: Radiation *IRF*
    $m_a(\omega_\infty)$ : Added mass

Evaluating the convolution integral in the radiation force is very expensive to compute at every time step of the simulation, and it is not possible on a real-time application. However, when the buoy is exposed to only harmonic waves of frequency $\omega_0$, the convolution in Equation (4) can be represented as:

$$F_R = c(\omega_0)\dot{z}(t) - m_a(\omega_\infty)\ddot{z}(t) \tag{5}$$

where $c(\omega_0)$ is the linear damping constant.

### 2.3. Froude-Krylov Forces

The Froude-Krylov forces $F_{FK}$ are the hydrostatic force $F_{FK_{st}}$ and the hydrodynamic force $F_{FK_{dy}}$. The hydrostatic force is the difference between the gravity force $F_g$ and the force caused by the hydrostatic pressure over the wetted surface of the buoy, and the hydrodynamic force is the integral of the unsteady pressure field over the wetted surface of the floating buoy caused by the incident waves.

$$F_{FK} = F_g - \iint P(t)\, \mathbf{n}dS \tag{6}$$

In linear models, the integration is done over the constant mean wetted surface, while in the nonlinear approach, the instantaneous wetted surface of the body is considered. This approach typically requires significant computational effort. However, Giorgi et al. proposed an algebraic solution that significantly reduces the computational effort for heaving axisymmetric geometries. The algebraic solution is found by defining the pressure

$P(t)$, the infinitesimal element of the surface $\mathbf{n}\, dS$, and the limits of integration. The total pressure for deep water waves can be found using Airy's wave theory.

$$P(t) = \rho g e^{\chi z} A \cos(\omega t) - \rho g z(t) \tag{7}$$

where:

ρ: Density of water
$g$: Gravity acceleration constant
$\chi$: Wavenumber

In Giorgi et al.'s work, the surface of the geometry of the buoy was defined in parametric cylindrical coordinates as:

$$\begin{cases} x(\sigma,\theta) = f(\sigma)\cos(\theta) \\ y(\sigma,\theta) = f(\sigma)\sin(\theta), \\ z(\sigma,\theta) = \sigma \end{cases} \qquad \theta \in [0, 2\pi) \wedge \sigma \in [\sigma_1, \sigma_2] \tag{8}$$

where $f(\sigma)$ is the profile of revolution of the point absorber. Using the radial $e_\sigma$ and tangent $e_\theta$ vectors canonical basis, and only the vertical component for heave motion restriction, the infinitesimal surface element, and the pressure becomes:

$$e_\sigma \times e_\theta\, dS = \mathbf{n}\|e_\sigma \times e_\theta\|\, d\sigma d\theta = \mathbf{n}\, f(\sigma)\sqrt{f'(\sigma)^2 + 1}\, d\sigma d\theta$$

$$P_z(t) = P(t)\cdot\langle \mathbf{n}, \mathbf{k}\rangle = P(t)\cdot \frac{f\prime(\sigma)}{\sqrt{f'(\sigma)^2 + 1}} \tag{9}$$

Combining these equations, the Froude-Krylov forces are:

$$F_{FK} = F_g - \iint P(t)f'(\sigma)f(\sigma)d\sigma d\theta \tag{10}$$

The paper continues defining the limits of integration based on the free surface elevation and the draft of the buoy $h_0$ at equilibrium as:

$$\begin{cases} \sigma_1 = z_d(t) - h_0 \\ \sigma_2 = \eta(t) \end{cases} \tag{11}$$

where $z_d(t)$ is the vertical displacement of the buoy from equilibrium. For a spherical point absorber, the profile of revolution derived in Giorgi et al. paper is $f(\sigma) = \sqrt{R^2 - (\sigma - z_d)^2}$, where $R$ is the radius of the buoy.

In this paper, the limits of integration and the profile of revolution have been redefined using calculus to derive the instantaneous wetted surface area of the sphere. Considering the bottom of the sphere as the lower limit of integration, the profile of revolution of the sphere at still water level (SWL) shown in Figure 2a is $f(\sigma) = \sqrt{R^2 - (-R + h_0 + \sigma)^2}$, where the SWL was used as a reference and the instantaneous wetted surface area is shaded in blue. In this case, the lower limit of integration is $-h_0$ and the upper limit is 0. When the buoy is displaced by $z_d(t)$, and the free surface elevation $\eta(t)$ is not zero as shown in Figure 2b the profile of revolution of the sphere becomes:

$$f(\sigma) = \sqrt{R^2 - (-R + h_0 - z_d(t) + \eta(t) + \sigma)^2} \tag{12}$$

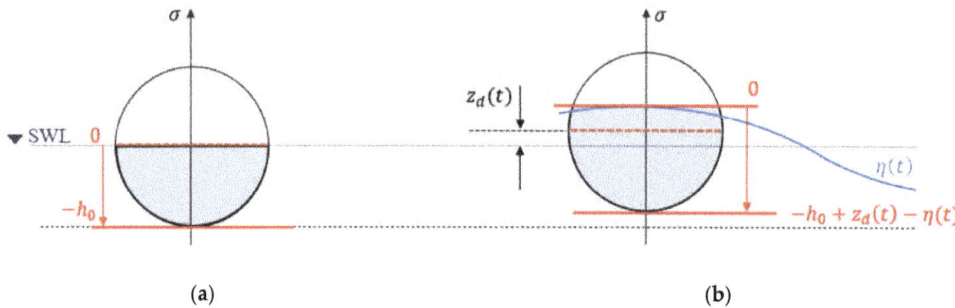


**Figure 2.** Heaving spherical point absorber at the still water level (SWL) on the left (**a**); to the right (**b**) is shown the displacement from resting position $z_d(t)$ and the wave elevation $\eta(t)$.

The limits of integration are: 0 for the upper limit, and $l_b = -h_0 + z_d(t) - \eta(t)$ for the lower limit. Combining Equations (8)–(12), the integral of the Froude-Krylov forces of a heaving spherical buoy can be expressed as:

$$F_{FK_z} = F_g - 2\pi\rho g \int_{l_b}^{0} (e^{\chi z} A\cos(\omega t) - \sigma)(-R + h_0 - z_d(t) + \eta(t) + \sigma)\, d\sigma \tag{13}$$

Solving the integral and separating the Froude-Krylov forces, the resulting nonlinear hydrostatic, and hydrodynamic forces of the system are:

$$F_{FK_{st}} = F_g + \frac{\pi\rho g}{3}(\eta(t) + h_0 - z_d(t))^2(3R - \eta(t) - h_0 + z_d(t)) \tag{14}$$

$$F_{FK_{dy}} = \frac{2\pi\rho g\eta(t)}{\chi^2}\left(e^{-\chi(\eta(t)+h_0-z_d(t))} + \chi\left(-R + \eta(t) + h_0 - z_d(t) + Re^{-\chi(\eta(t)+h_0-z_d(t))}\right) - 1\right) \tag{15}$$

The complete nonlinear model of a single degree of freedom WEC in heave mode can then be described as:

$$M\ddot{z}(t) = F_d + F_{FK_{st}} + F_{FK_{dy}} - c(\omega_0)\dot{z}(t) - F_c \tag{16}$$

where $M = (m + m_a(\omega_\infty))$.

For the subsequent control design, it will be convenient to write Equation (16) as:

$$\ddot{z}(t) = f_t(z(t), \dot{z}(t), \eta(t)) - \overline{F}_c \tag{17}$$

where $f_t$ is the true model of the system, and $\overline{F}_c = F_c M_t^{-1}$ where $M_t$ is the true mass. In contrast, the model used for control design is approximated as:

$$\ddot{z}(t) = f(z(t), \dot{z}(t), \eta(t)) - \hat{F}_c \tag{18}$$

$$f(z(t), \dot{z}(t), \eta(t)) = \left[F_d + F_{FK_{st}} + F_{FK_{dy}} - c(\omega_0)\dot{z}(t)\right]M^{-1}$$

$$\hat{F}_c = F_c M^{-1}$$

The calculation of the absorbed energy is [21]:

$$W = \int_{o}^{t} F_c \dot{z}(t)\, dt \tag{19}$$

### 3. Control Design

This section presents a sliding mode control (SMC) design, illustrated in Figure 3, for a heaving WEC based on the nonlinear differential equation model of Equation (16). Motivated by linear, complex conjugate control, the SMC reference trajectory is selected to match the frequency and phase of the excitation force. Since the WEC model is not linear, as assumed by complex conjugate control, it is recognized that this reference trajectory will not be optimal. Creating a real-time, energy extraction optimal reference trajectory is an open topic and well-suited for MPC applications. The reference velocity is then integrated to produce the reference displacement profile $z_r(t)$.

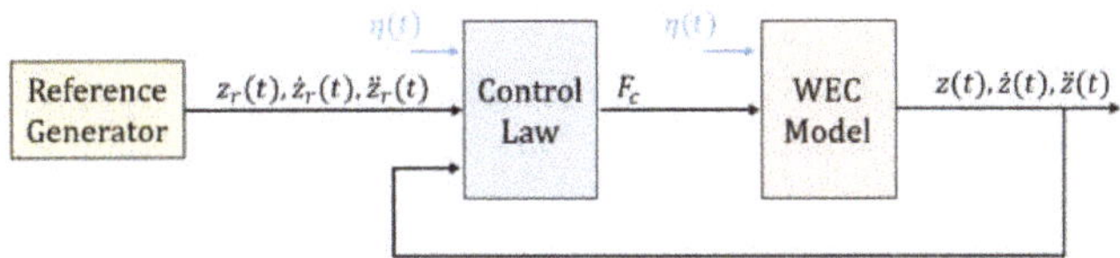


**Figure 3.** Configuration of proposed control schemes.

The sliding mode control is derived by first defining the sliding surface as:

$$s(z;t) = \dot{e}(t) + w\,e(t) \tag{20}$$

where $z$ is the state vector, $z = [z; \dot{z}]$ and $w$ is a real-positive value that sets the rate at which the system converges to the sliding surface. The problem of tracking a given profile becomes equivalent to maintaining the stationarity conditions:

$$\dot{s}(t) = \ddot{e}(t) + w\,\dot{e}(t) = \ddot{z}_r(t) - \ddot{z}(t) + w\,\dot{e}(t) = 0 \tag{21}$$

Substituting Equation (18) into Equation (21), and solving for $\hat{F}_c$ gives the control law:

$$\hat{F}_c = f\big(z(t), \dot{z}(t), \eta(t)\big) - \big(\ddot{z}_r(t) + w\dot{e}(t)\big) \tag{22}$$

This control law will only work if the assumed form of the model is perfect and there are no disturbances. To account for these inevitable situations, Lyapunov's Direct method is used to augment the control law of Equation (22) to ensure stability.

The Lyapunov candidate function is selected as:

$$V(t) = \frac{1}{2}s^2(t) \tag{23}$$

For stability,

$$\dot{V} = s(t)\dot{s}(t) = s(t)\big(\ddot{z}_r(t) - \ddot{z}(t) + w\dot{e}(t)\big) < 0 \tag{24}$$

Substituting the true model, Equation (17) into Equation (24) gives:

$$\dot{V} = s(t)\big(\ddot{z}_r(t) - f_t\big(z(t), \dot{z}(t), \eta(t)\big) + \overline{F}_c + w\dot{e}(t)\big) < 0 \tag{25}$$

Next, substitute a modified version of the control law of Equation (22) where we have added a new term, $A_c sgn(s(t))$, that does not affect the stationary condition imposed earlier:

$$\hat{F}_c = -\ddot{z}_r(t) + f\big(z(t), \dot{z}(t), \eta(t)\big) - w\dot{e}(t) - A_c sgn(s(t)) \tag{26}$$

Furthermore, note that $\hat{F}_c \neq \overline{F}_c$ are related by $\overline{F}_c = \frac{M}{M_t}\hat{F}_c$. The time-dependency notation will be omitted for the following equations for brevity. We now have the stability condition:

$$\dot{V} = s\left(\ddot{z}_r - f_t(z, \dot{z}, \eta) + \frac{M}{M_t}(-\ddot{z}_r + f(z, \dot{z}, \eta) - w\dot{e} - A_c sgn(s)) + w\dot{e}\right) < 0 \qquad (27)$$

or

$$A_c|s| > \ddot{z}_r - f(z, \dot{z}, \eta) - w\dot{e} + \frac{M_t}{M}(-\ddot{z}_r + f_t(z, \dot{z}, \eta) - w\dot{e}) \qquad (28)$$

As long as $A_c$ is sufficiently large to dominate the uncertainties on the right side of Equation (28), the system will be stable.

The discontinuous terms in the control of Equation (26) $A_c\,sgn(s(t))$ can cause chattering. This is eliminated by replacing the signum function with a hyperbolic tangent [22]. Then, the control law is interpolated by replacing $tanh(s(t))$ function by $s(t)/\Phi$ as:

$$F_c = F_d + F_{FK} - c(\omega_0)\dot{z}(t) - (\ddot{z}_r(t) + w\dot{e}(t))M + A_c\,tanh(s(t)/\Phi) \qquad (29)$$

where $\Phi$ is the boundary layer thickness.

The reference signal generator outputs the desired velocity profile of the WEC plant that is then integrated to obtain the desired displacement profile $z_r$. The control law takes as inputs the formed error signals, the desired acceleration profile $\ddot{z}_r$, the FK forces and the actual buoy displacement as in Equation (29). The control parameters $A_c$ and $\Phi$ were chosen during the simulation process. As aforementioned, the reference velocity profile is designed to match the frequency and phase of the excitation force. As described in Reference [5], the optimum velocity condition requires:

$$\dot{z}_r(t) = \frac{F_{exc}(\omega_0)}{2\,c(\omega_0)} \qquad (30)$$

To achieve this condition, phase control is needed so that the oscillation velocity $\dot{z}(t)$ is in phase with the excitation force and amplitude control to get a velocity amplitude of $|\dot{z}_r(t)| = |F_{exc}|/2c(\omega_0)$. However, this optimal condition leads to excessive motions of the device that would be infeasible in real-world applications. Therefore, the reference velocity profile is defined to be in phase with the excitation force with a suboptimal amplitude that varies for each sea state condition. Even though the WEC plant is nonlinear, in this study, the calculation of the reference velocity profile is based on the linear excitation force obtained from WAMIT [23].

To compare the results of the SMC, an approximate CCC [24] is used. Referring to the impedance matching principle [25], the optimal control force may be written as:

$$F_C^{opt}(\omega) = -Z_c^{opt}(\omega)\dot{z}(\omega) = -Z_i^*(\omega)\dot{z}(\omega) \qquad (31)$$

where the optimal control impedance $Z_c^{opt}(\omega)$ must equal the complex-conjugate of the intrinsic impedance $Z_i = R_i(\omega) + iX(\omega)$ composed by the intrinsic resistance $R_i(\omega)$ and the intrinsic reactance $X_i(\omega)$. For monochromatic incident waves, this optimal force may be written in the time-domain as:

$$F_C^{opt}(t) = -M_c\ddot{z}(t) + c_c(\omega_0)\,\dot{z}(t) - k_c\,z(t) \qquad (32)$$

For the optimal control, the controller values must equal the intrinsic values $M_c = M$, $c_c(\omega_0) = c(\omega_0)$, $k_c = k$, where $k$ is the linear approximation of the static Froude-Krylov force. By cancellation, this control force gives the optimal response velocity shown in Equation (30) when the WEC model is assumed to be linear, and $k$ is used instead of the nonlinear hydrostatic FK force. Since in this paper the WEC model is nonlinear, the optimal control force using CCC was modified to obtain a perfect cancellation of the $F_{FK_{st}}$ as:

$$F_C^{opt}(t) = -M\ddot{z}(t) + c(\omega_0)\dot{z}(t) + F_{FK_{st}} \qquad (33)$$

This control strategy used for comparison will be referred to as nonlinear complex-conjugate control $(NL - CCC)$ from this point on. The configuration of this control scheme is similar to that of the SMC shown in Figure 3. However, this control law only takes as inputs $z(t)$, $\dot{z}(t)$, $\ddot{z}(t)$, and, $\eta(t)$.

## 4. Simulation

As shown in previous studies [11], the significance of the Froude-Krylov nonlinear forces for a spherical body is great due to its varying cross-sectional area. The buoy parameters used in this simulation are listed in Table 1 and were chosen to resemble the parameters of a real device such as the WAVESTAR [26] device. The density of the body was chosen to be 500 kg/m$^3$ to make the draft of the buoy $h_0$ coincide with the center of gravity of the sphere, and the frequency-dependent hydrodynamic parameters were found by solving the radiation problem in WAMIT.

**Table 1.** Simulation parameters.

| Parameter | Values |
|:---:|:---:|
| Buoy radius $R$ | 2.5 m |
| Buoy mass $m$ | 32,725 kg |
| Buoy draft $h_d$ | 2.5 m |
| Added mass $m_a(\omega_\infty)$ | 14,019 kg |
| Radiation damping $c(\omega_0)$ | 11,208 N/(m/s) |
| Water density $\rho$ | 1000 kg/m$^3$ |
| Gravity constant $g$ | 9.81 m/s$^2$ |
| Wave amplitude $A(\omega_0)$ | 0.5 m |
| Wave frequency $\omega_0$ | 1.05 rad/s |
| Wavenumber $\chi(\omega_0)$ | 0.112 |
| SMC convergence rate $w$ | 8 |
| SMC coefficient $\phi$ | 1000 |
| SMC coefficient $A_c$ | 10 $kN$ |

The results showed in the following section correspond to the wave period of $T = 6$ s and $\omega_0 = 1.05$ rad/s. Therefore, the frequency-dependent parameters shown in Table 1 are based on this wave frequency.

Under the deep water assumption, the simulation was run for different monochromatic wave periods ranging from 3 to 9 s. In addition, the steepness of the wave $S$ was kept constant at 0.018, which is the highest wave steepness under the linear waves assumption [18]. To maintain the wave steepness constant, the amplitude of the wave is related to the wavelength as $S = 2A/\lambda = 0.018$.

Table 2 shows the range of wave periods and their corresponding wavelength and wave amplitude.

**Table 2.** Wave characteristics and control parameters.

| Wave Period (s) | 3 | 4 | 5 | 6 | 7 | 8 | 9 |
|:---|:---:|:---:|:---:|:---:|:---:|:---:|:---:|
| Wavelength (m) | 15 | 25 | 39 | 56 | 77 | 100 | 127 |
| Wave Amplitude (m) | 0.13 | 0.22 | 0.35 | 0.50 | 0.69 | 0.90 | 1.14 |
| SMC Reference Amplitude (m) | 0.6 | 1.1 | 1.8 | 2.19 | 2.14 | 2.05 | 1.9 |

The reference velocity profile in Equation (30) has been proven to be the optimum velocity [5] when using a linear approximation of the WEC, enabling maximum energy absorption for a given buoy shape. However, this is not the case for this nonlinear WEC model where the $F_{FK_{dy}}$ is dependent on the incident wave elevation and the buoy displacement. Additionally, it is well-known that CCC magnifies the buoy's motion, driving the device to be either fully submerged or completely out of the water for certain wave periods. This undesirable behavior can be eliminated with the SMC by keeping the amplitude of the desired displacement bounded:

$$h_0 + S_f R - 2R > z_r > h_0 - S_f R \tag{34}$$

where $S_f$ is a safety factor used to determine the maximum and minimum buoy height covered by water, as shown in Figure 4. Therefore, the amplitude of the desired displacement of the device was chosen to be a value within those limits. The SMC reference amplitude for each wave period is listed in Table 2.

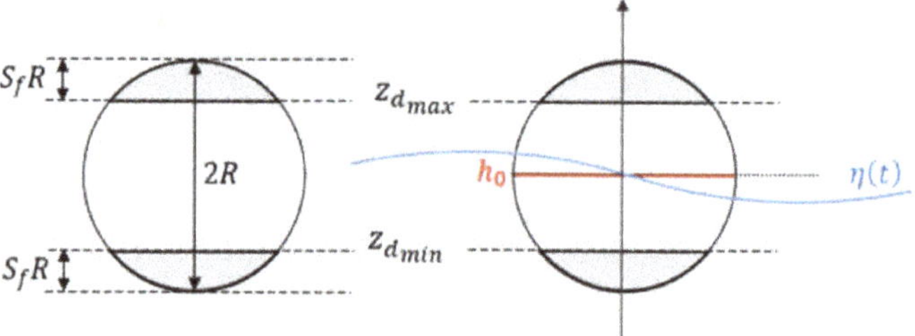


**Figure 4.** Spherical buoy limits of displacement.

## 5. Results

As aforementioned in Section 3, this paper compares the performance of the SMC against a complex-conjugate control (CCC) and a complex-conjugate control that includes the nonlinear static FK force in the control law (NL-CCC). The reference velocity used for the SMC is a cosine wave of the same frequency and phase as the linear hydrodynamic Froude-Krylov force obtained from WAMIT. To reduce the risk of driving the buoy completely out of the water or completely submerged into the water, a safety factor of $S_f = 10\%$ was used in the simulation, leaving 25 cm of clearance at the top and bottom of the spherical body.

The case shown in the results corresponds to a case where the CCC leads to a stable steady-state solution within the physical limits of the buoy. For higher wave periods, the solution with CCC would not reach steady-state. The maximum wave period at which the WEC controlled with the complex-conjugate control reached a steady-state solution within the physical limits of the buoy was at $T = 6s$. Therefore, that is the case that is shown in this paper. Figure 5 shows the relative displacement $z_d(t) - \eta(t)$ of the system using the different control strategies and the displacement limits to keep the buoy safe.

The relative displacement is kept within limits for the three control strategies used. The relative displacement when SMC is implemented is very well-behaved, and it takes advantage of the full range of motion available for the WEC. When CCC is implemented, there is a transient part for the first 15 s due to a 20-s ramp used to attenuate the overshoot caused by the control force. The relative displacement then reaches steady-state within the allowable range of motion. Note that the relative displacement with CCC is not in phase with the SMC. If the WEC plant were linear, the CCC and SMC would be in the phase. However, since the plant is nonlinear and the controller is a linear approximation, there is no perfect cancellation of the static FK force leading to a mismatch in relative displacements. Once the nonlinear static FK force is added to the CCC to form the NL-CCC, the relative displacement is in phase with the SMC. However, the amplitude of the relative displacement does not cover the full range of motion available since the minimum point at steady-state is already very close to the lower limit.

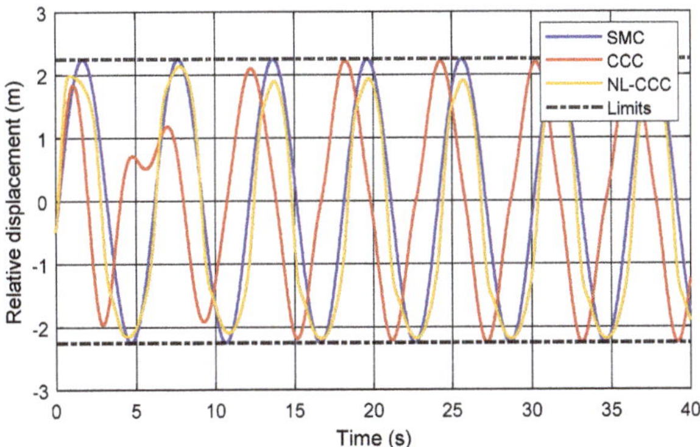

**Figure 5.** Relative displacement $z_d(t) - \eta(t)$ of the WEC when implementing SMC, CCC, and NL−CCC at $T = 6$ s.

The energy absorbed by the system when using the three control strategies and the control force used is shown in Figure 6. From this figure, it can be observed that the CCC performance is very poor compared to the SMC and the NL-CCC performance. This low energy absorption is caused by the very limited control force needed to maintain the WECs motion bounded. On the contrary, the energy absorbed when using SMC and NL-CCC are similar, with the SMC being slightly higher. However, the NL-CCC required a higher magnitude control force than the SMC to achieve slightly less energy absorption.

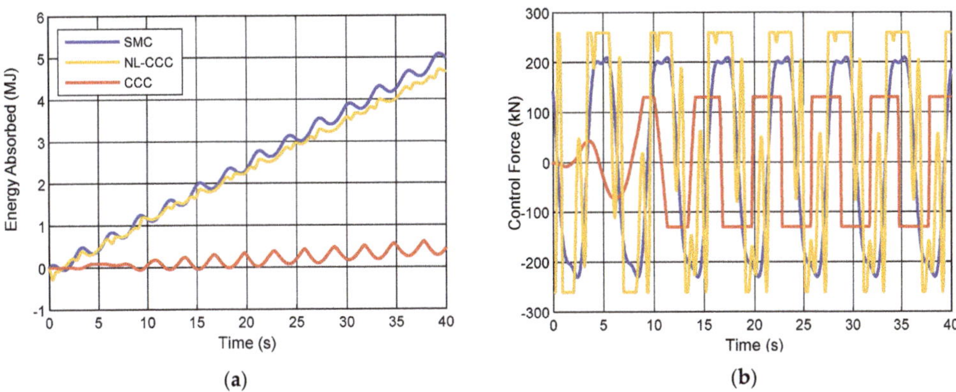

**Figure 6.** Absorbed energy (**a**) and control force (**b**) when implementing SMC, NL−CCC and CCC at $T = 6$ s.

Additionally, the SMC control force is a continuous function that does not need to be clipped in order to keep the WEC safe. On the other hand, the control force of CCC and NL-CCC are both saturated, causing discontinuities in the force. Furthermore, in order to maintain the WEC bounded in the safe range of motion, the control force of CCC and NL-CCC has to be limited at a different value for each sea state. While, for the SMC, this is not necessary, since the motion of the WEC is determined by the reference signals.

Although, in this case, the difference between the energy absorbed by the SMC and NL-CCC is very close, that is not the case for higher wave periods. The power absorbed by the WEC when using the three control strategies at different wave periods is shown in Figure 7. As the wave period increases, because the wave steepness is kept constant, the wave height increases. With increasing the wave amplitudes, the WECs displacement is significantly different from the wave elevation. Therefore, the mean wetted surface is significantly different from the instantaneous wetted surface used in this nonlinear approach, making the nonlinear Froude-Krylov forces very significant. Hence, when using a linear control strategy such as the CCC for higher wave periods, the control force must be highly limited in order to keep the WEC safe, reducing the power absorption significantly. Similarly, when using the NL-CCC, even though the nonlinear static FK force is incorporated in the controller, the nonlinear dynamic FK becomes very relevant; therefore, the control force must be limited to keep the WEC bounded. The hydrodynamic parameters used for the wave periods shown in this figure are shown in Appendix A.

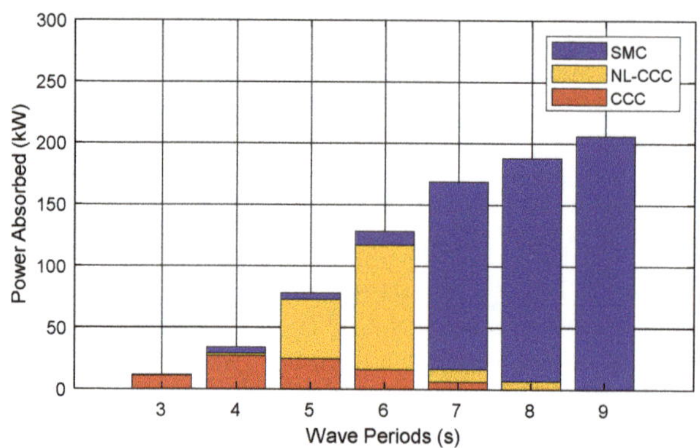

**Figure 7.** Absorbed Power for the different control strategies with varying wave periods and constant steepness of regular waves.

The difference in power absorption between the three strategies is very small in the shorter wave periods range, particularly between the SMC and the NL-CCC. Except for the WECs natural frequency of $\omega_n = 2.43 \frac{rad}{s}$, where, again, the CCC and NL-CCC control strategies need to be very limited to avoid extreme excursions of the WEC. Due to the SMC's ability to track the reference signal and assuming that there are no constraints in the control force for the longer wave periods, the power absorption outperforms that of CCC and NL-CCC.

The sliding mode controller is additionally very robust against modeling uncertainties or disturbances experienced by the system. From the parameters in Equation (29), a source of uncertainty could be the terms in the radiation force ($m_a(\omega_\infty)$, $c(\omega_0)$), and the wavenumber $\chi(\omega_0)$ in the dynamic Froude-Krylov force. These parameters can vary due to an error in the system identification, or the device is biofouled, the system is disturbed, etc.

The exact model assumes 100% certainty of the three parameters mentioned above and the perturbed model assumes 50% uncertainty of the three parameters simultaneously. In Figure 8 is shown a line with slope $-w$ in blue, and the sliding surface $s(t)$ in red for the perturbed model. In Figure 8a, the SMC coefficient $A_c$ of the term $A_c \tan h(s/\phi)$ in Equation (29) was kept the same as the value used in the exact model 10 $kN$. However, as seen from the figure, the initial loop starts at (0;0) due to matching initial conditions between the system states and the reference states, then the sliding surface keeps looping around the blue line, increasing in magnitude. This indicates that the error is not bounded, and therefore, the system could go unstable. Therefore, $A_c$ was increased to 100 $MN$ to handle the uncertainty errors. The sliding surface shown in red in Figure 8b forms a closed contour and loops around it, meaning that the error is controlled and will not go unbounded. Hence, when the model is not exact, the $A_c$ needs to be increased. For different wave periods, and depending on the modeling error or disturbances, the SMC coefficient $A_c$ might have to be adjusted to ensure the WEC is tracking the reference signals without much deviation.

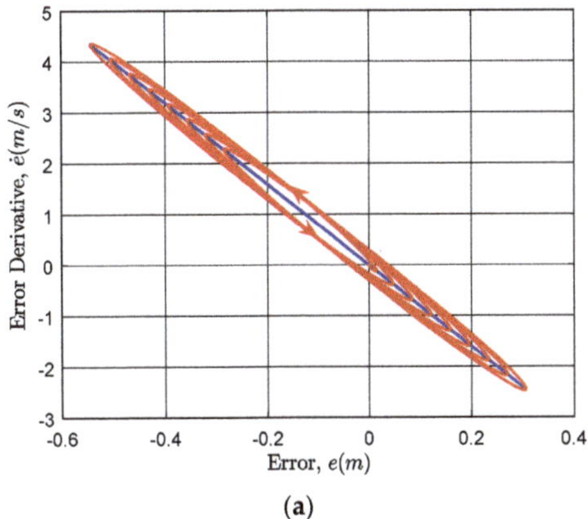 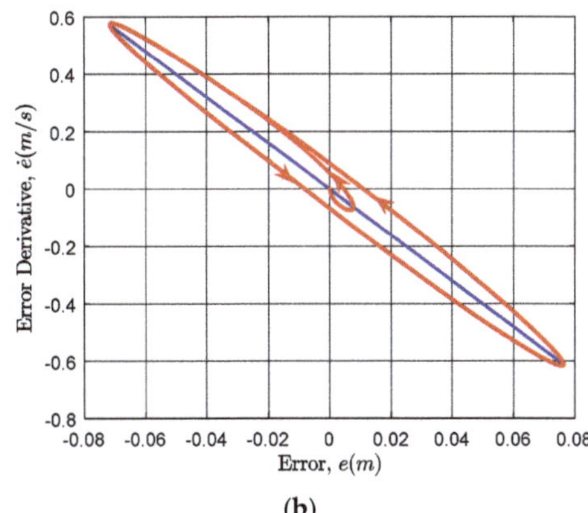

**(a)**  **(b)**

**Figure 8.** Phase−plane plot of sliding surface $s(t)$ with SMC coefficient of $A_c = 10\,kN$ on the left (**a**), and $A_c = 100\,MN$ on the right (**b**).

Figure 9a shows the error between the reference displacement and the WEC displacement for the exact and perturbed model with $A_c = 100\,MN$. The error for the exact model is close to 0, while, for the perturbed model, it oscillates between $\pm 7.76 10^{-2}$ m which is approximately a 3.5% error. To keep this error between a 5% error margin, the SMC coefficient $A_c$ had to be increased. As mentioned in Section 3, the higher the magnitude of this coefficient, the more robust the SMC is. However, as seen from the control force in Figure 9b, the increase in its magnitude barely affected the control force. This behavior is because the term $A_c \tan h(s(t))$ of the control law offsets the difference in the terms with uncertainties between the exact and perturbed model.

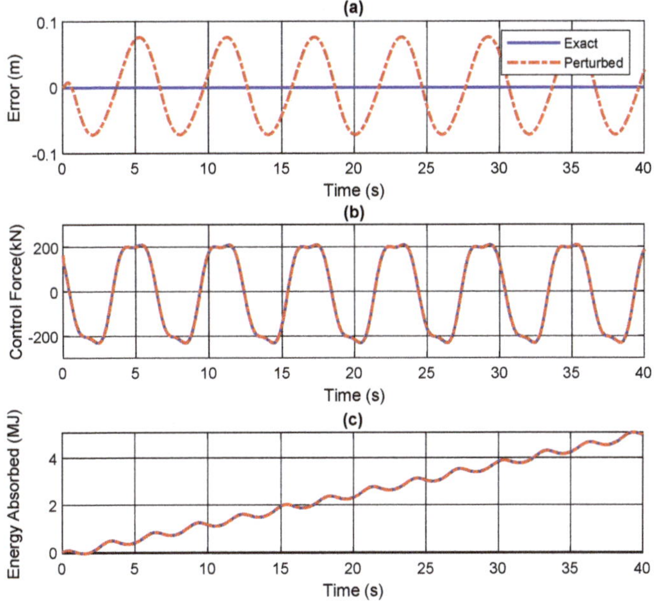

**Figure 9.** (**a**) Displacement error between reference and WEC displacement, (**b**) Control Force, and (**c**) Energy Absorption of the Exact Model and the Perturbed Model.

The energy absorption shown in Figure 9c does not seem to be influenced by the slight error in the WEC's displacement and the control force. Both the exact and perturbed models needed almost the same control force to deliver equivalent energy absorption. Therefore, the SMC has proven to work very well in systems prone to modeling errors and environmental disturbances such as wave energy converters.

## 6. Conclusions

Three different control strategies have been presented and applied to a spherical heaving point absorber wave energy converter. These control strategies have shown promising results when used in linear models of WECs. However, it has been shown that linear model approximations lead to inaccurate models when there is a significant difference between WEC displacement and wave elevation. Therefore, this paper presented a sliding mode controller applied to a nonlinear WEC model. This plant incorporates the algebraic nonlinear static and dynamic Froude-Krylov forces.

The SMC was compared to a complex-conjugate control, and a variation of this complex-conjugate control with a nonlinear static FK force term. The controllers aimed to obtain the higher power possible of all control strategies tested while maintaining the WEC within a safe range of motion. The simulation was done using linear and regular waves at a constant wave steepness of 0.018.

It was found that the proposed SMC successfully tracked the reference signals given at all of the wave periods tested. This was achieved by a continuous control force without constraints in magnitude, which would reduce the chance of failure in real-world applications by making the device more reliable. Furthermore, it was shown that the control force does not need to be limited at different values for each sea-state in order to keep the WEC in the safe range of motion, as long as the reference signals fall within those safe bounds.

The sliding mode controller allowed the maximum energy absorption between the control strategies tested at all of the wave periods. At lower wave periods, this was done by requiring less control force than the NL-CCC, and at higher wave periods, where the nonlinearities become very significant, the energy absorbed when using SMC proved to be substantially greater than when using CCC and NL-CCC. In addition, the sliding mode controller allows for the use of any desired reference signal, meaning that if there already exists a path that would deliver the optimal absorbed energy, it could just be used as a reference in the SMC and achieve optimal performance.

Furthermore, the SMC was simulated with modeling errors of 50% in the radiation force, and 50% error in the wavenumber used to calculate the nonlinear dynamic FK force. Despite these uncertainties, the SMC kept the wave energy converter in the tracking reference while using a continuous control force and delivering the same energy absorbed as the exact system. This robustness characteristic of sliding mode controllers is ideal for marine devices where the system parameters may vary over time due to aging, biofouling, corrosion, etc.

Future work includes extending the proposed sliding mode controller to work in irregular waves to test for more real-world scenarios. In addition, the designed controller can be experimentally validated by using a scaled-down point absorber hardware in the wave tank.

This paper used purely heaving point absorbers due to the computationally efficient algebraic nonlinear Froude-Krylov forces derived from Giorgi et al.'s work. However, this could be extended to more degrees of freedom in the future, where numerical approaches need to be used to solve the Froude-Krylov integrals, as explained in Reference [13].

**Author Contributions:** Methodology, T.D.G. and G.G.P.; Resources, G.G.P. and E.A.; Software, T.D.G.; Supervision, G.G.P.; Validation, T.D.G.; Visualization, T.D.G.; Writing—original draft, T.D.G.; and Writing—review and editing, E.A. and W.W.W. All authors have read and agreed to the published version of the manuscript.

**Funding:** This research received no external funding.

**Institutional Review Board Statement:** Not applicable.

**Informed Consent Statement:** Not applicable.

**Data Availability Statement:** Publicly available datasets were analyzed in this study. This data can be found here: https://github.com/tsdemont/Results_JMSE_1349687 (accessed on 25 August 2021).

**Conflicts of Interest:** The authors declare no conflict of interest.

### Appendix A

The hydrodynamic parameters used in the simulation for each wave period shown in Figure 7 are listed in Table A1. These parameters were obtained from WAMIT [23] for the spherical body with parameters listed in Table 1.

In addition, the amplitude of the SMC displacement reference used for each wave period is shown in Table 2. The energy absorbed by the WEC shown in Figure 7 was obtained using all of these parameters.

**Table A1.** Hydrodynamic parameters for different wave periods.

| Wave Period (s) | 3 | 4 | 5 | 6 | 7 | 8 | 9 |
|---|---|---|---|---|---|---|---|
| Wave frequency $\omega_0$ (rad/s) | 2 | 1.57 | 1.26 | 1.05 | 0.89 | 0.78 | 0.7 |
| Wavenumber $\chi(\omega_0)$ | 0.42 | 0.25 | 0.16 | 0.11 | 0.082 | 0.063 | 0.05 |
| Radiation damping $c(\omega_0)$ (N/(m/s)) | 16.19 | 16.81 | 14.35 | 11.21 | 8.51 | 6.64 | 5.02 |

## References


1. Gunn, K.; Stock-Williams, C. Quantifying the potential global market for wave power. In Proceedings of the 4th International Conference on Ocean Engineering (ICOE 2012), Dublin, Ireland, 17–19 October 2012; pp. 1–7. Available online: https://www.icoe-conference.com/publication/quantifying_the_potential_global_market_for_wave_power/ (accessed on 25 August 2021).
2. LiVecchi, A.; Copping, A.; Jenne, D.; Gorton, A.; Preus, R.; Gill, G.; Robichaud, R.; Green, R.; Geerlofs, S.; Gore, S.; et al. *Powering the Blue Economy: Exploring Opportunities for Marine Renewable Energy in Maritime Markets*; U.S. Department of Energy Office of Energy Efficiency and Renewable Energy: Washington, DC, USA, 2019; pp. 158–163. Available online: https://www.energy.gov/sites/prod/files/2019/03/f61/73355.pdf (accessed on 25 August 2021).
3. Faÿ, F.-X.; Henriques, J.C.; Kelly, J.; Mueller, M.; Abusara, M.; Sheng, W.; Marcos, M. Comparative assessment of control strategies for the biradial turbine in the Mutriku OWC plant. *Renew. Energy* **2020**, *146*, 2766–2784. [CrossRef]
4. Maria-Arenas, A.; Garrido, A.J.; Rusu, E.; Garrido, I. Control Strategies Applied to Wave Energy Converters: State of the Art. *Energies* **2019**, *12*, 3115. [CrossRef]
5. Falnes, J.; Kurniawan, A. *Ocean Waves and Oscillating Systems: Linear Interactions Including Wave-Energy Extraction*; Cambridge University Press: Cambridge, UK, 2020; ISBN 978-1-108-48166-3.
6. Babarit, A.; Clément, A.H. Optimal latching control of a wave energy device in regular and irregular waves. *Appl. Ocean Res.* **2006**, *28*, 77–91. [CrossRef]
7. Todalshaug, J.; Bjarte-Larsson, T.; Falnes, J. Optimum Reactive Control and Control by Latching of a Wave-Absorbing Semisubmerged Heaving Sphere. In Proceedings of the International Conference on Offshore Mechanics and Arctic Engineering, Oslo, Norway, 23–28 June 2002; Volume 4, pp. 415–423. [CrossRef]
8. Cretel, J.; Lewis, A.W.; Lightbody, G.; Thomas, G.P. An Application of Model Predictive Control to a Wave Energy Point Absorber. *IFAC Proc. Vol.* **2010**, *43*, 267–272. [CrossRef]
9. Cummins, W.E. *United States Department of the Navy. The Impulse Response Functions and Ship Motions*; David Taylor Model Basin; United States Department of the Navy: Bethesda, MD, USA, 1962.
10. Windt, C.; Faedo, N.; Penalba, M.; Dias, F.; Ringwood, J.V. Reactive control of wave energy devices—The modelling paradox. *Appl. Ocean Res.* **2021**, *109*, 102574. [CrossRef]
11. Merigaud, A.; Gilloteaux, J.; Ringwood, J. A Nonlinear Extension for Linear Boundary Element Methods in Wave Energy Device Modelling. In Proceedings of the International Conference on Offshore Mechanics and Arctic Engineering, Rio de Janeiro, Brazil, 1–6 July 2012; Volume 4. [CrossRef]
12. Giorgi, G.; Ringwood, J.V. Nonlinear Froude-Krylov and viscous drag representations for wave energy converters in the computation/fidelity continuum. *Ocean Eng.* **2017**, *141*, 164–175. [CrossRef]
13. Giorgi, G.; Ringwood, J.V. Computationally efficient nonlinear Froude–Krylov force calculations for heaving axisymmetric wave energy point absorbers. *J. Ocean Eng. Mar. Energy* **2017**, *3*, 21–33. [CrossRef]

14. Levant, A.; Levantovsky, L. Sliding Order and Sliding Accuracy in Sliding Mode Control. *Int. J. Control NT J CONTR* **1993**, *58*, 1247–1263. [CrossRef]
15. Gaebele, D.T.; Magaña, M.E.; Brekken, T.K.A.; Henriques, J.C.C.; Carrelhas, A.A.D.; Gato, L.M.C. Second Order Sliding Mode Control of Oscillating Water Column Wave Energy Converters for Power Improvement. *IEEE Trans. Sustain. Energy* **2021**, *12*, 1151–1160. [CrossRef]
16. Vu, M.T.; Le, T.-H.; Thanh, H.L.N.N.; Huynh, T.-T.; Van, M.; Hoang, Q.-D.; Do, T.D. Robust Position Control of an Over-actuated Underwater Vehicle under Model Uncertainties and Ocean Current Effects Using Dynamic Sliding Mode Surface and Optimal Allocation Control. *Sensors* **2021**, *21*, 747. [CrossRef] [PubMed]
17. Wahyudie, A.; Jama, M.A.; Assi, A.; Noura, H. Sliding mode and fuzzy logic control for heaving wave energy converter. In Proceedings of the 52nd IEEE Conference on Decision and Control, Firenze, Italy, 10–13 December 2013; pp. 1671–1677.
18. Mehaute, B.L. *An Introduction to Hydrodynamics and Water Waves*; Springer Study Edition; Springer: Berlin/Heidelberg, Germany, 1976; ISBN 978-3-642-85567-2.
19. Korde, U.A.; Ringwood, J. *Hydrodynamic Control of Wave Energy Devices*; Cambridge University Press: Cambridge, UK, 2016.
20. Todalshaug, J.H. Hydrodynamics of WECs. In *Handbook of Ocean Wave Energy*; Pecher, A., Kofoed, J.P., Eds.; Ocean Engineering & Oceanography; Springer International Publishing: Cham, Switzerland, 2017; pp. 139–158; ISBN 978-3-319-39889-1.
21. Falnes, J. A review of wave-energy extraction. *Mar. Struct.* **2007**, *20*, 185–201. [CrossRef]
22. Slotine, J.-J.E.; Li, W. *Applied Nonlinear Control*; Prentice Hall: Englewood Cliffs, NJ, USA, 1991; ISBN 978-0-13-040890-7.
23. Wamit, Inc. The State of the Art in Wave Interaction Analysis. Available online: https://www.wamit.com/ (accessed on 24 July 2021).
24. Todalshaug, J.; Falnes, J.; Moan, T. A Comparison of Selected Strategies for Adaptive Control of Wave Energy Converters. *J. Offshore Mech. Arct. Eng.* **2011**, *133*, 031101.
25. Nebel, P. Maximizing the Efficiency of Wave-Energy Plant Using Complex-Conjugate Control. *Proc. Inst. Mech. Eng. Part J. Syst. Control Eng.* **1992**, *206*, 225–236. [CrossRef]
26. Wavestar. Available online: http://wavestarenergy.com/ (accessed on 24 July 2021).

MDPI

St. Alban-Anlage 66

4052 Basel

Switzerland

Tel. +41 61 683 77 34

Fax +41 61 302 89 18

www.mdpi.com

*Journal of Marine Science and Engineering* Editorial Office

E-mail: jmse@mdpi.com

www.mdpi.com/journal/jmse